Communications
in Computer and Information Science 238

Gong Zhiguo Xiangfeng Luo Junjie Chen
Fu Lee Wang Jingsheng Lei (Eds.)

Emerging Research in Web Information Systems and Mining

International Conference, WISM 2011
Taiyuan, China, September 23-25, 2011
Proceedings

Volume Editors

Gong Zhiguo
University of Macau, China
E-mail: fstzgg@umac.mo

Xiangfeng Luo
Shanghai University, China
E-mail: luoxf@shu.edu.cn

Junjie Chen
Taiyuan University of Technology, China
E-mail: chenjj@tyut.edu.cn

Fu Lee Wang
Caritas Institute of Higher Education, Hong Kong, China
E-mail: pwang@cihe.edu.hk

Jingsheng Lei
Shanghai University of Electric Power, China
E-mail: jshlei@126.com

ISSN 1865-0929 e-ISSN 1865-0937
ISBN 978-3-642-24272-4 e-ISBN 978-3-642-24273-1
DOI 10.1007/978-3-642-24273-1
Springer Heidelberg Dordrecht London New York

Library of Congress Control Number: 2011936662

CR Subject Classification (1998): H.4, H.3, H.2, C.2.4, I.2.6, D.2

Typesetting: Camera-ready by author, data conversion by Scientific Publishing Services, Chennai, India

Printed on acid-free paper

Springer is part of Springer Science+Business Media (www.springer.com)

Preface

The 2011 International Conference on Web Information Systems and Mining (WISM 2011) was held during September 23–25, 2011 in Taiyuan, China. WISM 2011 received 472 submissions from 20 countries and regions. After rigorous reviews, 112 high-quality papers were selected for publication in the WISM 2011 proceedings. The acceptance rate was 23%.

The aim of WISM 2011 was to bring together researchers working in many different areas of Web information systems and Web mining to foster the exchange of new ideas and promote international collaborations. In addition to the large number of submitted papers and invited sessions, there were several internationally well-known keynote speakers.

On behalf of the Organizing Committee, we thank Taiyuan University of Technology for its sponsorship and logistics support. We also thank the members of the Organizing Committee and the Program Committee for their hard work. We are very grateful to the keynote speakers, session chairs, reviewers, and student helpers. Last but not least, we thank all the authors and participants for their great contributions that made this conference possible.

September 2011

Gong Zhiguo
Xiangfeng Luo
Junjie Chen
Fu Lee Wang
Jingsheng Lei

Organization

Organizing Committee

General Co-chairs

Wendong Zhang	Taiyuan University of Technology, China
Qing Li	City University of Hong Kong, Hong Kong

Program Committee

Co-chairs

Gong Zhiguo	University of Macau, Macau
Xiangfeng Luo	Shanghai University, China
Junjie Chen	Taiyuan University of Technology, China

Steering Committee Chair

Jingsheng Lei	Shanghai University of Electric Power, China

Local Arrangements Co-chairs

Fu Duan	Taiyuan University of Technology, China
Dengao Li	Taiyuan University of Technology, China

Proceedings Co-chairs

Fu Lee Wang	Caritas Institute of Higher Education, Hong Kong
Ting Jin	Fudan University, China

Sponsorship Chair

Zhiyu Zhou	Zhejiang Sci-Tech University, China

Program Committee

Ladjel Bellatreche	ENSMA - Poitiers University, France
Sourav Bhowmick	Nanyang Technological University, Singapore
Stephane Bressan	National University of Singapore, Singapore
Erik Buchmann	University of Karlsruhe, Germany
Jinli Cao	La Trobe University Australia
Jian Cao	Shanghai Jiao Tong University, China
Badrish Chandramouli	Microsoft Research, USA
Akmal Chaudhri	City University of London, UK
Qiming Chen	Hewlett-Packard Laboratories, USA
Lei Chen	Hong Kong University of Science and Technology, China
Jinjun Chen	Swinburne University of Technology, Australia
Hong Cheng	The Chinese University of Hong Kong, China
Reynold Cheng	Hong Kong Polytechnic University, China
Bin Cui	Peking University, China
Alfredo Cuzzocrea	University of Calabria, Italy
Wanchun Dou	Nanjing University, China
Xiaoyong Du	Renmin University of China, China
Ling Feng	Tsinghua University, China
Cheng Fu	Nanyang Technological University, Singapore
Gabriel Fung	The University of Queensland, Australia
Byron Gao	University of Wisconsin, USA
Yunjun Gao	Zhejiang University, China
Bin Gao	Microsoft Research, China
Anandha Gopalan	Imperial College, UK
Stephane Grumbach	INRIA, France
Ming Hua	Simon Fraser University, Canada
Ela Hunt	University of Strathclyde, UK
Renato Iannella	National ICT, Australia
Yan Jia	National University of Defence Technology, China
Yu-Kwong Ricky	Colorado State University, USA
Yoon Joon Lee	KAIST, Korea
Carson Leung	The University of Manitoba, Canada
Lily Li	CSIRO, Australia
Tao Li	Florida International University, USA
Wenxin Liang	Dalian University of Technology, China
Chao Liu	Microsoft, USA
Qing Liu	CSIRO, Australia
Jie Liu	Chinese Academy of Sciences, China
JianXun Liu	Hunan University of Science and Technology, China

Peng Liu	PLA University of Science and Technology, China
Jiaheng Lu	University of California, Irvine
Weiyi Meng	Binghamton University, USA
Miyuki Nakano	University of Tokyo, Japan
Wilfred Ng	Hong Kong University of Science and Technology, China
Junfeng Pan	Google, USA
Zhiyong Peng	Wuhan University, China
Xuan-Hieu Phan	University of New South Wales (UNSW), Australia
Tieyun Qian	Wuhan University, China
Kaijun Ren	National University of Defense Technology, China
Dou Shen	Microsoft, USA
Peter Stanchev	Kettering University, USA
Xiaoping Su	Chinese Academy of Sciences, China
Jie Tang	Tsinghua University, China
Zhaohui Tang	Microsoft, USA
Yicheng Tu	University of South Florida, USA
Junhu Wang	Griffith University, Australia
Hua Wang	University of Southern Queensland, Australia
Guoren Wang	Northeastern University, USA
Lizhe Wang	Research Center Karlsruhe, Germany
Jianshu Weng	Singapore Management University, Singapore
Raymond Wong	Hong Kong University of Science and Technology, China
Jemma Wu	CSIRO, Australia
Jitian Xiao	Edith Cowan University, Australia
Junyi Xie	Oracle Corp., USA
Wei Xiong	National University of Defence Technology, China
Hui Xiong	Rutgers University, USA
Jun Yan	University of Wollongong, Australia
Xiaochun Yang	Northeastern University, China
Jian Yang	Macquarie University, Australia
Jian Yin	Sun Yat-Sen University, China
Qing Zhang	CSIRO, Australia
Shichao Zhang	University of Technology, Australia
Yanchang Zhao	University of Technology, Australia
Sheng Zhong	State University of New York at Buffalo, USA
Aoying Zhou	East China Normal University, China
Xingquan Zhu	Florida Atlantic University, USA

Table of Contents

Applications of Web Information Systems

Applications of Web Mining

E-Government and E-Commerce

Geographic Information Systems

Information Security

Intelligent Networked Systems

Management Information Systems

Multi-agent Systems

Semantic Web and Ontologies

Web Information Processing

Web Information Retrieval

Web Intelligence

Web Interfaces and Applications

Web Services and E-Learning

Web Usage Mining

Simulation and Modeling for Centroid Algorithm Using OPNET in Wireless Sensor Networks

Hua Wang[1], Haiqing Cheng[2], and Huakui Wang[2]

[1] College of Computer Science and Technology, Taiyuan University of Technology
79 West Yingze Street Taiyuan, Shanxi, 030024 P.R. China
whchqlily@163.com

[2] College of Information Engineering, Taiyuan University of Technology
79 West Yingze Street Taiyuan, Shanxi, 030024 P.R. China
chqwhlily@163.com, huakuiw@sohu.com

Abstract. Location information of sensor node is necessary for many wireless sensor networks' applications. Simulation is a major method to analyze the performance of the location algorithm. This paper introduces the detailed process of modeling system using OPNET Modeler which is a software tool for network simulation. It is intended to provide an introduction to a common way of simulating for researchers working on positioning technique. By modeling for Centroid localization algorithm in wireless networks, structure of OPNET is described. Both behavior and performance of the modeled systems had been analyzed by performing simulations. And the experience of building a wireless sensor network in real world will be obtained through model designing, simulating, data collecting, and analyzing.

Keywords: OPNET Modeler; Wireless Sensor Networks (WSNs); Localization Technique; Simulation; Centroid Algorithm.

1 Introduction

In wireless sensor networks (WSNs), node location information is essential for wireless sensor networks. Localization technology is one of the key technologies. In the process of researching on positioning technology, the hardware experimental platform through building physical network can not achieve the aim of evaluating the performance of the algorithm. On the one hand, it is too expensive; on the other hand, after the test structure is built, the topology is difficult to change, especially for the one containing a large number of nodes in large-scale WSNs. Network simulation platform can be effectively used to overcome the above shortcomings through providing a strong lively simulation basis for the node localization. Numerous localization simulation systems have been developed and deployed during the past few decades such as [1-3]. While most of these systems rely upon NS2, GloMosim and OMNET^{++}, fewer use OPNET Modeler [4]. In the wireless transmission environment where there are many kinds of interferences, such as noise and multi-path and so on and time-varying characteristics of the channel is more obvious [5]. Thus, building a reasonable simulation for a wireless channel is more difficult. The

G. Zhiguo et al. (Eds.): WISM 2011, CCIS 238, pp. 1–10, 2011.

wireless module of OPNET Modeler provides a special fourteen stages transceiver pipeline to evaluate the characteristics of wireless communication. And user-supplied procedures interface is provided for user to modify the default procedures for special cases, if necessary. The main focus of this paper is to give a detailed overview of the process of the modeling of location algorithm in wireless environment and evaluate its performance using simulation-based analysis.

The rest of this paper is organized as follows. The next section describes the principle of Centroid algorithm. Section 3 provides the detail modeling process for nodes. Section 4 represents modeling for network and the conclusion is made in the section 5.

2 Centroid Algorithm

There are many positioning techniques which can be divided into two categories: range-based and range-free. [6] The first need the point-to-point distance or angle to calculate location, and the second evaluate the positions of unknown nodes without any information about distance or angle. The Centroid algorithm as a cost-effective alternative to more expensive range-based approaches is a typical kind of the range-free localization mechanisms. And Centroid algorithm is widely used in WSN due to its inherent advantages such as easy realization and lightweight. This section is focus on illustrating the principle of the Centroid algorithm.

As we all known, Centroid is the geometric center of a polygon and it can be regarded approximately as the center of mass. Thus, the average (geometric mean) of all points of the polygon is employed to represent the Centroid. In Centroid algorithm, unknown nodes which do not know their own location information estimate their coordinates by using a small set of nodes with known positions, called beacon-nodes. In the beginning of the location, the beacon nodes send out the location references including its location information to the unknown nodes. The unknown node calculates its own coordinate as soon as there are at least three beacons' packets with different identity are received.

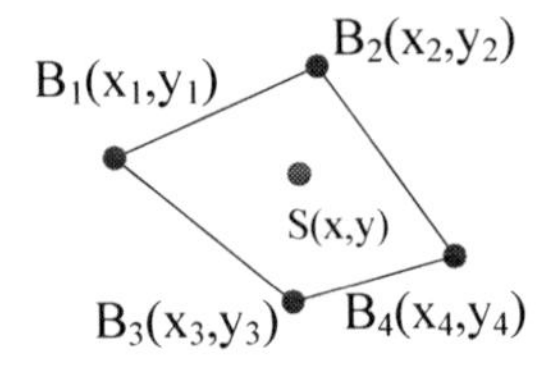

Fig. 1. Centroid location algorithm

As shown in Fig.1, a polygon is formed by beacon nodes B_1, B_2, B_3, and B_4. The coordinates of the node S can be calculated by:

$$(x, y) = \left(\frac{x_1 + x_2 + x_3 + x_4}{4}, \frac{y_1 + y_2 + y_3 + y_4}{4}\right) . \tag{1}$$

3 Node Modeling

OPNET Release 11.5A educational version is employed to describe the Centroid algorithm. Our WSNs will be created in the Project Editor which provides the resources needed to model all high-level components of a real-world network. With regard to the role playing in the course of positioning, the nodes in the WSNs can be divided into three categories: beacon nodes, unknown nodes and sink nodes. This section will introduce the functions of these three types of nodes and the detail modeling process.

3.1 Create Data Packet Model

Packet formats, which are referenced as attributes of receiver and transmitter modules within node models, define the internal structure of packets as a set of fields. For each field, the packet format specifies a unique name, a data type, a size in bits and so on. Packet formats are edited in the Packet format editor.

Beacon node sends packets to neighbor nodes by transmitting signals at a certain period to help the unknown nodes acquire the location references. The packet contains the beacon node's location information. Centroid algorithm requires that the packets must contain beacon node's *x*, *y* coordinate and *id* number which is exclusive in certain networks. To begin with, we must create packet format which is referenced as attributes of transmitter and receiver modules within node models. Packet model can be build as follows:

The packet format named "wsn_location_format" in this paper contains three fields: node id ("*id*") and beacon's coordinates ("*x_position*" and "*y_position*") And the "*size*" attribute is specified 32 bits.

id (8 bits)	x_position (8 bits)	y_position (8 bits)

Fig. 2. Packet format of a beacon message

It must be noted that suitable data type must be chosen for each field. For the node id field, the data type is set to "*object id*" and "*floating point*" for coordinates.

3.2 Create Beacon Node

In general terms, a node is terminal equipment or a circuit terminating equipment which will either originates data, or receives and processes data, or both. We create the node in the node editor. Beacon node generates a packet which is fit for the packet format defined as above at a rate and its own coordinates and id will be written in the new packet. At last, this packet will be sent out to its vicinity. From this definition, a beacon node has all of the following internal process models:

(1) Packet generating model "tx_gen": generate a packet.

(2) Packet processing model "tx_pro": write the location references in the new packet.

(3) Radio transmitter: configure the wireless pipe stage and channel attributes and allow packets to be sent outside of the beacon node via radio links.

(4) Antenna: can be used to exchange packets with other nodes when configure the suitable parameters (directionality or gain) for the antenna.

Fig.3 describes the node model for a beacon node. This node contains a packet generating module "tx_gen", a packet processing module "tx_pro", a wireless transmitter and an antenna module.

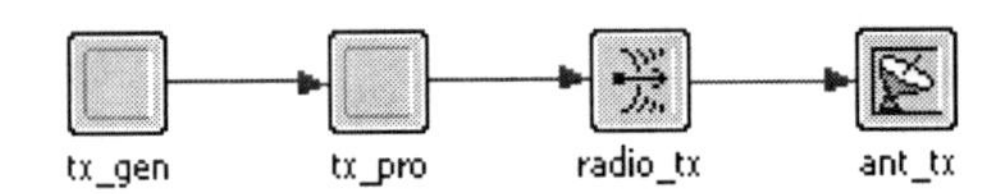

Fig. 3. Node model of a beacon

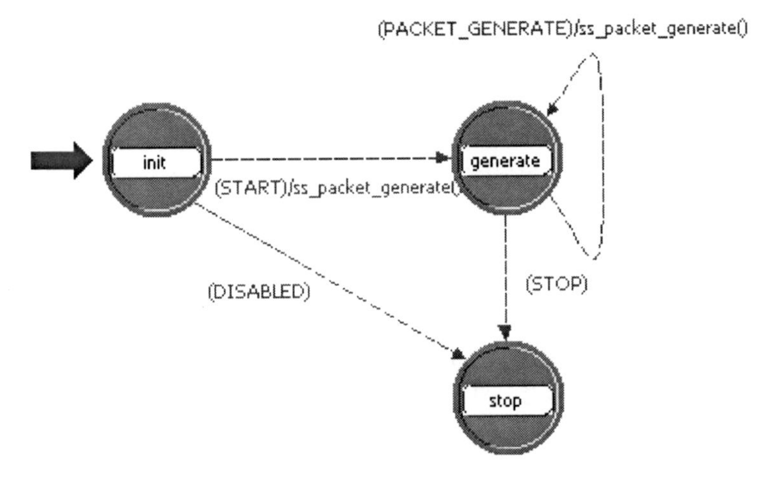

Fig. 4. The process model of module "tx_gen"

Each module's kernel is process model which is defined in the process editor. Process models describe the logic of real-world processes. The process model for the packet generating module named "tx_gen" is shown in Fig.4.

There are three states in this model. At the initial state called "init", a self interrupt that will indicate start time for packet generation is scheduled. When the interrupt of the start time arriving, the state transit to the "generate" state at which a packet is generated and the packet is sent to the lower layer. When the stop time configured in the model attribute arrived, the current state will transit to the next state named "stop".

The Packet processing module called "tx_pro" in our simulation includes only one state as shown in Fig.5.

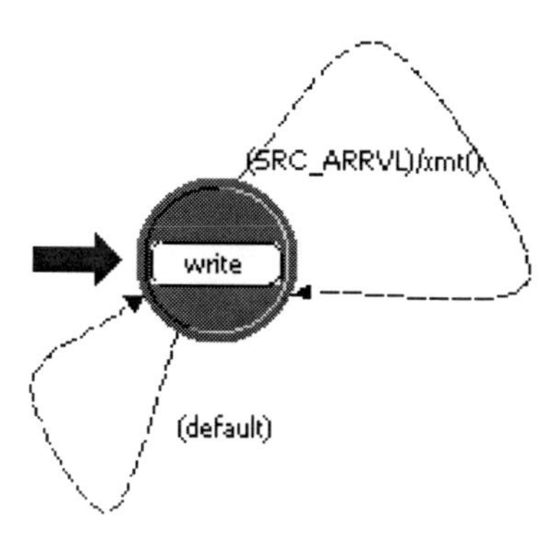

Fig. 5. The process model of module "tx_pro"

The node id number and beacon's coordinates are acquired at this state. Every node has a unique node number called node id number. Beacon's coordinates is the (x, y) position of the node which is acquired by calling the API function "*op_ima_obj_attr_get_dbl ( )*". The other function of this model is that the location references will be written in the new packet sent up by the model "tx_gen". All of these are carried out by the transition condition and the executive function. So long as the condition "SRC_ARRVL" is met, the function "xmt()" will be executed, when the interrupt type is stream interrupt the condition will be satisfied. The executive function is defined as:

```
FIN(xmt());
/*obtain beacon's id and get the packet from stream*/
    tx_node_id = op_topo_parent (op_id_self ());
    pkptr=op_pk_get(op_intrpt_strm());
/*get x and y position of the beacon node*/
op_ima_obj_attr_get_dbl (tx_node_id, "x position", &x);
op_ima_obj_attr_get_dbl (tx_node_id, "y position", &y);
/*write id, (x,y) in the packet*/
    op_pk_nfd_set(pkptr,"x_position",src_nx);
    op_pk_nfd_set(pkptr,"y_position",src_ny);
    op_pk_nfd_set(pkptr,"anchor_id",tx_node_id);
/*send out the pakcet*/
    op_pk_send(pkptr, 0);
FOUT.
```

The radio transmitter model and antenna are standard models in the OPNET library. We just need to configure the attributes according to the requirements.

3.3 Create Unknown Node

Unknown node receives the packets sent out by beacon nodes, and parses the received data packets. Utilizing the received location references from at least three different beacons, the node estimates its position according (1). Once its position information is obtained, the unknown node becomes a beacon node that is it will propagate location reference relating to itself as a beacon to help the other unknown nodes. The node model of unknown node is discribed in Fig.6.

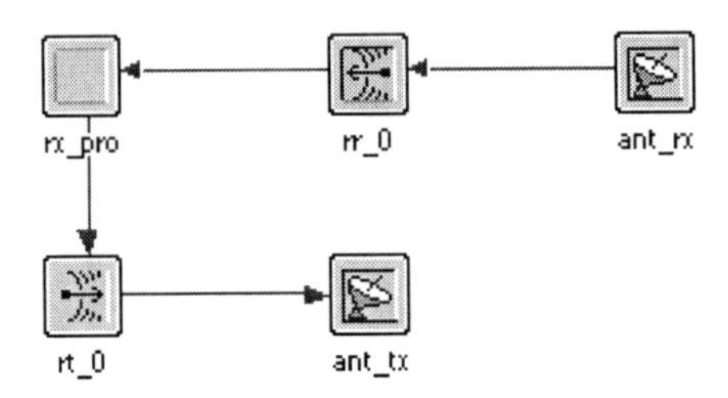

Fig. 6. Model of unknown node

The isotropic antenna "ant_rx" and radio receiver "rr_0" form the receiving part of the node. And the transmitter part consists of two parts: the radio transmitter model "rt_0" and antenna model "ant_tx". "rx_pro" is a management unit which manages the behavior of the other parts. Unlike the radio transmitter "rt_0", "rr_0" allows packets to be received from other nodes

Table 1. The function states

State	Function
"init"	Register statistic and create the list which is used to save the coming messages,
"wait"	Waiting for a beacon message carrying the location references
"restore"	Restore the coming message whose id number is different from the member of the list
"compute"	Evaluate the position of the node based on Centroid algorithm. And calculate the local location error.
"idle"	Schedule a self interrupt for generating a packet
"send"	Broadcast the location information as a beacon node.

By analyzing the characteristics of the model "rx_pro", it should contain six states in this model: "init", "wait", "restore", "compute", "idle" and "send" as shown in Fig.7. The function of each state is illustrated in TAB 1.

After initializing at state "init", the simulation kernel immediately transmit to state "wait" because "init" is a forced state. So long as arriving a packet (indicated by stream interrupt that is the condition "PACKET_STREAM-_ARRVAL" has be met), will the list which is employed to restore the packet information be inserted a new element at state "restore" by calling self-defined function "list_insert". Once the size of the list is more than 3, a scheduled self interrupt comes (satisfy "SELF_LOCATION_COMPLETE"). Entering into "compute", node's position has been calculated and the node completes the stage of positioning.

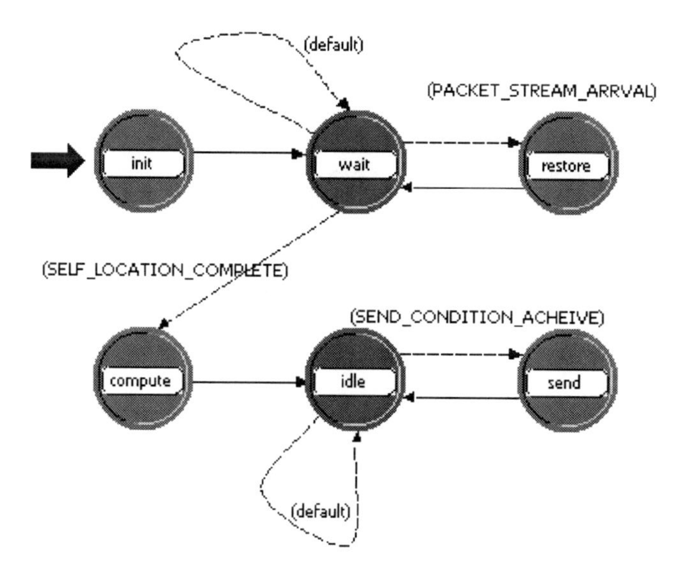

Fig. 7. The process model of module"rx_pro"

The other stage of unknown node is to propagate its own position information at a rate just like a beacon. At this time, simulation kernel is staying at "idle", until the time has arrived for sending. All the transition conditions are triggered by self interrupt scheduled in the process model. The function "list_insert" is defined as:

```
FIN(list_insert())
   pkptr=op_pk_get(op_intrpt_strm());
   op_pk_nfd_get(pkptr,"x_position",&temp_x);
   op_pk_nfd_get(pkptr,"y_position",&temp_y);
   op_pk_nfd_get(pkptr,"anchor_id",&anchor_node_id);
   if(the received beacon id is different from the
   previous)
   {
       temp_ptr =(TX_Cache*)op_prg_mem_alloc (sizeof
       (cache));
       temp_ptr->tx_node_objid=anchor_node_id;
       temp_ptr->tx_x=temp_x;
       temp_ptr->tx_y=temp_y;
       op_prg_list_insert(rxgroup_packet_lptr,temp_ptr,
       OPC_LISTPOS_TAIL);}
       op_pk_destroy (pkptr);
   FOUT.
```

3.4 Create Sink Node

To evaluate the performance of the location algorithm, average location error is used in this paper. Average location error indicates the average distance between the actual position and the estimated position of the unknown nodes.

$$e_{average} = \frac{\sum_{i=1}^{n} \sqrt{(x_{est} - x_a)^2 + (y_{est} - y_a)^2}}{n} . \tag{2}$$

The statement expressed in (2). Where, $e_{average}$ is average location error. (x_{est}, y_{est}) is the evaluating value of the (x_a, y_a). The sink node will collect the simulation statistics including the location error and the evaluation position of each node before the end of the experiment. And then, calculate the average location error. This node contains a data processing module, described by the process including only one unforced state, shown in Fig. 8.

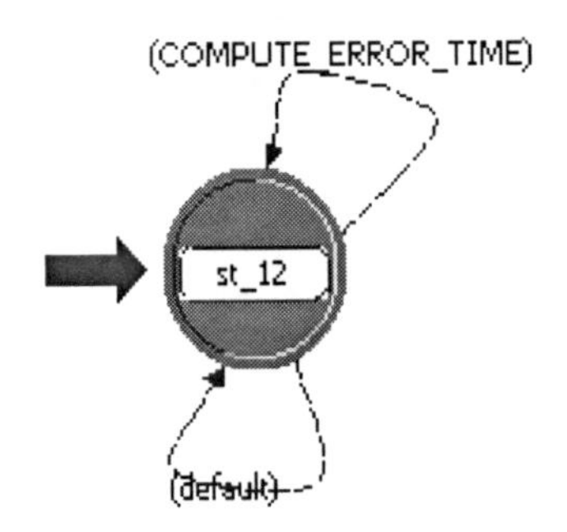

Fig. 8. The process model of sink node

At the end of the simulation, the transition condition "COMPUTE_ERROR_TIM E" is met; the sink node collects all the unknown nodes' location errors and writes it into the scalar statistics.

4 Network Model and Simulation Results Analysis

4.1 Building Network Topology

Topologies can be imported from vendor products or created manually in OPNET. Our choice is the latter technique. This section is focus on creating and editing network models in the Project Editor where we can also customize the network environment, run simulations and choose or analyze simulation results.

Find all kinds of the nodes we have created in the object palette and implement the node into the sensor field. We will get a wireless sensor network in which the unknown nodes can obtain their own location information automatically. In this study, sensor nodes are placed in a 50m×50m area. There are 30 unknown sensor nodes, 16 beacon nodes in this sensor field. Bird's eye view of the deployment of the sensor nodes is shown in Fig.9. The effective communication range of sensor nodes is set to 20m by using Receiver Group Configuration module.

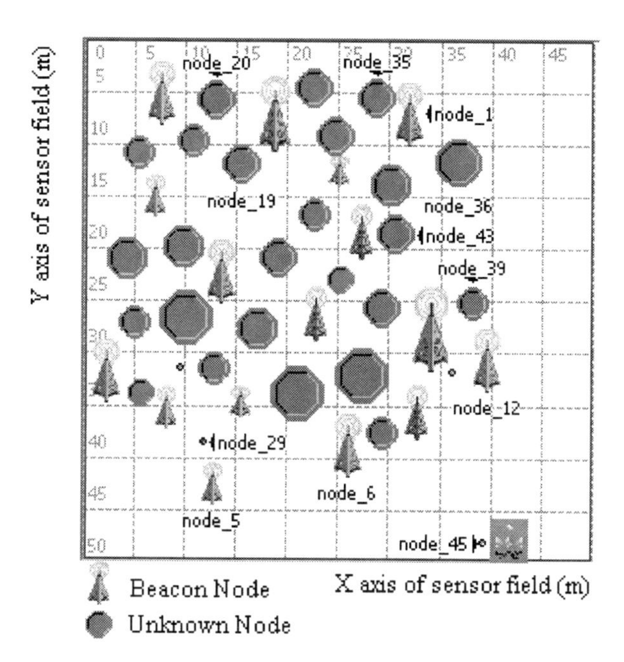

Fig. 9. The deployment of the sensor nodes

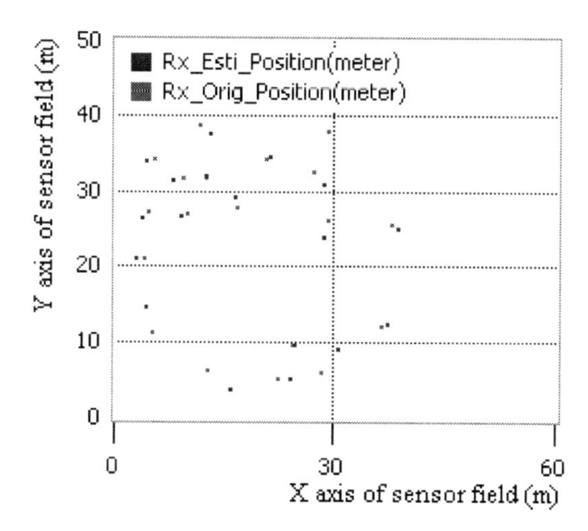

Fig. 10. Estimation results of the network

4.2 Data Collection and View Results

Two forms of data called vectors and scalar which are stored in distinct types of files can be used to store the results of the experiment. We set up several statistical measures in OPNET to study the performance of the location simulation. Fig.10 depicts the result of the positioning. The figure intuitively demonstrates the performance of the location mechanism. The red dots represent the nodes' actual physical locations and the blue ones are the estimation positions of the unknown nodes.

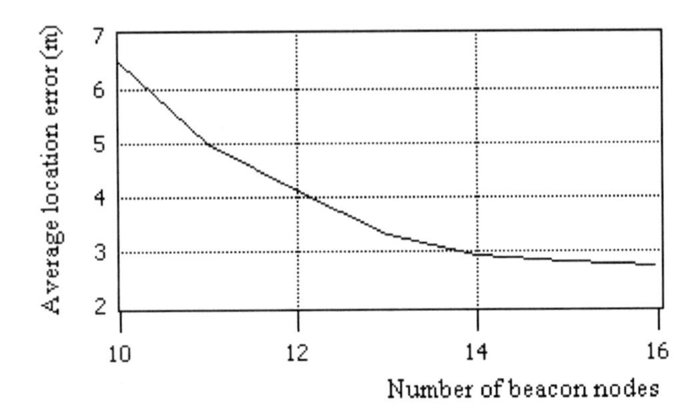

Fig. 11. Average localization error curve

Reduce the number of beacon nodes and run the simulation again. Open the analysis configuration tool, load the output scalar file in which the average location error is record and then select the scalar panel data for the display, we will get a curve that describes the location error varying the number of the beacons. The average error increases when the number of beacon nodes decreases, as shown in Figure 11.

5 Conclusion

In this paper, OPNET editors: project editor, node editor, process editor and packet editor are illustrated in examples. What's more, we describe the general process of the wireless modeling for localization in wireless sensor networks which might be helpful for some researchers who work on localization. OPNET can be used to analyze the location mechanism for the wireless sensor networks easily and perfectly. In addition, it provides a user-friendly graphical and text interface.

References

1. Molina, G., Alba, E.: Location discovery in Wireless Sensor Networks using metaheuristics. Applied Soft Computing 11, 1223–1240 (2011)
2. Caballero, F., Merino, L., Gil, P., Maza, I., Ollero, A.: A probabilistic framework for entire WSN localization using a mobile robot. Robotics and Autonomous Systems 56, 798–806 (2008)
3. Nezhad, A.A., Miri, A., Makrakis, D.: Location privacy and anonymity preserving routing for wireless sensor networks. Computer Networks 52, 3433–3452 (2008)
4. Koubaa, A., Alves, M., Tovar, E.: A Comprehensive Simulation Study of Slotted CSMA/CA for IEEE 802.15.4 Wireless Sensor Networks. In: 6th IEEE International Workshop on Factory Communication Systems (WFCS), pp. 183–192. IEEE Press, Torino (2006)
5. Ren, Z., Wang, G., Chen, Q., Li, H.: Modeling and simulation of Rayleigh fading, path loss, and shadowing fading for wireless mobile networks. Simulation Modeling Practice and Theory 19, 626–637 (2011)
6. Mao, G., Fidan, B., Anderson, B.D.O.: Wireless sensor network localization techniques. Computer Networks 51, 2529–2553 (2007)

An Interactive Modeling System Based on SIFT

Chengtao Yi[1], Xiaotong Wang[1], Xiaogang Xu[2], and Bangsheng Nie[2]

[1] Department of Navigation, Dalian Naval Academy,
116018 Dalian, China
[2] Department of Arming System and Automation, Dalian Naval Academy
116018 Dalian, China
{yctao1024,wangxiaotong}@163.com, xxgang@zju.edu.cn,
nbs1024@sina.com

Abstract. Aiming at the problem of low efficiency and low photorealistic models in the current navigation simulator, a new interactive system of image-based modeling (IBM) Based on SIFT was studied. The goal of this paper is 3D reconstruction from multiple Images of the scene or the object. To overcome the shortcoming of manual matching in IBM, a new algorithm based on SIFT was provided. To improve the third dimension of 3D reconstruction, a new way of texture image extraction and fusion were adopted. Experimental results show that the image matching algorithm is fit for images even under conditions such as move, rotation, slight scale and affine transformation, and the 3D models generated by this paper are sufficiently accurate and photorealistic to meet the requirements in the virtual reality and visualization applications. Compare with the previous system of image-based modeling, it is simpler, more convenient and easier to create photorealistic models.

Keywords: Image matching, SIFT, stratified reconstruction, texture mapping.

1 Introduction

At present, the visual system virtual 3d modeling problem can come true by GBM (Geometry-based modeling) or IBM (Image-based modeling) in the marine simulator. The GBM is traditional modeling methods, it has relatively complete theoretical framework, and its common modeling software have 3dmax, Multigen Creator and so on. But the GBM faces many problems: a) the structural geometry modeling is complex and trivial, and requires a lot of human and manual, such as section of a few minutes of 3d animation, the scene modeling may need one to two months. b) it is relevant between drawing speed and scene complexity. If the Complexity is the higher, Geometric strips which are deal with in the cutting, vanishing and illumination computation are more and more, and the efficiency is lower. c) the realistic is not better. Model performance is often for the price of the non-real calculation, and the existing algorithm is still unable to simulate or implementation of some special lighting and simulated atmosphere effect. IBM is 3d reconstruction technique, which was originated in the middle of 19s and tries to break traditional mode of the GBM fundamentally. Compared with these GBM, IBM technology has the following features: a) only shoot

G. Zhiguo et al. (Eds.): WISM 2011, CCIS 238, pp. 11–18, 2011.

(samples) through digital camera or video scene or object, the calculation of sample data generated the three-dimensional coordinates of scene or object, it reduce the complexity and intensity of the scene or object modeling, it has the advantage of modeling, faster and less system resources etc. b) Reconstruction algorithm for calculating resource requirement is not high, it improves the rendering demand and calculation of large-scale scene or complex scenes, it displays that hardware capabilities are limited. c) Reconstruction model can achieve real renderings of environmental photograph effect, provides more realistic and identify.

Recently, there are already some good commercial software and scientific software based on IBM technology, such as: PhotoModeler[1], ImageModeler[2], Interactive modeling system from multiple images[3], The plant modeling based on image [4], The tree model based on image[5], VideoTrace interaction model system from the video[6], etc. These software can come true semi-automatic modeling through method of artificial interaction in the absence of any information under the premise of geometric constraints, But the biggest deficiency in the software is the burden of user interaction. The main reason is that the image matching features with reconstruction algorithm is unable to maintain stability conditions caused by scaling or rotating sheltered viewpoint or noise in the software, it affects seriously the accuracy of model and could lead to fail to rebuild even. In 1999, Lowe put out SIFT(Scale Invariant Feature Transform), and improved method[7] in 2004.Feature point which is extracted By SIFT is invariable not only for image translation and rotation transformation scaling, but also for illumination change and complex affine transformation and projection transformations. It is widely concerned in target recognition and image retrieval and image mosaic[8]. But SIFT applied less in 3d reconstruction, and the important characteristic of SIFT operator is that the matching points are many and stable, it is useful in 3d reconstruction, it can effectively solve interaction burden in the image matching software. Therefore, this paper puts forward a method of the 3d reconstruction system based on the improvement SIFT feature matching, and does the relevant test.

2 Overview of the System

This paper realizes the 3d reconstruction system in following technical route: firstly, generate eigenvector descriptors using an improved algorithm of SIFT extracted features. Secondly, Get the initial matching points Based on the characteristics of image features, and get the accurate matching points based on an epipolar line constraint and the further improvement of relaxation algorithm. Then, get the projective reconstruction characteristic points using hierarchical reconstruction algorithm, make sure the upgrading matrix through the camera calibration technique, and make the feature points coordinates in projective space into the feature points coordinates in Euclidean space by the upgrade matrix. At the same time, using cluster to reconstruct the adjustment method optimized. Finally, Extract the texture images from the original image sequences and fuse them by texture fusion method, and rebuild model texture mapping. The Workflow of the system is shown in figure 1.

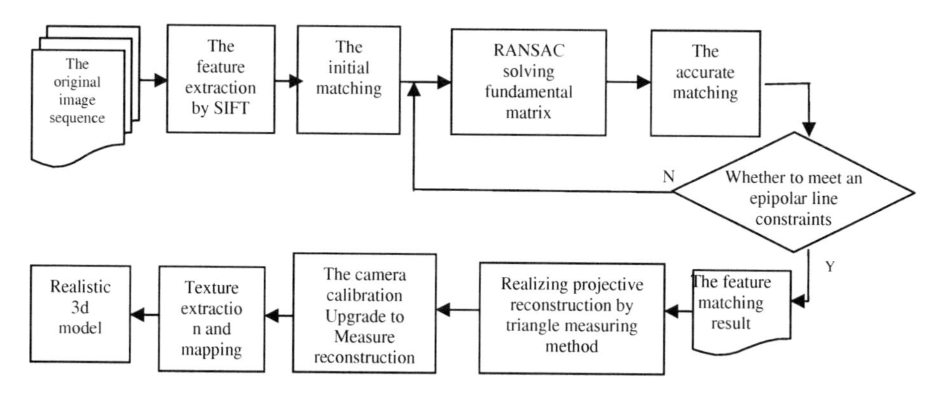

Fig. 1. System work flowchart

3 The Improved Algorithms of 3D Reconstruction System Based on Sift

3.1 The Feature Points Extraction and Matching Based on SIFT

1) multi-scale feature points extraction[9]

SIFT is actually a translation, which can translate the image into a collection of scale invariant feature. Firstly, detect the feature of scale-space, deciding the position and the scale-space of the feature. Secondly, appoint the main direction of the feature by calculating the gradient direction histogram of each pixels neighborhood of the feature. Lastly, construct the simplified feature descriptors based on the feature points. The main steps of feature point extraction using SIFT are as follows:

a) Detection of extremum of scale-space. Set up image pyramid by filtering the original image with a series of Gaussian filter with a certain multiple of scaling factor. Then do minus between every two adjacent Gaussian blur images to acquire Gaussian differential images pyramid. Remove the bottom and the top image of the Gaussian differential pyramid images, and then get a candidate scale and points of interest with invariant direction in the remaining images of each piece on different scales.

$$\begin{cases} G(x,y,\sigma) = \dfrac{1}{2\pi\sigma^2} e^{-(x^2+y^2)/2\sigma^2} \\ L(x,y,\sigma) = G(x,y,\sigma) \times I(x,y) \\ D(x,y,\sigma) = L(x,y,k\sigma) - L(x,y,\sigma) \end{cases} \tag{1}$$

here $G(x,y,\sigma)$ is a two-dimensional Gaussian function, σ is the standard deviation of Gaussian normal distribution. $I(x,y)$ is each image pyramid, while $L(x,y,\sigma)$ is the convolution of the images and the two-dimensional Gaussian function, named as the image scale space. $D(x,y,\sigma)$ is DOG operator, defined as the difference of adjacent Gaussian filtered images of each group, which is an approximation of normalized LOG

(Laplacian of Gaussian) operator. DOG compares their gray value of each pixel in its 3 × 3 with 8 pixels neighborhood and the two corresponding 9 pixels neighborhood of the adjacent upper and lower pyramid image. If the pixel is the extremum in the entire 26 pixels neighborhood, then this feature point should be choose to be a candidate key point.
b) Feature point positioning. In each extreme point, we used a detailed model (three-dimensional quadratic function) to fit in order to remove the instability points, including the removal of any point whose offset vector is greater than the threshold preset. In order to enhance the independency of the feature point and to improve performance of matching we remove the point whose contrast is below the preset threshold and the edge point whose principal curvature extremum does not meet certain conditions.
c) Appointment of the main direction. SIFT calculates the gradient direction histogram so as to get the rolling invariability of the detected features. The main steps are as follows: First, compute the gradient direction of the points belong to the neighborhood of the feature points; then, get the absolute value of the gradient, multiply the gauss function value to be the weighted value; at last, get the main direction of the character points by voting.
d) Construction of simplified feature descriptors based on the feature points. From the paper [7], we can know that for any stable feature point, in its scale space feature points the feature point as the center there are 4×4 sub-regions. In each sub-region there are 4×4 pixels. Form a seed point by calculation of the gradient direction histogram of each sub-region. By sorting eight vector direction information of each seed point, we can form a 4×4×8=128 dimensional feature vector. That is the SIFT feature descriptor. If normalize the 128-dimensional feature descriptor, you can enhance the robustness of illumination changes.

Lowe's paper [7] shows that we will get about 2000 stable feature points from a typical 500 × 500 image, so the cost of calculation is very large. In order to reduce time-consuming part of feature extraction, we should reduce the number of feature points as more as possible in the premise of ensuring the quality of the feature points. Therefore, we choose an 8 × 8 window in our experiments, which means we use a 2×2×8 = 32 dimensional feature vector to describe each feature point.

2) reversible matching algorithm based on SIFT feature vector
When computing SIFT feature vector, the same point may have a number of directions, so the points belong to different feature points, but in fact they may be the same point, and this will cause the repeating of the matching points. To inhibit this condition we used the reverse method of matching. To use two images as an example, the realization steps are as follows:

a) For the feature points set of two images, traverse each feature point of the first image, and calculate the smallest Euclidean distance between the corresponding eigenvector of this feature point with all the eigenvectors of another image. Generally we use exhaustive method to calculate the Euclidean distance of feature vectors of two images, but the computational complexity is $O(n^2)$[10]. It can't be real-time. k-d tree can quickly and efficiently find the nearest neighbor points. And complexity is only O(nlgn). However, when the feature point descriptor's dimension is greater than 10, the

efficiency of k-d tree will be much lower than the exhaustive method. Therefore, we use the approximate nearest neighbor algorithm[10]. Its high-dimensional space searching efficiency has markedly improved than k-d tree.
b) We choose nearest neighbor matching algorithm because nearest neighbor matching method[11] can elicit the non-relationship matching feature points due to image occlusion and background confusion. First, calculate the ratio(NN/SCN) r of nearest neighbor distance and the second nearest neighbor distance, then set the threshold t, and consider the correct matching points whose r is less than t. This is because for wrong match, similar distance may have a large number of other errors of matching in high-dimensional feature space, thus its value of r is relatively high. the threshold t is 0.8 in paper [7], while t generally equals to 0.6 in this paper.
c) Similarly complete the nearest neighborhood matching of all the feature points of the second image.
d) Because the above steps will unavoidably produce some error matching, it is necessary to eliminate the additional error matching in accordance with the geometric restrictions and other constraints to improve robustness. In this paper we use random sample consensus (*RANSAC*) method and epipolar constraint to eliminate the outliers.

3.2 Hierarchical Reconstruction Optimization Algorithm

This paper uses 3d model reconstruction algorithm in paper [3], Based on the feature points come true the projective reconstruction by matching triangle measuring method, and calculate the upgrade matrix through the camera calibration technique, and make the feature points coordinates in projective space into the feature points coordinates in Euclidean space by the upgrade matrix for meeting Visual requirements. In order to solve the accumulative error produced by the reason that the image sequence two images before to determine the reference coordinate system. The paper rebuilds result global optimization by using cluster adjusting method, Coherent adjusting is Calculation method that adjust the camera's perspective projection matrix and the spatial point coordinates, Get reconstruction results that making 2d image point projection minimum error is little. Objective function for:

$$\varepsilon = \sum_{i,j} \| \hat{x}_{ij} - x_{ij} \|^2 = \sum_{i=1}^{m}\sum_{j=1}^{n} d^2(x_{ij}, P_i X_j) \qquad (2)$$

Which P_i was the perspective projection matrix of the ith image, X_j was the jth 3D reconstruction point, $\hat{x}_{ij}$ was the reprojection homogeneous coordinates of the ith image, x_{ij} and was the real image coordinates of the corner. m was the number of the camera and n was the number of the 3D points.

3.3 Texture Mapping Technique and Extraction

Texture mapping scheme and extraction is, above all, extracting directly texture image from the original image scene or objects, then, processing by image fusion method, Finally the model is mapped to reconstruct the corresponding strips. Because object three-dimensional model of the same triangles are generally exists in multiple image texture, texture extraction must ensure two aspects: accuracy and optimization. Accuracy reflects that it not selects the blocked trigonal texture; optimization reflects that texture forming two triangles texture should not have apparent juncture .Therefore; this paper uses the perspectives of texture image fusion method, which can put two images into a seamless image. image fusion Generally divided into two steps, the first step is the image fusion, two images will be merged into the same space coordinates, make two images as an image; The second step is to eliminate flat-fell seam, two image can be fusion a real image after removing juncture. To eliminate the Mosaic, this paper uses fusion algorithm of the weighted average. Realizing method is, weighted fusion can be expressed as:

$$I(i,j) = \alpha I_1(i,j) + (1-\alpha) I_2(i,j) \tag{3}$$

which weighting coefficients α=[0,1],if α=0.5, it is equivalent to two images taken average; if αchanges, namely at different points of its value is different also, when α changes from 0 to 1, Image changes from $I_2(i,j)$ to $I_1(i,j)$, It can realize the smooth transition between images, thus it eliminates the stitching.

3.4 Manual Correct

Even SIFT ensure three-dimensional reconstruction result was not affected by the original image sequences, scaling transformation, the noise, color changing rotation and so on, there was always error matching when the distance was too long for taking picture or the condition for picture was different from each other very much. So, this paper firstly take matching automatically by SIFT, then modify, add or delete the matching feature points by manual style.

When Texture mapping, the texture images were processed by image fusion method, if the result processed was not good, user can select the material trigonal texture images by Interactive tools.

4 Experimental Analysis

In order to verify the effectiveness of this algorithm, with two images for example, we realize the algorithm by using OpenGL and OpenCV in the Pentium CPU (R) 4 GHz 2.4, vc + + 6.0 for WinXP. the features of SIFT extract parameters: in order to reduce time of establishing scale space and generating feature vector, when Image and Gaussian kernel convolute, No prior to image magnification. in the process of feature point positioning the contrast threshold is m = 0.03, neighborhood is 8×8 Child window, feature points are descript with 32 dimension characteristic vector, matching parameters of feature points are NN/SCN =0.6,the epipolar line searching width is 40, the tolerance of the epipolar line e=0.65. Other parameters determined by experience, and remain unchanged. Reconstruction results are shown in figure 2. Left, middle, right figures are original figure, reconstruction model figure and rebuild model side grid texture figure. Empty and deformation is due to the model parameters of the reconstruction of only two original figures in reconstruction figures, they could be eliminated by increasing number of original image. Thus, ①the improved SIFT algorithm can be matched, it is applicable to matching problem to the same object from different perspectives of image, and the matching veracity is 95%. ② This 3d reconstruction algorithm has good robustness and accuracy, reconstruction model is correct, it has "pictures" reality after adding texture

Fig. 2. Some reconstruction result examples based on two images

5 Conclusion

3d reconstruction is currently a hot spot in the field of computer vision and difficulties. This paper realizes an interactive 3d reconstruction system based on the research results. Import multiple images which bring different perspectives from the same object, The algorithm of this paper can identifies the matching relations of image, obtaining more stable matching points, and reconstructing successful 3d objects model based on the matching point. This algorithm has strong robustness. Improved SIFT feature points ensure three-dimensional reconstruction result was not affected by the original image sequences, scaling transformation, the noise, color changing rotation and so on. SIFT feature vector, established k -d tree and introducing an epipolar line constraint reduce the complexity of the matching feature points, and improve the real-time performance of the algorithm. The system uses stratified reconstruction and optimization method, it enhances the robustness of the reconstruction algorithm, and the method of the perspectives of texture image fusion makes the transition more natural among the textures. this system is simple interactive, convenient operation, more realistic than before, and satisfy modeling demand that most of the scene or object quickly lifelike model in virtual reality and other marine simulator.

Acknowledgment. This work is supported by the Science Development Foundation of Dalian Naval Academy (Grant #2010004).

References

1. http://www.photomodeler.com/
2. http://www.realviz.com/products/im/index.php
3. Liu, G., Peng, Q.S., Bao, H.J.: An Interactive Modeling System from Multiple Images. Journal of Computer–aided Design & Computer Graphic 16(10), 1419–1424 (2004)
4. Quan, L., Tan, P., Zeng, G.: Image-based plant modeling. ACM Trans. on Graphics (SIGGRAPH) 25(3), 772–778 (2006)
5. Tan, P., Zeng, G., Wang, J., Kang, S.B., Quan, L.: Image-based tree modeling. ACM Trans. on Graphics (TOG) and Proc. of SIGGRAPH 2007 (2007)
6. Antonvan, V.D.H., Anthony, D., Thorsten, T.: VideoTrace: Rapid interactive scene modelling from video. In: Proceedings of SIGGRAPH 2007 (2007)
7. Lowe, D.G.: Distinctive image features from scale-invariant keypoints. International Journal of Computer Vision 60(2), 91–110 (2004)
8. Moradi, M., Abolmaesumi, P., Mousavi, P.: Deformable Registration Using Scale Space Keypoints. In: Proc. of SPIE Medical Imaging, San Diego, CA, pp. 791–798 (2006)
9. Wang, G.M., Chen, X.W.: Study on a New Algorithm of Feature Matching-SIFT. Journal of Yancheng Institute of Technology Natural Science Edition 20(2), 1–5 (2007)
10. Arya, S., Mount, D., Netanyahu, N.: An optimal algorithm for approximate nearest neighbo searching in fixed dimensions. Journal of the ACM 45, 891–923 (1998)
11. Alkaabis, D.: A new approach to corner extraction and matching for automated image registration. In: 25th IEEE International Geoscience and Remote Sensing Symposium, GARSS (2005)

Research on Service-Oriented and Component-Based Simulation Platform

Yang Shi, Ming-hua Lu, Ming-yan Xiao, and De-sen Zhang

Navy Submarine Academy, Qingdao, 266071, China
sy59537@163.com

Abstract. The state of the art in distributed simulation is described and the current challenge and requirement for distributed simulation are discussed. A simulation framework based on service-oriented architecture is proposed for its virtues in achieving simulation composability. Three component-based software engineering technologies for constructing service-oriented simulation platform are compared. As a case, multi-tier model architecture for navy warfare simulation is proposed and the simulation platform is established according to Java EE/EJB specification. The implementation and experiment of this simulation platform are expatiated in details.

Keywords: distributed simulation, composability, service oriented, component-based software engineering, Java EE/EJB.

1 Introduction

Modeling and simulation is a field involved many disciplines and widely used in various domains. It plays the key role in solving complex problems before the actual physical systems are built. Distributed simulation have gained great achievements, but with the rapid increasing in problems' complexity and new requirements, the existing simulation technology faced some huge challenges, such as low efficiency and less flexibility, and can not meet the demand on reducing time and resource costs associated with development and validation of simulation models has also increased. The fundamental reason is that existent distributed simulation frameworks have shortcomings for these demands. In recent years, the concept of Service-Oriented Architecture (SOA) and Component-Based Software Engineering (CBSE) technology were proposed and practiced in software developing. These provide a way to achieve the goal of composability and flexibility in modeling and simulation through a set of lean and reusable components.

2 The State of the Art in Distributed Simulation

As we known, distributed simulation has gone through two main technology stages, DIS and HLA[1]. In each stage simulation has its own research topic which we called "simulation ability". In 1980s the topic is how to connect the physically distributed simulator by network and make them work together, the concrete project was

G. Zhiguo et al. (Eds.): WISM 2011, CCIS 238, pp. 19–27, 2011.

SIMNET (SIMulator NETwork) conducted by NATO. Based on the success of SIMUNET, IEEE propose a standard named DIS protocol which make the distributed have interconnectivity, the ability of connection between simulators.

In the last decades distributed simulation research focus on the interoperability, the ability which enable simulations developed by various fields to interactive at the application level. The most important result is High Level Architecture (HLA), a novel simulation framework conducted by DMSO. HLA guarantees consistency and interoperability between federates by provides the means of entity state management, time management, data distribution management and so on [2]. As the newest simulation framework, HLA has been widely used in various domains. But recent researches show that HLA has some shortcomings, some of them are listed as follows.

(1) Low efficiency and flexibility in modeling and simulation.

(2) Inability to provide for reusability of fine-grained simulation component in other simulation context without significant source code modifications.

The reason is that the RTI(Run Time Infrastructure), the underlying implementation of HLA, is not an open simulation architecture. RTI is not capable of combining the simulation components in a flexible way to meet various objectives without substantial modification, furthermore it can not support simulation integration continuously.

3 SOA-Based Simulation Framework

3.1 Figures SOA for Simulation Composability

With the problems' complexity increasing, one time simulation experiment always involved many resources, such as tools, data, models and applications. Preparing for one specific simulation experiment means to migrate or modify existing codes in addition to programming new codes, which takes far more time and cost than simulation running. It is a huge challenge that we faced and have to tackle. To achieve more flexibility and higher efficiency in simulation modeling, developing, deploying and integration, we need simulation have the feature of composability, the ability to create, select and assemble simulation components in various combinations. The *composability theory* can explain why HLA cannot provide this ability[3]. Some projects have been made to remedy this problem, such as HLA evolved[4] and webHLA[5].

With the advance in software technologies, there have been no significant breakthroughs in building composable simulation framework. Both theoretical and practical researches have provided the base of simulation composability. Service-Oriented Architecture (SOA) is a solution for simulation composability. SOA isolated the service definition with service implementation and achieved the business abstract by a standard of service interface, so all the service can be integrated and provide reconfigurable service process in a reasonable manner.

Enterprise Service Bus(ESB) is the infrastructure of SOA, which play the role of software bus, it glue all the components compliant with ESB standard together and make them communicate with each other. SOA is propitious to service integration from various domains and it improves the composability to great extent [6].

SOA is suitable for simulation framework not only because SOA is dominating thought for large scale software developing but also SOA has the virtues to meet the new simulation requirements as follows: SOA makes the connectivity possible between various simulation resources, legacy applications and new developed applications can be integrated together seamlessly, components developed by various domains can be deposited and rehearsed for specific simulation contest, so both flexibility and efficiency can be ensured. Recently, some researches based on SOA have being carried on such as[7],[8]. Fig. 1 shows the evolution history of distributes simulation technology.

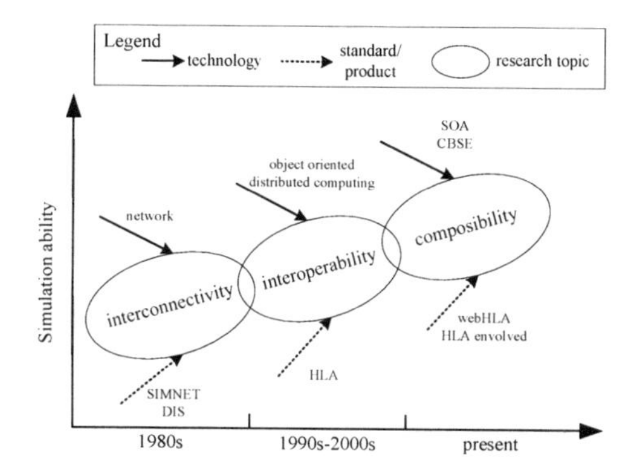

Fig. 1. Roadmap of distributed simulation. This shows that each stage of distributed simulation is characterized by contemporary application requirements and technology background.

3.2 CBSE for Composable Service Implementation

SOA is a thought or philosophy rather than a concrete technology, which guides building composable software architecture. To achieve this goal, CBSE is the best practice. Currently, there are three component-based technologies for distributed computing, that are CORBA, .NET and Java EE/EJB (Enterprise Java Beans).

Limited by the technology background of CORBA and .NET, both of them emphasize on implementation of transparent service and isolation the views of developer and customer, so little considerations are made on components management, combing and configuration. Furthermore, CORBA has the shortcoming of too complexity and .NET is lack of openness and ability to migration among different OS platform because it rely on Windows platform.

Java EE/EJB provides a novel specification for component developing, management, deployment that is in favor of simulation composability. A simulation platform built with Java EE/EJB is shown as Fig. 2.

In this platform, J2EE application server plays the core role which is constructed by three layers as follows.

(1) The underlying software layer includes simulation services and common service. Simulation services are group of specific simulation services, such as time management, event generation and management, entity management and data management. Common services include naming, security, message, data link, management and deployment service.

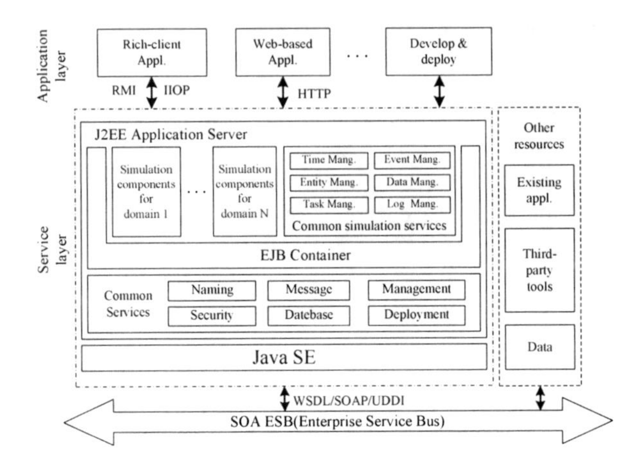

Fig. 2. Simulation framework based on SOA and Java EE/EJB

(2) The upper and the most important layer is the EJB container which is tailored and developed according EJB specification, it provides the life-cycle management for simulation components deployed into it.

(3) The top layer is a group of simulation components deployed into EJB container; these components are developed with EJB specification and provide a well defined service interface for users.

This simulation paradigm has several outstanding characteristics as follows:

(1) One component can be called by several different applications distributed on network at the same time and instances of these components are managed by EJB container automatically. This leads to binary level reusability without any source codes modifying and thus improves the efficiency of simulation resources.

(2) The structure makes a clear interface between developers and users and decreases the load and difficulty for simulation, some of complicated works such as component life-cycle management are moved to EJB container, some of work such as deploying and maintaining can also be done by software tools. So to achieve higher simulation efficiency is possible both for users and developers.

(3) This framework makes the transparent service possible, the style of simulation client applications can be vary from rich-client with GUI, command-line to web-based, furthermore, the location of simulation applications can be local or remote. This improves the flexibility and availability.

This structure provides a general scheme not only for platform independent service request and response but also for components deployment and management. From the users' perspective, simulation activity just means to select components from the deposit and recombine them together for specific instances without necessary to know internal details. From the developer's perspective, simulation activity means to develop the simulation components according to EJB specification.

4 Navy Warfare Simulation Platform: A Case Study

4.1 Model Architecture for Navy Warfare Simulation

Simulation modeling is fundamental work and preconditions for all simulation activities, so reasonable model architecture for warfare simulation must be established before construction of platform. To build this model architecture, two criteria must be followed: *completeness* and *fidelity* [9]. Completeness means that models should cover the domain problems; fidelity means that the simulation result of model should be consistent with real world. Fig. 3 shows the model architecture.

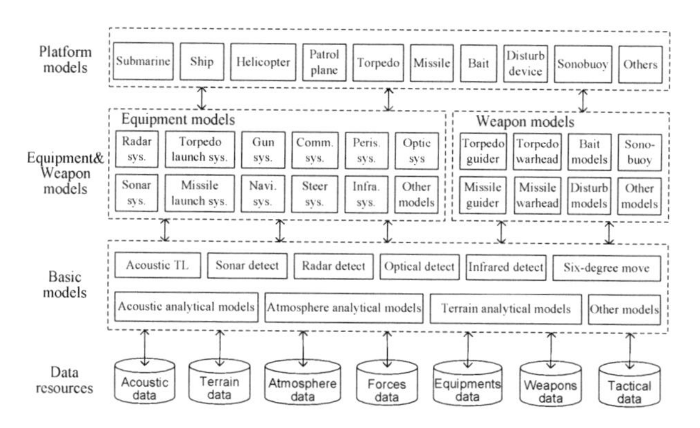

Fig. 3. Model architecture for navy warfare simulation

It is a multi-tier and modular model architecture which involves several model layers for environment, equipment, weapon, forces and others.

Basic models include several types of analytical model for environment, sensors and entity movement. These are fundamental and refined models, so upper layer models can called them and get the acceptable results.

Equipment and weapon models involve models of various devices and weapons equipped by forces. These models simulate the work process according to the performance of equipments and weapons. Each type of equipment and weapon can be parameterized by simulation data resources at the bottom of model architecture.

Platform models involve models of various forces' C4ISR system and status. Each type of force can be constructed by models at lower layer.

Two adjacent layers are isolated from each other, the lower layer provided services for the upper layer. This multi-tier and modular architecture decreases the complexity and is propitious to implementation according to Java EE/EJB.

4.2 Implementation of Simulation Platform

Simulation Components Developing and Deploying. Based on the model architecture, the models can be implemented as components. There are two types of works that are development of simulation service components at server side and development of simulation applications as client side.

According to EJB specification each simulation component must be developed as class which has two parts, one is component *interface class* which defines the service interface; the other is component *definition class* which describes the implementation details. e.g. the interface class of "torpedo" component likes this:

```
@Local
public interface TorpedoLocal { // interface class
  // methods for service declearing
  public boolean init(...);
  public void move(...);
  ...
}
```

The definition class which implemented the methods of "torpedo" component likes this:

```
@stateful
public class TorpedoBean implements TorpedoRemote,
TorpedoLocal {
  protected WEAPONIYPE typeName;     // attributes
  ...
  public boolean init(...) {...}     // methods imple.
  public boolean move(...) {...}     // methods imple.
}
```

The developed components are bundled into files as "*.jar" or "*.war" format and deployed to application server. Components which have been deployed in application server can provide service now. Glassfish version 3.1 is chosen as Java EE application server. Glassfish is a high performance application server in which an EJB container manages the simulation components deployed in it.

Calling Components from Client Application. After developing and deploying of simulation components client-side applications can use the simulation services. Client-side applications can be various and play different roles through calling and rehearsing different services at server side, for example, application can be simulation scenario editor, simulation monitor, data views and so on. There are two ways to call service from client applications.

(1) Lookup and call service using JNDI(Java Naming Directory Interface).

JNDI is a standard interface for component register and lookup, Using JNDI user can lookup and call components deployed in server. An example of component calling using JNDI is shown as follows.

```
Properties props = new Properties();
props.setProperty(...); // set JNDI property
props.setProperty(...); // set server port
InitialContext ctx;
try {
  ctx = new InitialContext(props);
  // lookup service using JNDI
  tor=ctx.lookup("XXX-ejb/Torpedo/remote");
  tor.init(...);  // call service method
  tor.move(...);  // call service method
} catch (NamingException e) {
  out.println(e.getMessage());
}
```

(2) Call service using dependency injection

JNDI is too fussy for programming, so EJB version 3 introduced the concept of dependency injection which uses annotation to simplify the lookup and calling of service components. A piece of code is shown how to use dependency injection to call service as follows.

```
public class Torpedo implements TorpedoRemote,
TorpedoLocal {
  // dependency injection
  @EJB
  private EngineLocal engine;
  @EJB
  private SensorLocal sensor;
  ...
  public boolean init(...) {
    boolen bE = engine.init(...);// call service method
    boolen bS = sensor.init(...);// call service method
    ...
  }
}
```

Using the annotation "@EJB", components are declared as an object instance, thereby the methods of components can be called and provide service. Comparing with JNDI mode, dependency injection can simplify the coding.

Simulation Execution. A general simulation execution process is shown as Fig. 4. It includes several steps as follows.

Step 1. Simulation prepare. In this stage, users edit or load simulation scenario through scenario editor. Scenario is represented as XML format, which include forces, tactical missions, battlefield environment, running conditions and so on [10].

Step 2. Simulation initialization. When simulation start, component instances for initial forces are generated by calling corresponding components and then each force instance creates its own equipments' instances.

Step 3. Simulation running. During this period, force instances run according to the programmed process and various weapon instances will be created when forces fire.

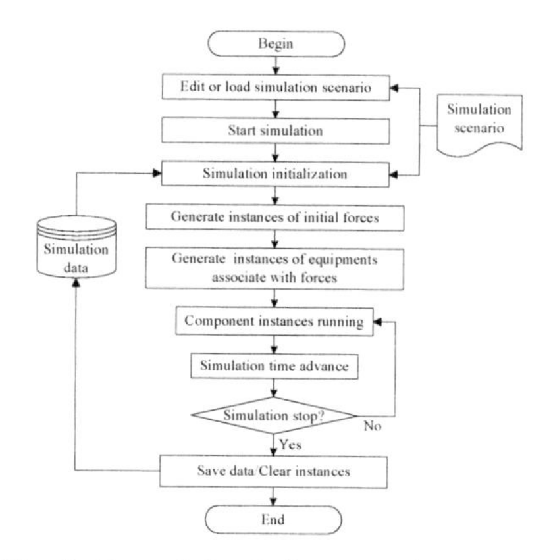

Fig. 4. General simulation execution process

Step 4. Simulation loop decision. When some specific conditions such as simulation timeout, loop times overflow, special events occurring (forces destroyed, weapon exploding) are met, simulation will stop and go to step 5. These conditions are set in simulation scenario in advance.

Step 5. Simulation post-process. When simulation stop some transactions such as data storing, component instances clearing and so on will be done for the next simulation experiment.

The simulation platform also supports multi-task, which means it can perform several simulation tasks either from one or more users during the same period. As mentioned above, client applications invoke request and application server creates instances of simulation components. During simulation execution clients call the services and the Java EE server takes charge of the component instances' life-cycle management such as activate, deactivate and so on automatically. Fig. 5 shows a classic runtime profile of both client and server during once simulation execution.

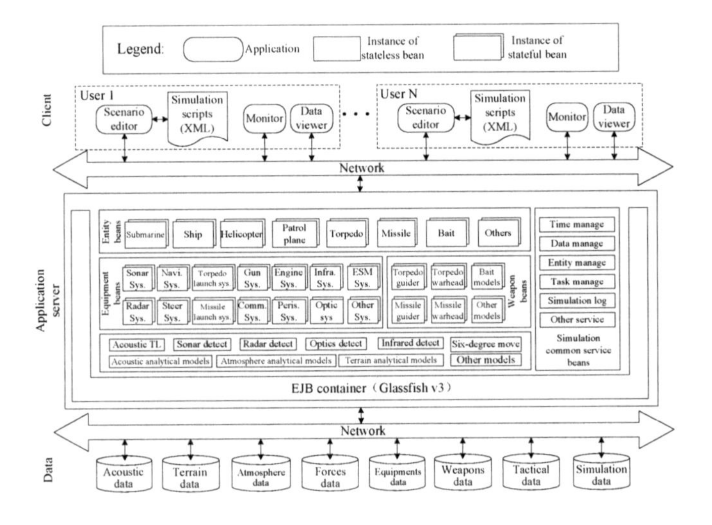

Fig. 5. Classic runtime profile of simulation execution

5 Conclusion

This paper proposes a simulation framework based on SOA, which gives an elegant paradigm for simulation composability. Under this framework existing applications, third party tools, data resources and new developed applications can be integrated seamlessly and constantly.

CBSE is the best practice for achieving simulation composability nowadays, this paper compares the advantages and disadvantages among three popular CBSE technologies. As the comparing result, Java EE/EJB is chosen as the specification for building navy warfare simulation platform. On this platform, both users and developers can do their work through a way of "toy bricks", developers make toy bricks for special simulation functions and users build their simulation scenes through these toy bricks.

Researches and experiments have shown that simulation platform based on CBSE provides the composability, flexibility and high efficiency required to operate in an on-demand simulation environment. Future researches will focus on the performance improvement of simulation platform.

References

1. Fullford, D.: Distributed Interactive Simulation: It's Past, Present, and Future. In: Proc. of 1996 Winter Simulation Conference, pp. 179–185. IEEE Press, Washington, DC (1996)
2. Davis, W.J., Moeller, G.L.: The High Level Architecture: Is There A Better Way? In: Proc. of 1999 Winter Simulation Conference, pp. 1595–1601. IEEE Press, Phoenix (1999)
3. Bartholet, R.G., Reynolds, P.F., Brogan, D.C., Carnahan, J.C.: In search of the Philosopher's Stone: Simulation composability versus component-based software design. In: Proc. of 2004 Fall Simulation Interoperability Workshop, Orlando (2004)
4. Moller, B., Morse, K.L., Lightner, M., Little, R., Lutz, R.: HLA Evolved - A Summary of Major Technical Improvements. In: Proc. of 2008 Fall Simulation Interoperability Workshop. 08F-SIW-064, Orlando (2008)
5. Fox, G.C., Furmanski, W.: WebHLA - An Interactive Programming and Training Environment for High Performance Modeling and Simulation. In: Proc. of 1998 Fall Simulation Interoperability Workshop. 98F-SIW-216, Orlando (1998)
6. Bell, M.: Service-Oriented Modeling: Service Analysis, Design, and Architecture. John Wiley&Sons Inc., New Jersey (2008)
7. Liu, L., Russell, D., Xu, J.: Modelling and Simulation of Network Enabled Capability on Service-Oriented Architecture. Simulation Modelling Practice and Theory 17, 1430–1442 (2009)
8. Tsai, W.T., Cao, Z., Wei, X., Paul, R., Huang, Q., et al.: Modeling and Simulation in Service-Oriented Software Development. SIMULATION:Transactions of the Society for Modeling and Simulation International 83, 7–32 (2007)
9. Shi, Y., Xiao, M.Y.: Design and Implementation of HLA-based Submarine Warfare Simulation Platform. Journal of System Simulation 22, 2106–2109 (2010) (in Chinese)
10. Li, W.B., Shi, Y.: Submarine Operation Simulation Suppostion Based on XML. Computer Simulation 24(3), 104–107 (2007) (in Chinese)

Research on Automatic Recommender System Based on Data Mining

Qingzhang Chen, Qiaoyan Chen, Kai Wang, Zhongzhe Tang, and Yujie Pei

Department of computer science and technology, Zhejiang University of technology, HangZhou 310000, China
Qzchen@zjut.edu.cn, Yujiepei1986@163.com

Abstract. By using ART neural network and data mining technology, this study builds a typical online recommendation system. It can automatically cluster population characteristics and dig out the associated characteristics. Aiming at the characteristics of recommendation system and users' attribute weights, this paper propose a modified ART algorithm for clustering MART algorithm. It makes recommendation system to set the weight value of each attribute node based on the importance of user attributes. The experiment shows that the MART algorithm has better performance than the conventional ART algorithm and can get more reasonable and flexible clustering results.

Keywords: the automatic recommender system, Adaptive Resonance Theory, data mining technology, association rules.

1 Introduction

With the rapid growth of network information, users need more and more time via the internet to obtain information that they are interested in. This gave birth to the information recommendation system. Based on the historical records of users, the recommended system will initiatively recommend the information to them for choosing according to their preferences.

View of the recommendation system [1] [2] to meet the needs of individual users, save users' time of searching information and also can receive timely updated information related with their own preferences, it has been widely used in e-commerce site [3] [4] [5] [6] [7]. These e-commerce sites help users find the goods matching their needs as soon as possible by the way of proposing recommendations product information.

2 The Research Status of Online Recommendation System

The main recommendation technology of online recommendation system can be divided into non-personalized recommendation, attribute-based recommendation, Item-to-Item correlation and People-to-People correlation. The main recommended way of recommendation technology can be divided into content-based filtering and collaborative filtering. Many scholars have conducted research on the recommendation

G. Zhiguo et al. (Eds.): WISM 2011, CCIS 238, pp. 28–35, 2011.

system. In China the more successful recommendation systems are Douban, Dangdang, Taobao, and Jingdong Mall and so on. Typical studies abroad are: Sarwar B. et al in 1998, the GroupLens research system [6] and L. Terveen et al in 1997 proposed PHOAKS system [8].

In this study, the recommended system uses a personalized recommendation mechanism. First, artificial neural network technology is used to produce the user group, then use data mining technology to mine the association rules between the user and items as well as between the items according to the existing transaction data in transaction database. Then get the information recommended rules. At last, based on users' input, determine the user type and return the product. We use ART (Adaptive Resonance Theory) algorithm of artificial neural network technology to form the user clustering. The foreign scholars on the application situation of ART clustering algorithm: Massey, L. proposed the use of ART clustering algorithm in document classification [9]. In 2009 Bhupesh Gour and Sudhir Sharma made the ART clustering algorithm applied in the fingerprint identification system [10]. ART clustering algorithm applied research situation of domestic scholars: BaiLin of Xidian University proposed intrusion detection algorithms based on ART clustering to solve network security problems [11]. Bai Yinglong and Li Chuntao of China Medical University proposed a cluster model of personality characteristics of patients based on ART clustering algorithm [12].

3 The Overall Framework of Recommender Mechanism

The framework of on-line automatic recommender mechanism is shown in Figure 1. This framework relies on the use of ART neural network technology to deal with user's personal information. By analyzing the user's personal properties information and classifying all online users, user type information will be obtained. Apply data mining technology to deal with historical transaction data and users' types, and identify the association rules between the users and the selected items as well as between items, and store them into the knowledge database. When a user is on-line or initiates a service request, the system by identifying the user type information to find the appropriate rules fitting for this user' type, mine the user's interest information and present personalized recommendations information to the user. The processing of the automatic recommender mechanism consists of two phases, namely preprocessing stage and on-line stage.

3.1 The Design of Preprocessing Stage

Preprocessing stage is the analysis of user profile, including the user's attributes and historical transaction data because the information extract from the database may exist format incompatibility. Figure 2 describes the operation flow of the preprocessing phase, it is: Step 1. Select the relevant user data, including user attributes (for example,the user's gender, the user's age, etc.) and historical transaction data (such as transaction ID, user ID and transaction items). Step 2. Design ART classification network: the user property is set to the input vector, classification task based on the ART network is created, and the ART algorithm for classification will be follow-up introduced. Step 3. Enter the experimental value, obtain classification results. Step 4. Use Apriori algorithm for data mining applications. Integrate ART classification results and historical transaction data to calculate the relationship between the user types and items as well as between items. Outputs are their association rules, and stored in the system's knowledge base. Step 5. Create Association Rules Knowledge Base: the results of data mining in the form of rules are stored in the Knowledge Base.

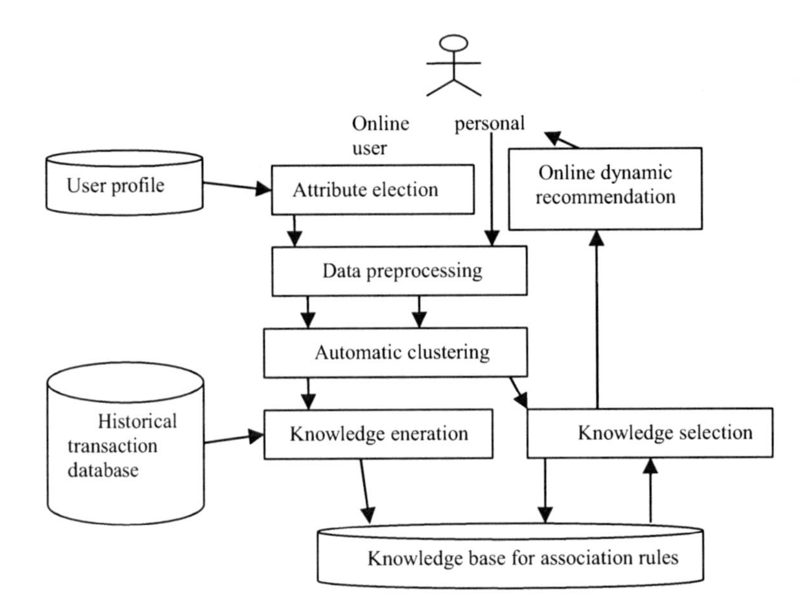

Fig. 1. The framework of on-line automatic recommender mechanism

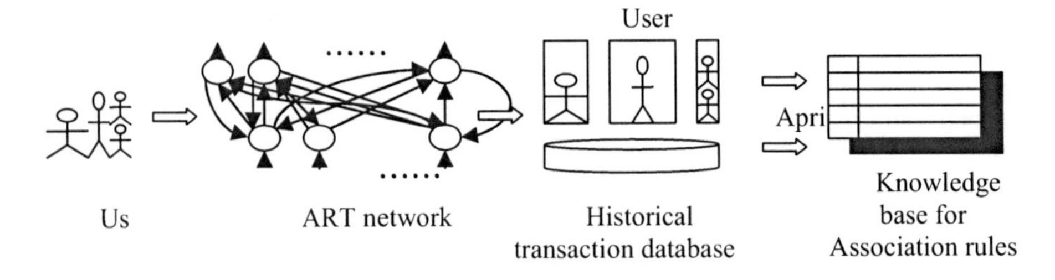

Fig. 2. Automatic recommending process: preprocessing flow

3.2 The Design of On-Line Stage

According to preprocessing, the recommended knowledge is stored in the Knowledge Base. As long as the online user enter the required items, automatic recommendation system will automatically process (shown in Figure 3.) Processes have the following points: (1) users are classified based on user attributes by ART network. (2) System read the knowledge base based on user type and required item. (3) Combine with association rules and extract related items for the recommended.

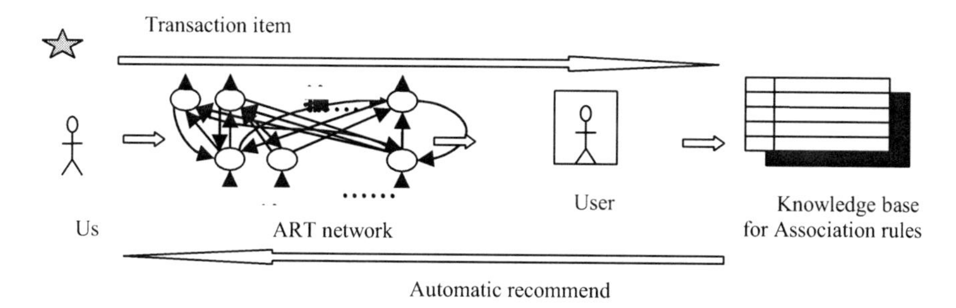

Fig. 3. Automatic recommending process: on-line processing flow

Automatic recommender system recommends the items to the users through the network interface. The number of recommended items is decided by the ranking of confidence level of the extracted rules.

3.3 The Design of Clustering with ART Network

ART technology is used for classifying users into proper clusters. There are three main components of ART network structure: A. input layer: the user attribute is designed as the input vector. Linear transfer function is used to translate the property into a binary vector. B. output layer: output the clustering results of the users. C. network connection layers: input layer and output layer are usually connected by adaptive weights link, which includes bottom-up connections with adaptive weights are called and top-down connections with adaptive weights are called. The clustering steps of ART are introduced as follows:

Step 1: Let $I_1, I_2, ..., I_n$ to be the n chosen attributes of user which is to be designed as the input vector. According to each attribute, the relevant input nodes of each attribute I_n are denoted as $I_n = \left[X_1^n, X_2^n, ..., X_{tn}^n\right]$, where t_n is the number of the network input nodes to the nth attribute and X_{tn}^n Î {0, 1}. The input vector to be used for ART network is, therefore, represented as this: $\left[X_1^1, X_2^1, ..., X_{t1}^1, X_1^2, X_2^2, ..., X_{t2}^2, ..., X_1^n, X_2^n, ..., X_{tn}^n\right]$, where t1+t2+... ,+tn=P and t1,t2,... ,tn≧1.

Step2: Set Nout=1 as initial value. This means only one initial output node is set up for ART network at the beginning.

Step 3: Initialize the weight matrix W, where $W^t[i][1]=1, W^b[i][1]=\frac{1}{1+P}$, P is the number of input nodes. $W^t[i][1]$ is the adjustable weight connected from output layer (one node at the beginning) down to input layer. $W^b[i][1]$ is the adjustable weight connected form input layer up to the output layer.

Step 4: Input one training vector to the input layer of ART network.

Step 5: Determine the activation output, i.e., to derive net [j]: $net[j]=\sum_{i=1}^{P} W^b[i][j] \bullet X[i], 1 \le j \le Nout$.

Step 6: Find the maximum activation net [j*]: $net\left[j^*\right] = \max_j net\left[j^*\right], 1 \le j \le Nout.$

Step 7: Calculate the similar value V j*,: $\|X\| = \sum_{i=1}^{p} X[i]$

(1) $\left\|W_{j^*}^t \bullet X\right\| = \sum_{i=1}^{p} W^t[i]\left[j^*\right] \bullet X[i]$

(2) $V_{j^*} = \dfrac{\left\|W_{j^*}^t \bullet X\right\|}{\|X\|}$

Step 8: Perform the vigilance test. If V_{j^*} is less than the pre-defined vigilance value r that means the pattern is not close related to the output category and, hence, next output node comparison is needed. If V_{j^*} < r, go to Step 9, else go to Step 10.

Step 9: Test whether there is another similar output unit.
If Icount < Nout, Try next maximum activation net [j] of output layer unit, set Icount = Icount + 1 and net [j*] = 0, go to Step 6. Else if there is no other output unit for testing, then

(1) Generate new category: Let Nout = Nout + 1, and create new connected weight

$$W^t[i][Nout] = X[i], W^b[i][Nout] = \frac{X[i]}{0.5 + \sum_{i=1}^{p} X[i]} \tag{1}$$

(2) Set the output value. If $j = j^*$, then $Y[j] = 1$, else $Y[j] = 0$.
(3) Go to Step 4, input a new vector X.
Step 10: If V_{j^*} > r, i.e., pass the vigilance test is passed, then

$$W^t[i]\left[j^*\right] = W^t\left[j^*\right] \bullet X[i], W^b[i]\left[j^*\right] = \frac{W^t[i]\left[j^*\right] \bullet X[i]}{0.5 + \sum_{i=1}^{p} W^t[i]\left[j^*\right] \bullet X[i]\left[j^*\right]} \tag{2}$$

(1) Adjust the weighted value with
(2) Set the output value. If j= j *, then $Y[j]$=1, else $Y[j]$= 0.
Step 11: If there is no more clustered generated and the learning cycle has converged, then output the result of classification. Otherwise go to Step 4 for next cycle learning.

3.4 The Modified ART Algorithm (MART)

As we discussed before, the conventional calculation of similar value only consider the "1" role, did not consider the "0" role in top-down connections adaptive weight. However, in practical application, 0 and 1, respectively, are two states and useful information in determining. They are equally important. Moreover, the importance of each attribute of the user is different, so add the weight of the user attributes as the calculation parameters; we will modify the similarity values calculated as:

$$V[j]=\sum_{i=1}^{p}W[i][j^*]\bullet x[i])\bullet\frac{M[i]}{\sum_{i=1}^{p}M[i]}+\sum_{i=1}^{p}1-W[i][j^*])\bullet(1-x[i])\bullet\frac{M[i]}{\sum_{i=1}^{p}M[i]}(4) \quad (3)$$

$M'[i]$ means the weights of input attribute, which is the importance of the ith node, and $M'[i]\in R^+$. This method can accurately compare the two vectors, but do not exist the pros and cons issue. We can use example to verify the improved similar values computing formula. Let p =10, the template vector be $W'_{j^*}=001111$, and the input vector be $x[i]=001000$. According to the formula (3), the similar value is"1", but the two vectors are clearly not similar. It will bring the error to determine the actual classification. The calculation of similar value using the formula (4) and three different sets of weights respectively is shown as below:

a) Let $M'[i]=[1,1,1,1,1,1]$, and the V is: $V=1*\frac{1}{6}+1*\frac{1}{6}+1*\frac{1}{6}=0.5$.

b) Let $M'[i]=[1,1,1,0,0,0]$, and the V is: $V=1*\frac{1}{3}+1*\frac{1}{3}+1*\frac{1}{3}=1$

c) Let $M'[i]=[1,1,1,4,4,4]$, and the V is: $V=1*\frac{1}{15}+1*\frac{1}{15}+1*\frac{1}{15}=0.2$

Through the above three cases, we find the calculated result of similar values is not the same. In the first case, all nodes have the same importance, get the similar value 0.5. This similar value is more reasonable than the similar value 1 that the original formula gets. In the second case, the first three nodes are set the same weight, and are different with the other nodes. The output valuc of the similar value is 1, which means that the input pattern and the template vector are similar. In the third case, the last three nodes will be set to a higher weight. This means that the last three nodes are more important than the other nodes. Because the input vector and the three nodes of template vectors are the same, so the similar value is smaller than that in the first cases.

The implementation steps of new algorithm MART (Modified ART) is similar to the ART algorithm, but firstly need to set the importance of each node (attribute),i.e., $M'[i]$. According to the similar user properties, form user groups using MART algorithm. This result can be used in the course of the next mining process.

3.5 Generate Strong Association Rules

Firstly, obtain frequent item sets using the Apriori algorithm, then according to frequent item sets to generate strong association rules. The thumb of rule is $X\Rightarrow Y$, and so we can calculate the confidence Level event that when X occurs, Y also occurs. If it has the set minimum confidence Level, then the strong association rules will be set up. The confidence Level is calculated by: $confidence(X\Rightarrow Y)=P(Y|X)=support(X\cup Y)/support(X)$ where $X\cap Y=\phi$. The specific process is: Let each frequent item set be l_i. X is the subset of this frequent item sets and $(l_i-X)\cap X=\varnothing$. $support(X)$ means that the transaction history contains the number of item X. $support(l_i)$ is the number of l_i included in the transaction history, so the confidence Level of $X \Rightarrow (l_i - X)$ is $support(l_i)/support(X)$, if its value is equal to or greater than the set minimum confidence Level, it is the strong association rules. And so on, until all of the strong association rules are generated, which are the recommended rules between the user types and items as well as between items, and deposit them into the knowledge base, which are used for online recommendation stage of the system. Online recommendation stage is: when a user is online, the system process and obtain user type information by the ART algorithm, then preprocessing stage draw strong association rules, which are recommended rules, recommend items associated with the user types to the user.

4 Conclusions and Further Work

This study combines artificial neural network technology and data mining, and presents a personalized automatic recommendation system. Aim at the "XNOR" state of property vector to be considered and the weight problem of properties, the classical ART algorithm will be modified and get MART algorithm. And create a user-oriented book recommendation system, which produce the user clustering using MART algorithm and each user's type, then use the Apriori algorithm for mining strong association rules between the user type and books as well as between books, i.e., book recommended rules, it contains each user's interest information. At last according to the entered information and the user type, recommend the appropriate related items or information to the user. The advantages of this book recommended mechanism are: 1. Community features: use ART network to cluster user communities, and then recommend appropriate book information to each different community. 2. Association feature: using Apriori algorithm of data mining can find the strong association rules between the user community and books, and between books and books. 3. High efficiency of user clustering: when the ART network clustering is completed, the online system can recommend the appropriate information to the user, so it is very fit for dynamic changes and rapid response environment of network.

In addition, this book recommended mechanism is a prototype framework; the data mining algorithms is the typical Apriori algorithm. In order to improve the performance of Apriori algorithm, there are many variants of further improvement and expansion of Apriori [13]. Therefore, the performance optimization of Apriori algorithm can be identified for another future research direction of this article.

References

1. Sarwar, B., Karypis, G., Konstan, J., Riedl, J.: Analysis of Recommendation Algorithms for E-commerce. In: ACM Conference on Electronic Commerce, pp. 158–167 (2000)
2. Yu, P.S.: Data Mining and Personalization Technologies. In: The 6th International Conference on Database Systems for Advanced Applications, pp. 6–13 (1999)
3. AltaVista (2002), `http://www.altavista.digital.com`
4. Amazon (2002), `http://www.amazon.com`
5. Hill, W.C., Stead, L., Rosenstein, M., Furnas, G.: Recommending and evaluating choices in a virtual community of use. In: Proceedings of CHI 1995, pp. 194–201 (1995)

6. Konstan, J., Miller, B., Maltz, D., Herlocker, J., Gordon, L., Riedl, J.: GroupLens Applying Collaborative Filtering to Usenet News. Communications of ACM 40(3), 77–87 (1997)
7. Shardanand, U., Maes, P.: Social Information Filtering: Algorithms for Automating 'Word of Mouth. In: Proceedings of the Computer-Human Interaction Conference, CHI 1995 (1995)
8. Editmax (2009), http://www.editmax.net/n1229c15.aspx
9. Massey, L.: On the quality of ART1 text clustering. Neural Networks, 771–778 (2003)
10. Gour, B., Bandopadhyaya, T.K., Sharma, S.: High Quality Cluster Generation of Feature Points of Fingerprint Using Neutral Network. EuroJournals Publishing, 13–18 (2009)
11. Bailin: Research on intrusion detection system based on neural computing and Evolution Network. Xi'an University of Electronic Science and Technology (2005) (Chinese)
12. Bai Y.-l., Li C.-T.: Design of characteristics in patients with cluster model. Chinese General Medical (2007) (Chinese)
13. Zheng, L.-y.: Trie-based algorithm of mining association rules. Journal of Lanzhou University of Technology 30(5), 90–92 (2004) (Chinese)

AVS Fast Motion Estimation Algorithm Based on C64x+ DSP

Liu Yan-long, Li Fu-jiang, and Zhang Gang

College of Information Engineering, Taiyuan University of Technology,
Taiyuan, China
ly10050@163.com

Abstract. Motion search is one of the most computationally complex modules for AVS. The algorithms with high degree of computing regulation and parallel potential are good for DSP optimization. Based on the C64x+ DSP characters, the paper proposes square-step early termination search algorithm. The Very-Long-Instruction-Word (VLIW) architecture and signal instruct multiple data (SIMD) is widely used for DSP. Based on the VLIW and SIMD, the paper analyzes the optimization methods to accelerate the speed of motion search. The AVS MAD modules are optimized with assembly instruction set and Cache optimization. The experiments show that the total instruction cycle of assembly code is 38.53% to 50.53% of the one consumed by the original c code with compile optimization for different block MAD modules.

Keywords: AVS, motion search, VLIW, SIMD.

1 Introduction

AVS is the latest video coding standard established by China [1]. It employs several powerful coding techniques such as intra coding with several different spatial prediction directions, variable block-size motion estimation, 8x8 integral discrete Cosine transform, context-adaptive variable length codes, loop filter, and so on[2].

Based on the block matching video coding system, motion search is one of the most computationally complex modules. The full search algorithm is best if measured in the predicting error, and has features such as fixed length step, which is easily designed into parallel lines processing ASIC. But the computation quantity of full search algorithm is large, it calculate matching result of all the points within the search area to choose the best matching point. Therefore there are many fast motion search algorithm such as lozenge search algorithm (DS) [3], hexagon search algorithm (HS) [4], the cross lozenge search algorithm (CDS) [5] [6]and cross lozenge hexagon search[CDHS] [7], which greatly reduce the search complexity, the accuracy of motion estimation and video quality.

The adjacent blocks mostly belong to the same moving object with similar movement, and the algorithms with high degree of computing regulation and parallel potential are good for DSP optimization, so the paper proposes square-step early termination search algorithm. Based on the VLIW and SIMD of DSP, the

G. Zhiguo et al. (Eds.): WISM 2011, CCIS 238, pp. 36–43, 2011.

paper uses the method assembly instruction set and Cache optimization to accelerate the speed of motion search.

2 The Square-Step Early Termination Search Algorithm

From the features of the motion stadium in video, adjacent blocks mostly belong to the same moving object with similar movement that is relevant motion vector (MV). Table 1 and Table 2 shows motion vector residuals (MVd) distribution probability obtained by the full search algorithm in the typical video sequences of the CIF and D1 resolution respectively. From the table 1, for the CIF sequence, about 73% of the adjacent block MVd less than 1, while about 74% of the adjacent block MVd for the D1 less than 1; For CIF sequences, about 90% of the adjacent block MVd less than 5, while about 91% of the adjacent block MVd for D1 less than 5. MV has a strong correlation in the space. Therefore, in the process of motion search, select the motion vector prediction value firstly, and then search around the corresponding points of the predictive value to get the final motion vector.

In addition to predict mode has a strong correlation between adjacent macroblocks, the rate distortion optimization also has a strong correlation between adjacent macroblocks. Based on the current encoding macroblock, the modes of top, upper right, left is respectively modA, modB, modC, and the RDO of top, upper right, left is respectively JA, JB, and JC. Jmax is Maximum for the JA, JB, JC, Jmaxab is Maximum for JA and JB, Jmaxac is Maximum for JA and JC, and Jmaxbc is Maximum for JB and JC. JD is the RDO for the current macroblocks. All-zero blocks of the decision threshold is set to Jmax0. Set the decision threshold TH as follow

$$TH = \begin{cases} J_{\max} + 256, \bmod_A = \bmod_B = \bmod_C \\ J_{\max ab} + 128, \bmod_A = \bmod_B \neq \bmod_C \\ J_{\max ac} + 128, \bmod_A = \bmod_C \neq \bmod_B \\ J_{\max bc} + 128, \bmod_B = \bmod_C \neq \bmod_A \\ J_{\max 0}, \bmod_B \neq \bmod_C \neq \bmod_A \end{cases} \quad (1)$$

To improve accuracy and reduce the decision error rate, the decision threshold set threshold TH as follow

$$TH = \begin{cases} \alpha_0 + \lambda\beta_0, TH > \alpha_0 + \lambda\beta_0 \\ \alpha_1 + \lambda\beta_1, TH < \alpha_1 + \lambda\beta_1 \\ TH, others \end{cases} \quad (2)$$

λ is the Lagrange multiplier. By different sequences for testing with different QP to get experience value, α_0 is 1024, α_1 is 480, β_0 is128 and β_1 is 32.

VLIW describes instruction set idea, in which the compiler combines much simple, independent instruction into a single instruction word [8]. VLIW extract the highly parallel instruction data from the application program, and allocate these machine instructions evenly to the many execution units in chip. The characteristic of the DSP support VLIW is that an instruction execute several operations which can be realized as one function (such as integer add, subtract, multiply, shift, read and write memory, etc.) of RISC instructions. C64 + has two data processing data path A and B, each channel has four functional units (L, S, M and D) and the registers which contains 32 32-bit registers. Functional unit execute logic, shift, multiplication, addition, and data addressing and other operations. Two data addressing units (.D1 and .D2) transfer data between registers and memory. 4 functional units for each data path connect a single data bus to the registers on the other side of the CPU in order to exchange data for registers on both sides. Each 32-bit instruction takes a functional unit. C64x + CPU can generally combine 8 instructions into a single instruction word. There is no doubt that fully parallel execution of 8 instruction is most efficient, so try to get the eight functional units of A, B group execute in parallel as far as possible.

In signal processing and other compute-intensive code, packaging data processing is a powerful method of internal parallelism while maintaining the code density. Many signal processing functions perform the same operation on many data elements and these operations are usually independent [9]. For packaging data processing, programmers can operate on the data in a single compact instruction stream which reduce the code size and improve efficiency of the implementation. DSP packaging data processing is the same type of processing on multiple independent data using a single instruction. C64x + provides rich handling instructions for the packaging data processing and supports double word read and store instructions, which you can visit the 64-bit data once. Per clock cycle issue up to two dual-word load or store instructions. C64x + gives the peak bandwidth of 128 bits per cycle access to on-chip memory. The packaging data type is the cornerstone of C64x + packaging data processing, and each packaging data type packs more data unit into a 32-bit general purpose registers which support the highest level of 8-bit unsigned and 16-bit signed number.

From DSP features above, the fast algorithm by reducing the search points is not fit for the DSP optimization, while algorithms which have high degree of computing regulation and parallel potential is good for DSP optimization algorithm. This paper proposes an early termination of the square-step search algorithm and specific steps as follow.

- Initialize the number of search steps N0, according to the MV situation around the block, predicts the current block MVP, according to the scope of the corresponding point (-4, +3) of the predictive value determine the reference block.
- Calculate the value of rate distortion optimization within the scope of$\pm$ 4 reference block, to determine the minimum value of SAD, and the best MVd at the same time
- According to the formula (1) and (2) to determine the threshold TH
- Judge the size of MVd, if MVd = 3 or MVd =- 4, skip to Step 5); if MVd $\leq$ 3 and MVd $\geq$ -4 and SAD $\leq$ TH, skip to Step 6), otherwise go to Step 5 .)
- Begin a starting forecast point with the new best MVd, in the range of the corresponding point (0,8) to the predictive value determine reference block, then jump to step 4); and number of search steps N increase1, if N $\geq$ 8, skip to Step 6)
- Terminate the motion search, this is the best MVd.

Table 1. Probability of MV(CIF)

Sequence		0≤ MVd < 1	1≤ MVd < 2	2≤ MVd < 3	3≤ MVd < 4	4≤ MVd < 5	MVd≥5
foreman	MVx	65.38	15.79	5.92	3.17	1.73	8.00
	MVy	66.83	17.42	5.26	2.64	1.43	6.42
bus	MVx	75.04	4.01	0.76	1.21	1.43	17.54
	MVy	85.37	10.27	0.99	0.34	0.20	2.82
ice	MVx	78.04	5.54	2.58	2.09	1.08	10.66
	MVy	76.92	6.75	2.87	2.06	1.07	10.31
soccer	MVx	65.73	14.57	4.93	2.70	1.26	10.79
	MVy	73.48	11.62	2.14	1.47	0.90	10.39

Table 2. probability of MV(D1)

Sequence		0≤ MVd < 1	1≤ MVd < 2	2≤ MVd < 3	3≤ MVd < 4	4≤ MVd < 5	MVd≥5
foreman	MVx	67.94	12.81	5.79	3.53	1.69	8.23
	MVy	69.72	14.33	5.14	2.97	1.33	6.49
bus	MVx	77.61	5.51	1.30	1.57	1.56	4.45
	MVy	74.60	4.35	0.84	1.29	1.41	17.50
ice	MVx	84.65	10.55	1.05	0.47	0.21	3.06
	MVy	81.13	4.73	2.34	1.75	0.87	9.18
soccer	MVx	65.73	14.57	4.93	2.70	1.26	10.79
	MVy	73.48	11.62	2.14	1.47	0.90	10.39

3 Cache Optimization

DM6446's store memory includes chip inside and outside two parts: C64x + DSP apply an advanced internal memory structure with two levels: program memory (L1P) of first level is 32KB and the data memory (L1D) is 80KB; internal memory (L2) of second level is 64KB[10]. Internal memory can also be divided Cache and common data space SRAM according to our need. The access to on-chip memory can be up to 600MHz (the same to the CPU), but the total size is only 176KB. The maximum DDR2 memory out of chip is 256MB, but the access rate is only 160MHz, so this is a key point in real time that storage space of DSP must be allocated reasonable to make the program running quickly.

The Cache reduces costs while improving the processing speed. The CPU starts to look up data you need from the Cache, remove the data directly from the Cache [11] if it does. Cache miss mean that the data of the current processor is not in Cache, start the access to memory at the same time. During the Cache miss for pause time, CPU stands still and can not be executed instructions. Similar to the background independent of the data processing operation [12], the Cache continues to load the follow-up data of the Cache block after the CPU resumes.

Cache miss includes L1D Cache miss, L2 Cache miss and L1P Cache miss. These cache miss can be divided into three categories: one is compulsory miss, that is the failure due to access to a data for the first time; the other is capacity miss, that is the data visited is not in Cache because the workload of the CPU outride the size of the Cache, the last is conflict miss, that is the visited data in Cache is replaced by another data CPU cross-refers.

The performance of cache is greatly dependent on the re-use of cache lines, which is also the main purpose of optimization. We generally layout the data and code memory appropriately, and adjust the order of the CPU's memory access to achieve this goal, to decrease the cache miss and increase rate of Cache hit. Cache optimization includes procedures and data Cache Optimization.

1) Program Cache Optimization

The AVS coding algorithm code can be divided into two parts: the key execution code and initialization / rare event code. When cache optimized, only the key execution code is frequently referred, thereby remove the rare events among them to reduce the size of the dynamic storage area and the L1P conflict miss. According to the order of key execution function in program to adjust its position, and put the code frequently used together, Make the code assigned to consequent storage space to improve rate of instruction Cache hit. For the relatively large function code, the algorithm can be divided into several small pieces suitable for the size of program cache to eliminate the capacity miss in L1P.

2) Data Cache Optimization

Improving data Cache hit rate is actually improving the reuse of data, using the four methods.

- All data of the working set fits into cache (no capacity misses by definition), but conflict misses occur. It can be eliminated by allocating data contiguously in memory.
- Thrashing is caused if more than two read misses occur to the same set evicting a line before all of its data was accessed. Provided all data is allocated contiguously in memory, this condition can only occur if the total data set accessed is larger than the L1D capacity. These conflict misses can be completely eliminated by allocating the data set contiguously in memory and pad arrays as to force an interleaved mapping to cache sets.
- In this read miss scenario, data is reused, but the data set is larger than cache causing capacity and conflict misses. These misses can be eliminated by splitting up data sets and processing one subset at a time. This method is referred to as blocking or tiling.

The L1D write buffer can be the cause for additional stalls. Generally, write misses do not cause stalls since they pass through the write buffer to the lower level memory. However, the depth of the write buffer is limited to four entries. To make more efficient use of each 128-bit wide entry, the write buffer merges consecutive write misses to sequential addresses into the same entry. If the write buffer is full and another write miss occurs, the CPU stalls until an entry in the buffer becomes available. Also, a read miss causes the write buffer to be completely drained before the miss is serviced. This is necessary to ensure proper read-after-write ordering.

4 Assembly Instruction Set Optimization for MAD

The Very-Long-Instruction-Word (VLIW) architecture developed by Texas Instruments (TI) is supported by the C64/C64x+ DSP. It is very helpful to process data more efficiently. On the C64x CPU a fetch packet will always be eight instructions. On the C64x+ CPU this may be as many as 14 instructions due to the existence of compact instructions in a header based fetch packet. Fetch packets are always 256 bits wide. In most cases, the CPU fetches VLIW (256 bits wide) to supply up to eight 32-bit instructions to the eight functional units during every clock cycle. But not all of the eight instructions can be executed in parallel. The first bit of every 32-bit instruction determines if the next instruction belongs to the same execute packet as the previous instruction, or whether it should be executed in the following clock as a part of the next execute packet. So there are three patterns of execution sequences for the eight instructions: fully serial, fully parallel, and partially serial. Absolutely, the second pattern is the best. Pipeline resolves this problem.

Single instruction multiple data (SIMD) is another important technique that supported by the C64/C64x+ DSP. On the TMS320C64x DSP, there are eight highly independent functional units. Six ALUs (32-/40-bit), and each supports single 32-bit, dual 16-bit or quad 8-bit arithmetic per clock cycle. Two multipliers support four 16 x16-bit multiplies (32-Bit Results) per clock cycle or eight 8 x 8-bit multiplies (16-bit results) per clock cycle. There are many SIMD instructions in the C64/C64x+ DSP instruction set. For example, LDDW can load double word from memory one time. Suppose the data are 8-bit wide, it can load four data one time. Another example, ADDSUB2 let two sets of two 16-bit operands do addition and subtraction simultaneously. So we can use the SIMD instructions to operate multiple data according to programmers' needs flexibility and these data can do the same or different operation at the same time.

AVS motion search is the most time-consuming module for AVS encoder. It needs to use assembly instruction set optimization. We make use of VLIW and SIMD of C64x+ DSP to improve motion search module. The purpose of motion search is to obtain motion vector MV and the minimal value of rate distortion optimization, and the assembly instruction set MAD is implemented. The MAD assembly code of 16 × 16 block is finished with two circular strategies. The inner loop in parallel finds the minimal RDO of two lines and its coordinates. The outer loop controls the number of loop for the horizontal and vertical search. Note that, because the inner loop is based on two lines parallel search, so the number of rows MAD must be a multiple of 2. Because the location of reference data can not be determined, so it can only use non-aligned reading instruction LDNDW. LDNDW read out eight refImg and 8 srcImg, and then calculate the absolute value of residuals between refImg and srcImg with SUBABS4. Do the dot product operation between the absolute value of residuals and 0x010101 with the instruction DOTPU4. So the sum of the absolute value of residuals for the total line can be obtained. Finally, the sum of the absolute value of residuals for the eight lines can be added to the current block. The smallest RDO and its coordinates can be obtained by constantly shifting the search range and compare the different RDO. The optimization for block 16x16, block 16x8 and block 8x16 is similar to block 8x8. The key code is implemented as follow.

```
mad8x8_loop:
   ADD    .D2  B_hvl, 1, B_hvl
|| DOTPU4 .M1  A_err0l, A_k_one, A_row_0
|| ADD    .S2  B_row_2, B_mad_3, B_mad_2
|| DOTPU4 .M2  B_err3h, B_k_one, B_row_3
|| SUBABS4 .L1  A_src5l, A_ref5l, A_err5l
|| SUBABS4 .L2X  B_src5h, A_ref5h, B_err5h
|| LDNDW  .D1  *A_ref_d(A_p7), A_ref7h:A_ref7l
|| MV     .S1  A_p1, A_f
loop_1:
   ADD    .D2X  A_row_0, B_mad_0, B_mad_0
|| ADD    .S2   B_row_1, B_mad_2, B_mad_1
|| ADD    .S1  A_row_3, A_mad_4,  A_mad_3
|| SUBABS4 .L1X  A_src4l, B_ref4l, A_err4l
|| SUBABS4 .L2   B_src4h, B_ref4h, B_err4h
|| DOTPU4 .M2   B_err5h, B_k_one, B_row_5
|| DOTPU4 .M1   A_err6l, A_k_one, A_row_6
|| LDNDW  .D   *A_ref_d(A_p6), B_ref6h:B_ref6l
loop_2:
  [A_i] BDEC .S1  mad8x8_loop, A_i
|| ADD    .S2X  B_mad_0, A_mad_2, B_mad
|| SUBABS4 .L2  B_src2h, B_ref2h, B_err2h
|| SUBABS4 .L1  A_src3l, A_ref3l, A_err3l
|| DOTPU4  .M2 B_err4h, B_k_one, B_row_4
|| DOTPU4  .M1 A_err5l, A_k_one, A_row_5
|| LDNDW   .D *A_ref_d(A_p3), A_ref3h:A_ref3l
|| SUB    .D2  B_v1, 1,  B_v1
```

5 Experimental Results

The experiment is carried out through simulation on CCS3.3. All tests in the experiment are run on the Pentium Dual E2140 1.60GHZ with 1.93GB DDR2 and the operation system is Microsoft Windows XP. The experiment results are listed in the following tables:

Table 3. MAD module optimization experiment result

module	Original C	*assembly code optimization*
16x16	5754	*2452*
16x8	3706	*1428*
8x16	3514	*1428*
8x8	*1134*	*573*

As can be seen from above Table, after the optimized assembly-level, the MAD module cycles of 16x16 block is 42.61% of clock cycles required for C-level optimization; the MAD module cycles of 16x8 block is 38.53% of clock cycles required for C-level optimization; the MAD module cycles of 8x16 block is 40.64%of clock cycles required for C-level optimization; the MAD module cycles of 8x8 block is 50.53% of clock cycles required for C-level optimization.

6 Conclusions

The paper studies the optimization method for AVS motion search which is time-consuming module. Combing the feature of VILW and SIMD on C64x+ DSP, the paper proposes square-step early termination search algorithm. The AVS MAD modules are optimized with assembly instruction set and Cache optimization. The experiments show that the total instruction cycle of assembly code is 38.53% to 50.53% of the one consumed by the original c code with compile optimization for different block MAD modules.

Acknowledgments. It is a project financed by Chinese National Natural Science Foundations (60372058, 60772101).

References

1. Audio Video coding Standard Workgroup of China, GB/T20090.2-2006. Information technology - Advanced coding of audio and video - Part 2:Video (2006) (in Chinese)
2. Fan, L., Ma, S., Wu, F.: Overview of AVS Video Standard. In: Proc. 2004 IEEE Intl. Conf. Multimedia & Expo., pp. 423–426 (2004)
3. Zhu, S., Ma, K.K..: A new diamond search algorithm for fast block matching motion estimation. In: Proceedings of International Conference Information, Communication and Signal Processing, vol. 1, pp. 292–296 (1997)
4. Zhu, C., Lin, X., Chau, L.: Hexagon-Based Search Pattern for Fast Block Motion Estimation. IEEE Transactions on Circuits and Systems for Video Technology 12, 349–355 (2002)
5. Zhu, C., Lin, X., Chau, L.: Enhanced Hexagon Search for Fast Block Motion Estimation. IEEE Transactions on Circuits and Systems for Video Technology 10, 1210–1214 (2004)
6. Cheung, C.H., Po, L.M.: A novel cross-diamond search algorithm for fast block motion estimation. IEEE Transactions on Circuits and Systems for Video Technology 12(12), 1168–1177 (2002)
7. Cheung, C.H., Po, L.M.: Novel cross-diamond-hexagonal search algorithms for fast block motion estimation. IEEE Transactions on Multimedia 7(1), 16–22 (2005)
8. TMS320C64x/C64x+DSP CPU and Instruction Set Reference Guide, SPRU732g (February 2008)
9. TMS320C6000 Assembly Language Tools User's Guide, SPRU186n (April 2004)
10. Texas Instruments Incorporated, TMS320C64x DSP Two-Level internal Memory (SPRU610) (2002)
11. Texas Instruments Incorporated TMS320C64x+ DSP Cache User's Guide (SPRU862) (October 2006)
12. Texas Instruments Incorporated, Using Cache Tune to Improve Cache Utilization on TMS320C6000 Targets, SPRAA01 (2004)

Intelligent Decision Support System for Exercise Prescription: Based on Rough Sets

Fing Beng*

Shandong University of Finance. 38, Shungeng rode, Jinan, China
fbfyh@163.com

Abstract. This paper presents Intelligent Decision Support System (IDSS) for exercise prescription to cure obesity, through literature review, field survey with the application of rough sets theory. The gross structure and functional structure of this system are addressed in this paper. 300 samples are collected and are classified into two groups with 220 samples being training sample and 80 samples being test sample. The experiment is conducted based on these two sample groups and results are acquired. The IDSS for exercise prescription can analyze treatment protocols of various obese patients and propose some practical models and offer references to perfect exercise prescriptions for different obese patients.

Keywords: rough set, exercise prescription, obesity, decision support, system.

1 Introduction

Newly developed learning and reasoning technologies in computer science have long been experimented in the area of sport. This is not only because there are many practical application systems in that area but also because abundant factors in the sports area need to be taken into special consideration, which make it very challenging technologically.

Traditional models for artificial intelligence are knowledge-intensive, so much so that domain knowledge in kinesiology, health science and sports physiology can only be obtained from specialists in those domains. Furthermore, those knowledge when processed, needs decoding artificially, and this top-down way of modeling construction is extremely complicated, laborious and those models are hard to maintain. The greatest obstacle of this modeling lies in the acquisition of trivial knowledge. With the wide application of computer technology in sports rehabilitation, a large quantity of corresponding information has been collected, and people begin to prefer a bottom-up way of modeling through which models are generalized and synthesized from the underlying data rather than transcendental realms. However, manual inspection and analysis are no longer applicable in large data base processing; hence it has become our top priority to develop tools and technologies that are helpful in obtaining undiscovered interesting models from data bases. Nowadays query tools in database management system can accomplish simple tasks such as acquiring data about the

* Corresponding author.

G. Zhiguo et al. (Eds.): WISM 2011, CCIS 238, pp. 44–50, 2011.

number and the average age of people who suffers obesity, yet intelligent data analysis has to be applied to some advanced queries such as what are the main factors influencing obesity treatment outcome and how these factors interact with each other.

At present, application of computer technology in sports medicine mainly exists in detection, laboratory test, and diagnosis as well as information management systems. But computer technology hasn't been employed for rehabilitation which is of great importance in sports medicine. As a matter of fact, even though some patient has been diagnosed with certain disease, the treatment outcome varies with differences in age, physique, rehabilitation degree, medical history and even sex. Some rehabilitation project which is effective for certain patient may be less effective or ineffective at all for other patients. Therefore it is worthwhile to study how to choose rehabilitation projects according to different patients so as to control illness quickly and efficiently.

Obesity is a medical condition in which body fat especially triglyceride has accumulated to such a great extent that it leads to apparent overweight and excessively thick adipose layer. Obesity is caused by a combination of excessive dietary calories and change of metabolism. It has an adverse effect on health increasing the risk of cardiovascular disease and cancer; affecting digestion and internal secretion functions; leading to soft tissue injuries, decreased fertility and mental disorders.

Obesity is a common disease severely impairing human health and according to the data base of in a rehabilitation center, the majority of patients there suffer from this disease. The present paper chooses this case hoping to analyze various obesity treatment projects according to the given original data, acquire some valuable models and offer reference so as to optimize various obesity patients' exercises prescriptions.

2 Gross Structure of the System

Traditional decision support system consists of human-computer interaction system, data base, pattern store, methods and its corresponding management system. Knowledge base and inference mechanism, the core component of IDSS, can only be introduced by decision support system. Because intelligent decision support system of rough sets data analysis (RSDA-DSS) changed the traditional way of acquiring knowledge in IDSS, its structure also varied accordingly.

As is illustrated in figure 1, RSDA-DSS is composed of five parts, namely customer interface module, DBADM, rough sets analysis model, problem solving model and background knowledge acquisition model.

The functions of these five models are as follows. Customer interface module guides interaction between customers and the system. It can also decide which mining task to take, submit system data argument and supply decision makers with mining results. The function of DBADM is connecting application program with the data base, accessing data in the data base, and generating training sets and information tables. Rough sets analysis model can discover knowledge from a great number of data, and it is the main source of knowledge in knowledge base. This model is made up of data prep and role generation model. The former is in charge of reduction, discretization, and generalization analysis of data as well as null value estimation. The latter being the core of data mining part, includes attribute reduction, value reduction

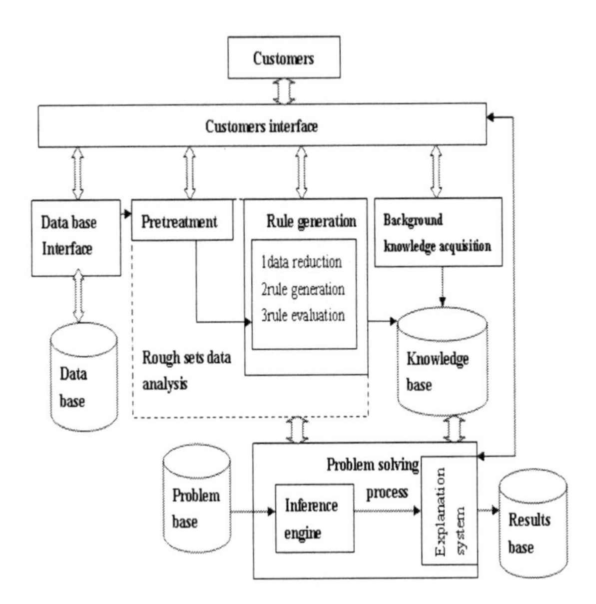

Fig. 1. RSDA-DSS structure

and solution of minimal rule sets. It also evaluates the final mined rule and stores rule satisfying certain support and confidence in the knowledge base for later use. For data mining can not discover all the expert knowledge, customers can make use of background knowledge acquisition model to make necessary supplement to knowledge base. Problem solving model will firstly analyze the issue including discrete segmentation of condition attribute, conception division and then employ inference engine and interpreter to offer theory and its explanation to customers. The inference engine is to obtain the solution to the issue according to given facts and using knowledge in the knowledge base and then presents the solution to customers. The interpreter is used to record solving process of the inference engine, synthesize the foundation of this process and supply it to customers.

3 Functional Structure of the System

Functional structure of the system is shown in figure 2.

4 Data Collection and Presentation

There is a great amount of information in data base about obesity patients, with a rather complicated structure. Under the guidance of some experts, we simplify the original data with propriety and select some major attributes for analysis.

Basic information of the obesity patients: sex, age and blood types.

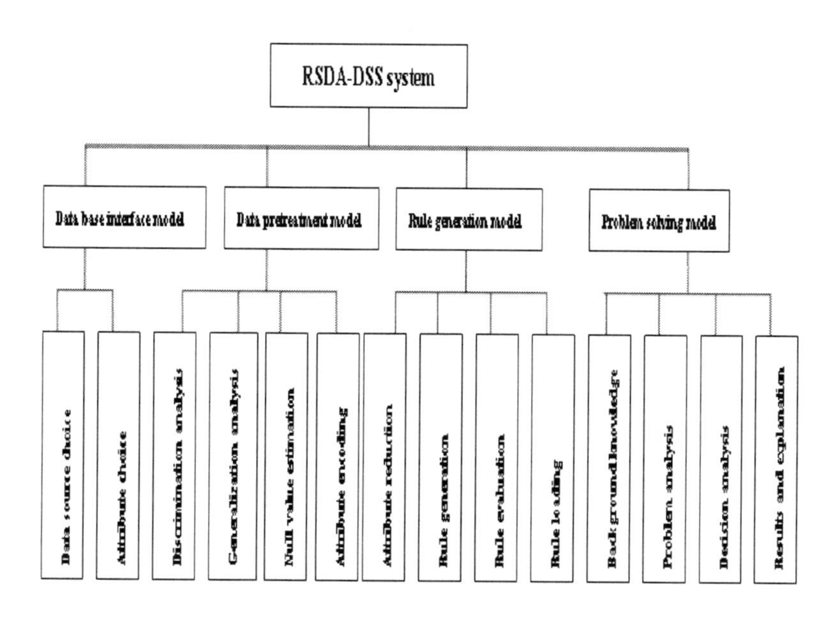

Fig. 2. Structure figure of the system function

Physical condition: consciousness, breath mood, and physical quality which is the comprehensive reflection of the patients' nutrition status.

Patients' condition: obesity can be classified into 5types such as obesity caused by overeating, press, edema, anemia, and fatigue according to site of fat, body shape, developing speed, scope and degree of obesity. Those five types of obesity often coexist, making it hard to differentiate among them clearly. So we simplify patients' condition into three degrees, slight, serious and critical on the basis of major diagnosis given by doctors and the initial condition of illness in the medical report.

Treatment protocol: treatment protocols for obesity are very complex, and they can resort to rehabilitation guidance, aerobic exercise, medication, diet regulation, massage treatment and operation treatment. Excessive varieties and usages involved in the medication make the structural analyses impossible, thus we simplify the medication and take the purpose of medicine instead of medicine. The purposes can be classified into reduction of diet and energy, enhancement of metabolism and energy consumption, reduction of material absorption and fat reduction.

Due to the frequent changes of the treatment protocols of patients, we deal with the medium protocol as follows: the apparent recorded recovery with the protocol change as a result of economic reasons is regarded as effective protocol and the effect is marked as 'recovery'; those changes of protocol for the worsening illness are regarded ineffective protocol and the effect is marked as 'no recovery'; changes of protocols for the same person are regarded as the reception of the second treatment. The conservative patients for the steadiness of illness will retain the former protocol, abandoning the conservative treatment.

Curative Effect refers to the effect achieved through corresponding recovery treatment protocols. The recovery records given by the fitness center can be categorized into healing, recovery, no recovery, worsening and the records will be seen as decision attribute.

The final data selected are shown in table 1. In the table, the treatment number as the object marking, the curative effect as the decision attribute, and the conditional attribute set $C=\{C_{01},C_{02},C_{03},C_{04},C_{05},C_{06},C_{07},C_{08},C_{09},C_{10},C_{11},C_{12},C_{13},C_{14},C_{15},C_{16}\}$. The collected samples in the decision table of conditional attribute add up to 300.

Table 1. Data structure of treatment protocol

attributes	code	assignments	attributes	code	assignments
Patients number	C_{00}	Primary code , patient ID	Rehabilitation guidance	C_{09}	First-class/routine
Sex	C_{01}	male/female	Aerobic exercise	C_{10}	T/P
Age	C_{02}	0～100	Dieting regulation	C_{11}	T/P
Blood type	C_{03}	O/A/B/AB	medication	C_{12}	T/P
Mood	C_{04}	normal/low	Massage Therapy	C_{13}	T/P
Breath	C_{05}	normal /hard	Reduce fat	C_{14}	T/P
Action	C_{06}	normal /hard	Acupuncture therapy	C_{15}	T/P
Physical quality	C_{07}	excellent/good/ fair/poor	Operation	C_{16}	T/P
Patients' condition	C_{08}	serious/slight	Curative effect	D	Cured /better / not better / serious

5 Experiment Results

Among 2 groups of experiment data, 200 data in the first group as exercise samples and the 80 data as the test samples. The experiment process and the results are listed below:

5.1 Discretization Analysis

The attribute of age in the example is consecutive, thus through employing the discretization of the rough sets and genetic algorithm, 6 phases are set and 5 division points are （49,64,68,77,82）. Though the reduction of redundant division points, the final discretization results are （*， 49］，（49, 64］，（64, 77］，（77， *）.

5.2 Attribute Reduction and Generation and Appraisal of Regulations

Through contrast between the several models and reduction, appraise the results by the measurement offered in literature 5, see table 2. In table 2, C for set of conditional attributes, RED for reduction results, N_r, E_a, E_i and E_w respectively for numbers of regulations, attribute evaporation, instance evaporation and data evaporation.

Table 2. Contrast between the several models and reduction

β		0	0.2	0	0.2
δ_0		0	0	0.2	0.2
ADBARK algorithms	RED	$C\{C_{01},C_{04},C_{06}\}$	$C\{C_{01},C_{04},C_{05},C_{06}\}$	$C\{C_{01},C_{06}\}$	$C\{C_{01},C_{05},C_{06}\}$
	N_r	102	107	82	85
	E_a	17.65	23.53	11.76	17.65
	E_i	66.00	64.33	72.67	71.67
	E_w	83.31	83.56	86.04	86.92
MIBARK algorithms	RED	$C\{C_{01},C_{04},C_{06}\}$	$C\{C_{01},C_{04},C_{05},C_{06},C_{10}\}$	$C\{C_{01},C_{04},C_{06}\}$	$C\{C_{01},C_{04},C_{05},C_{06}\}$
	N_r	102	96	102	107
	E_a	17.65	29.41	17.65	23.53
	E_i	66.00	68.00	66.00	64.33
	E_w	83.31	85.38	83.31	83.56
CSBARK algorithms	RED	$C\{C_{03},C_{04},C_{05},C_{06},C_{10},C_{11}\}$	$C\{C_{03},C_{04},C_{05},C_{06},C_{09},C_{10},C_{11},C_{12}\}$	$C\{C_{03},C_{04},C_{05},C_{06},C_{10},C_{11}\}$	$C\{C_{01},C_{03},C_{04},C_{05},C_{06},C_{10},C_{11}\}$
	N_r	109	73	109	96
	E_a	35.29	47.06	35.29	41.18
	E_i	63.67	75.67	63.67	68.00
	E_w	85.17	90.06	85.17	88.04

Results show:
The bigger β is, the fewer the regulations, the higher attribute evaporation, instance evaporation and data evaporation are. The bigger δ_0 is, the more the regulations, the lower attribute evaporation, instance evaporation and data evaporation are.

Little difference exists between the data reduction of ADBARK and MIBARK algorithms. CSBARA enjoys higher data reduction.

5.3 Test Results

$\beta=0.2$, $\delta_0=0.2$ and CSBARK algorithm are employed to test the results, and table 3 shows the outcome

Table 3. Test results

Group	Test sample	Test accuracy (%)
Serious	9	66.7
Cured	7	57.1
Better	43	81.4
Not better	21	61.9
Total	80	

5.4 Decision Analysis

As this application does not directly forecast results according to condition attributes, but give suggestions about treatment protocol based on patients' physical condition; therefore we should make corresponding alternation to determination algorithm.

Proposition I: if rule$\varphi_1 \wedge \varphi_2 \rightarrow \psi$, then$\varphi_1 \rightarrow (\varphi_2 \rightarrow \psi)$.

Prove: $\varphi_1 \wedge \varphi_2 \rightarrow \psi <=> \neg(\varphi_1 \wedge \varphi_2) \vee \psi <=> \neg\varphi_1 \wedge (\neg\varphi_2 \vee \psi) <=> \neg\varphi_1 \vee (\varphi_2 \rightarrow \psi) <=> \varphi_1 \rightarrow (\varphi_2 \rightarrow \psi)$.

Therefore, we classify antecedent rules into two parts, the first describing physical condition and illness state of the obesity patients, the second describing treatment protocols. In that way, we can transform the mined rule according to this proposition and apply the decision analysis algorithm.

6 Discussion

RSDA-DSS system can also be applied to other fields to make decisions; for instance, it can be used to make forecast and diagnose malfunctions according to customers' requirements. The obtained rules will be better in quality with more data and more precise description of data's features. The case in the present paper involves too much complicated factors, many protocols vary gradually in treatment and there exists great connection among the original and latter protocols. Given those difficulties, we process treatment protocols separately in order to simplify data description and mining process, which however has greatly influenced accuracy of the results. It follows that the core problem of the RSDA-DSS system's application to optimizing treatment protocols exists in finding a better method to collect sample data.

References

1. Yang, S., Liu, Y.: Attribute Reduction Algorithm in Rough Sets Model: Based onβ-δ_0. Management Science of China 10 (2003)
2. Yao, Y.Y.: A comparative of Rough fuzzy sets and rough sets. International Journal of General Systems 109, 227–242 (1998)
3. Ziarko, W.P.: Rough sets, fuzzy sets and knowledge discovery, pp. 32–44. Springer, New York (1994)
4. Pawlak, Z.: Rough set—Theoretical aspects of reasoning about data, pp. 9–51. Kluwer Academic Publishers, London (1991)
5. Yu, W., Renji, W.: "Data Reduction" Based on Rough Set Theory. Journal of Computer Science and Technology 21(5), 393–400 (1998)

The Study on the Methods of Identifying Ice Thickness in the Capacitive Sensor Measuring System

Chang Xiaomin[1] and Dou Yinke[2]

[1] Measurement and Control Technology Institute of Taiyuan university of technology, 030024, TaiYuan, China
douyk8888cn@yahoo.com.cn
[2] College Electric and Power Engineering of Taiyuan university of technology, 030024, TaiYuan, China
douyk8888cn@yahoo.com.cn

Abstract. Due to wild values or singular values existing in each set of data sampled with the capacitive ice layer thickness detection system, bigger errors always occur in determining ice layer thickness. To improve the accuracy, denoising methods for data based on Segmented Average Value (SAV) and Singular Spectrum Analysis (SSA) are designed. By comparison, the latter is better in term of effect, filtering wild values and singular values and improving the accuracy.

Keywords: capacitive sensing, ice thickness, sensor, noise treatment.

1 Introduction

Ice thickness is an important parameter for river ice and sea ice. Using electromagnetic induction, capacitive sensors to detect ice thickness are a new real-time measurement method [1]. Its fundamental principle as shown in Figure 1 is that it decodes under the control of the single-chip computer, and turns the multi-way switch on to connect each induction electrode to the corresponding sine wave generator while other electrodes to the ground. The sine signal from the generator flows through a voltage divider resistor and then the capacitor of diffuse electric field formed by the metal electrode and its surrounding medium and other electrodes, finally to the "ground surface"[2] where the power supply on. The signal from two terminals of the capacitor flows through the rectifying detector and then the filter, where the voltage on each electrode is converted into DC signal [3], which is processed by an A/D converter later. The same signal collection is applied to each electrode once. After completing this, it writes data to the mass storage SD-card via the SPI bus for storage. Due to electrodes being at different positions in the sensor, the medium each of them inducts can only be ice, water or air. Therefore, the ice layer thickness can be figured out based on data measured with each electrode.

The essential of capacitive ice layer thickness sensor is that the capacitance formed by an induction electrode with the "virtual ground" varies when a medium with certain dielectric constant is approaching the electrode[4]. Since electrodes inside the

G. Zhiguo et al. (Eds.): WISM 2011, CCIS 238, pp. 51–58, 2011.

capacitive sensor being in different media, under ideal conditions, voltages measured with different electrodes shall be on different straight-line segments and the data curve got shall be a ladder diagram composed by three smooth and jitter-free lines as shown in Figure 2(a). But in practice, the curve drawn with data got in each collection isn't a smooth ladder diagram as shown in Figure 2(b), due to complexities inside ice &water and influences of temperature and other parameters of the sensor circuit

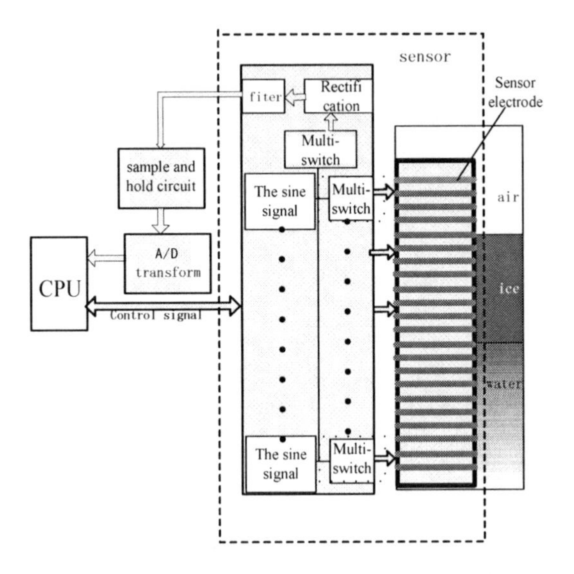

Fig. 1. The principle of the capacitive sensor measurement system for detecting ice thickness

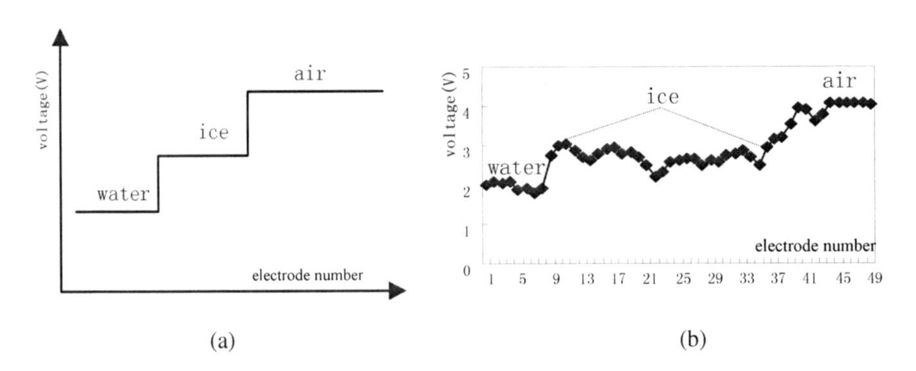

Fig. 2. Voltage curves of ice thickness under ideal conditions and a set of actual sampling data

The ice layer thickness calculated with the single-chip computer is one of the threshold voltages set by the author respectively at ice-water and air-ice interfaces[5]. For example, the ice-water interface threshold voltage in Figure 2 (b) may be set as 2.5V, for any electrode whose voltage below 2.5V, it may be deemed in the water; similarly, that of the air-ice interface may be set as 3.5V, so the electrode whose voltage above 3.5V may be deemed in the air. However, threshold setting is often interfered by voltage wild values, causing big error in ice thickness calculation are acceptable.

2 The Research Method of the System

In order to calculate ice thickness based on measurement data accurately, two methods of mathematics algorithm were studied. The same basic idea is taken in the two methods, that is designing a mathematics algorithm to "denoise" the curve shown in Figure 2 (b) as far as possible, i.e. filtering any wild value and singular point (the point that is significantly different from the line's average value) to convert the curve to a 3-step ladder curve as smooth as possible, selecting two threshold voltages, and then calculating the ice thickness with a program. In this paper, the two mathematics algorithms designed are analyzed and compared.

2.1 Segmented Average Value Method

Let a set of sampling data have n numbers, make 5 numbers into a segment from 1st, so get (n-5) segments. For instance, numbers 1-5th forms Segment 1, 2-6th Segment 2, 3–7th Segment 3 … (n-5)-n Segment (n-5). After completing the segmentation, average each segment firstly.

$$\overline{E} = Ave(Bi) = \frac{1}{5}\left[A(i) + A(i+1) + A(i+2) + A(i+3) + A(i+4)\right]$$

Then average absolute values of differences between the 5 numbers in each segment and the said average value, i.e.

$$Di = \frac{1}{5}\left[abs(A(i) - \overline{E}) + abs(A(i+1) - \overline{E}) + \ldots\ldots.. + abs(A(i+4) - \overline{E}\right]$$

Then compare each value with the average value, let the value minus Di when it is bigger than the average value $\overline{E}$, and vice versa. This is the simple segmented average value method.

Then program with Matlab to "denoise" the raw data and smooth the curve. Two curves before and after treatment with the Matlab program are shown in Figure3.

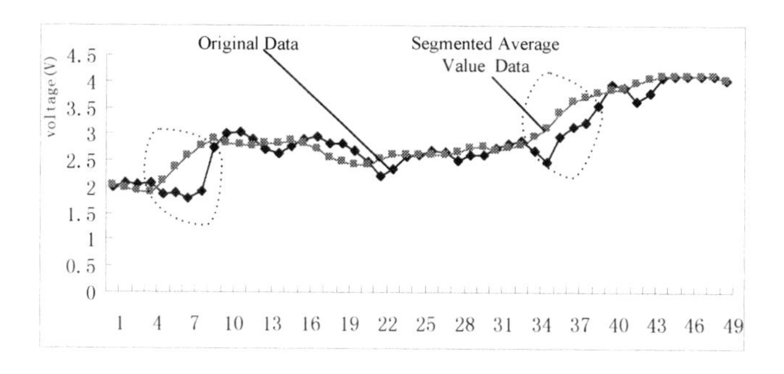

Fig. 3. Comparison between data processed based on SAWM and raw data

It's indicated in Figure 5-16 that the processed data curve becomes much smoother than the raw data curve, and the three steps keep distinct. However, the voltage at jump point on the processed curve is "raised" obviously (two sections enclosed with dashed lines in the figure), which would cause inconvenience to threshold voltage setting during post-treatment, or else improper setting would cause big error of calculation results.

2.2 Singular Spectrum Analysis Method

The Singular Spectrum Analysis (SSA) Method is a generalized analytical method for power spectral density [6, 7], which can denoise and extract features of unstable time series and nonlinear time series, as well as implement dynamic reconfiguration of time series for not subject to sine wave assumption and adopting time-domain frequency domain analysis for identification and description of signal.

Singular Spectrum Analysis Theory:

For a $N \times M$ real matrix A, it would be presented as

$$A = V \times S \times U^T$$

Here U^T represents the transpose of U, V, U and S are $N \times M$ orthogonal matrixes, and S is an N× N matrix. For matrix S, its element at (i,i) is λ_i while others are zero. So λ_i is the singular value of matrix A.

The specific method for denoising smooth curve fitting of singular spectrum is as follows:

Let the time series of length N measured from a system be $\{x_i\}$, $i = 1, 2, \cdots, N$, give its embedded space dimension M, $M < \frac{N}{2}$, and construct an attractor track matrix,

$$X = \begin{bmatrix} x_1 & x_2 & \cdots & x_{N-M+1} \\ x_2 & x_3 & \cdots & x_{N-M+2} \\ \cdots & \cdots & \cdots & \cdots \\ x_M & x_{M+1} & \cdots & x_N \end{bmatrix}$$

X shows evolution features of reconstructed attractor in phase space. If there's no noise in an original signal or S/N of it is very high, and the track matrix is singular, X can be decomposed into K $(K < M)$singular values $\lambda_1 \geq \lambda_2 \geq \cdots \geq \lambda_K > 0$ arranged in non-increasing order; on the contrary, it can also be decomposed into K $(K = M)$ singular values $\lambda_1 \geq \lambda_2 \geq \cdots \geq \lambda_K > 0$ arranged in non-increasing order, therefore the singular value K is related to the nature of the system.

Take $S_i = \log\left(\lambda_i / \sum_{j=1}^{k} \lambda_j\right)$, then S_1, S_2, $\cdots$, S_M is referred to as the singular spectrum of the system. From the above analysis, it can be seen that whether the system has noise is up to whether the track matrix has any zero singular values.

From this angle, it can be seen that a singular spectrum represents relationship of energy taken in the entire system by each state variable. Generally, several values in the front of spectrum values are bigger and others smaller, and bigger ones are corresponding to feature elements in the signal while smaller ones to noise elements. So keep bigger values and set the smaller ones to zero artificially, and then transform $\{x_i\}$ reversely into a new time series $\left(x_i'\right)$, which would reduce interfering noise and smooth measurement data to a certain extent.

Then expand X based on Empirical Orthogonal Function (EOF) into M feature vectors E^k ($1 \le k \le M$), referred to as empirical orthogonal function, and define the k-th principal component p_i^k as orthogonal projection coefficient of original time series $\{x_i\}$ on E^k.

$$p_i^k = \sum_{j=1}^{M} x_{i+j} E_j^k, \quad 1 \le i \le M$$

If each principal component and empirical orthogonal function is known, then each component series can be reconstructed according to the following formula:

$$x_i^k = \begin{cases} \frac{1}{M} \sum_{j=1}^{m} p_j^k E_j^k & (M \le i \le N - M - 1) \\ \frac{1}{i} \sum_{j=1}^{i} p_j^k E_j^k & (1 \le i \le M - 1) \\ \frac{1}{N - i + 1} \sum_{j=i-N+M}^{M} p_j^k E_j^k & (N - M + 2 \le i \le N) \end{cases}$$

Then x_i, the i-th element in the series reconstructed with the k-th principal component, can be shown as

$$x_i = \sum_{i}^{k} x_i^k$$

In smoothing based on SSA, it requires considering interception of principal components, namely choosing which principal components to reconstruct signals. If choosing too few principal components, it would loss feature information of part signals; but on the contrary, it would include excessive interference components. Therefore, the optimum of principal components could be selected according to the experimenter's estimates or repeated selection by each measurement.

According to the above analysis, it is indicated that iterative smooth treatment of measurement data based on SSA will obtain a smoother data curve. Here take our sensor data for example to analysis and process.

The following is a set of raw data collected with the capacitive ice layer thickness sensor in experiments on the Inner Mongolia section of the Yellow River, as shown in Table 1.

Table 1. A set of sampling experimental data

1	2	3	4	5	6	7
2	2.08	2.04	2.06	1.85	1.9	1.75
8	9	10	11	12	13	14
1.58	1.91	1.99	1.96	2.35	2.47	2.54
15	16	17	18	19	20	21
2.58	2.75	2.78	2.67	2.67	2.54	2.32
22	23	24	25	26	27	28
2.02	2.16	2.36	2.38	2.44	2.45	2.23
29	30	31	32	33	34	35
2.38	2.38	2.52	2.57	2.62	2.34	2.07
36	37	38	39	40	41	42
2.39	2.59	2.64	2.99	3.56	3.63	3.31
43	44	45	46	47	48	49
3.48	3.74	3.77	3.83	3.84	3.85	3.72

Based on the 49 data, a time series $\{X_i\}$, $i = 1,2,\cdots,49$ of length N=49 is obtained. Select an embedded space dimension M to construct the attractor track matrix, and then perform the artificial zero-setting and inverse operations on singular values, thus obtain a curve after denoising based on SSA. Finally, the curve of $M = 20$ and 3 singular values kept is selected as the optimum, and the fitting curve and original curve are shown in Figur 4.

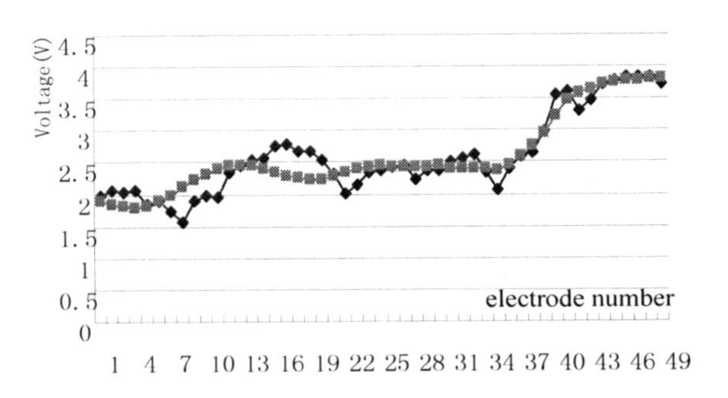

Fig. 4. Curve denoised based on SSA and original curve

It can be seen that the denoised fitting curve is much smoother than the raw data curve and eliminates bigger noise signals from Figure 4.

The data fitting method of denoising treatment based on SSA can eliminate bad values (noise signals) effectively and correct sensor data largely. Comparing with the smooth treatment based on the least square method (LSM), LSM can process a single bad value more effectively with smaller errors, but it can't detect a cascade of continuous bad values and only can correct the first one to a certain extent. On the contrary, not only a single bad value, but also a cascade of ones can be corrected by the smooth treatment based on SSA more accurately. Therefore, the method is more effective for data curve fitting generally. Although it can find a wild value accurately and correct it, it would change normal values neighboring to the wild value at the same time, which is disadvantageous to keeping original information and follow-up process of measurement data. In addition, it is more difficult to select the embedded space dimension M and eliminate singular values representing noise. Either of the improper selections would cause data loss and corruption, so it is difficult to grasp.

To compare curve fitting effects of the segmented average value method and singular spectrum analysis method, the author processed the same set of data with the two methods, obtaining curves as shown in Figure 5.

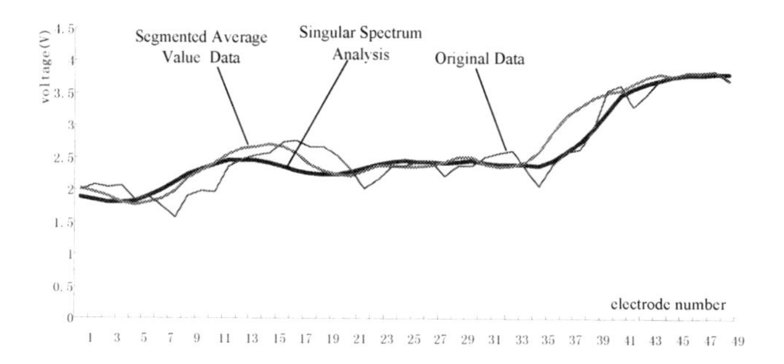

Fig. 5. Comparison between curves processed with the two methods and original curve

Obviously, it can be seen from the three curves that the effect of curve after "denoising" with SSA method is the best, which is the closest to and smoother than the original curve, and with three distinctive steps. Whereas the curve processed with the segmented average value method is smoother than the original one, but its effect is worse than that with SSA method. Therefore, it adopts the SSA method for the data "denoising" treatment algorithm in this design at last.

After "denoising" the original curve, it also requires calculating ice layer thickness based on threshold setting method. When setting threshold with the curve processed based on SSA, it is set as 85% of the highest voltage of each step on the curve. For example, air-ice interface threshold voltage is set as 85% of 3.81V, the highest voltage of electrode in the air, i.e. 3.23V is the threshold voltage. By this, the error of ice thickness calculation with the computer is reduced to 1cm.

3 The Conclusion

The ice thickness identification algorithm is the difficulty in designing a connect ice thickness sensor. As the capacitor in the capacitive ice layer thickness sensor is affected largely by environmental changes, wild value or singular value occurs more easily in the data measured with it. It can realize "denoise" of data and smoothing of curve with both of the two methods, designed by the author, the segmented average value method and singular spectrum analysis method. However, the latter produces better effect than the former, and improves accuracy of ice thickness identification greatly.

Acknowledgment. This paper was sponsored by Shanxi youth natural science foundation (No. 2010021018) and Shanxi Science and Technology research project (No. 20110321026-02). I would like to thank the Office of Science and Technology in Shanxi Province in this paper to give financial support.

References

1. Haas, C.: Evaluation of ship-based electromagnetic-inductive thickness measurements of summer sea-ice in the Bellingshausen and Amundsen Sea Antarctica. Cold Regions Science and Technology 27, 1–16 (1998)
2. Liu, Y., MingHao, Y., YanHong, L.: Using fringe field capacitive grain moisture sensor. The Journal of China Agricultural University, 58–61 (December 2007)
3. Dong, X., Changxi, L.: Research on Uniplanar Scattering-Field Capacitive Sensors for Measuring the Water Content in Sand. Chinese Journal of Sensors and Actuators 21, 2000–2004 (2008)
4. Meng, W., Yang, Y.: The Principle and application of capacitive sensor. J. Modern Electronic Technology, 49–52 (July 2003)
5. Dou, Y.-k., Chang, X.-M., Qin, J.-M.: The study of a capacitance sensor and its system used in measuring ice thickness, sedimentation and water level of a reservoir. In: Error Theory and Data Processing, Harbin, China, November 2002, pp. 88–89. International Harbin Institute of Technology Press, Ding zhenLiang (2009)
6. Qi, z., Mi, D., Xu, Z., et al.: Singular spectrum analysis in mechanical malfunction diagnosis. Noise and Vibration Control 1, 82–88 (2008)
7. Forum on Information Technology and Applications, ChengDu,China, pp. 616–619 (May 2009)

Research on User Defined Mapping Techniques Based on Geographic Information Center

Liu Rong, Cheng Yi, Ge Wen, and Huang Ruiyang

Zhengzhou Institute of Surveying and Mapping,
Zhengzhou, China
chxycy@126.com

Abstract. Design and make their own map is a dream of many people. Supporting by SOA techniques and geographic information services, people can achieve their dream online. In the paper a user defined mapping system is put forward and key techniques are analyzed. The system includes GI provider, requester and geographic information center. GI provider can encapsulate geographic data into OGC data services or encapsulate GIS software components into processing services and register these services to geographic information center. GIC in charge of geographic service description, classification, service metadata storage, service assessment, maintenance and inquires. GI requester can search services via GIC, get dada service from others, processing data with processing services online and select proper symbols to display it.

Keywords: user defined mapping, geographic information service, geographic information center, services discovery, services matching.

1 Introduction

The modern society has stepped into information age, while the computer network technology has changed the living state of people in some sense, and has formed a completely new society organization, which we call "internet" society. With the supporting of internet, GIS servers such as Google Earth, Microsoft Terra Server etc can provide flexible, online services to internet users. And users can define individual maps to fit their own demands by these services which come from various servers. The key techniques in this process focused on how to publish geographic information resources as internet services, how to discover services in internet environment and how to integrate different services into one map view. In this paper, we put forward a method to enable this process.

2 The Framework of UDM Based on GIC

As Fig. 1 shows, the framework includes geographic information services provider, geographic information requester and geographic information center.

G. Zhiguo et al. (Eds.): WISM 2011, CCIS 238, pp. 59–66, 2011.

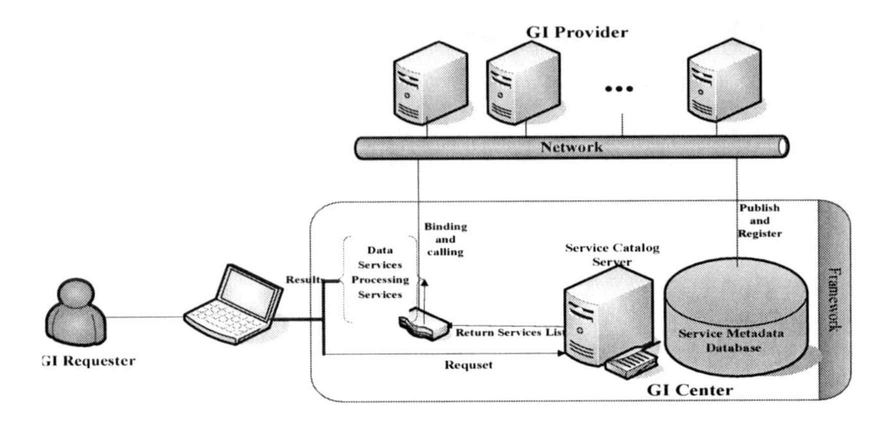

Fig. 1. Framework of UDM based on GIC

2.1 Geographic Information Center

Geographic web service is the basic form in Geographic Information Center (GIC), includes not only all kinds of geographic data services, but also geographic processing services, display services and etc. All these services can register into GIC. GIC is a core component in the system. It is in charge of geographic service description, classification, service metadata storage, service assessment, maintenance and inquires.

2.2 Geographic Information Provider

There are two different with traditional Web GIS systems. The first one, everyone can be geographic information provider in the system. Geographic data in system includes two parts: infrastructure data and thematic data. Infrastructure data provided by system builder and thematic data can be provided by both system builder and users. The Second one, the way of geographic data provide is not transfer their data to system server, but encapsulates their data into GI Services according to OGC standards and in charge of update of their own services.

2.3 Geographic Information Requester

Geographic information requester in the system can use web data services resource as local data files. And also they can use processing service to process and display their data. Of course, they need to understand basic knowledge of relative standards.

3 The Key Techniques of User Defined Mapping Based on Geographic Information Center

3.1 Geographic Information Center Building

Summarizing advantages and disadvantages of current geographic information center, Orcale 9 is used to build geographic information catalog database. A flexible and appropriate classification system is built and a detailed storage scheme of geographic information metadata is designed to describe the furthest of geographic information.

(1) Establishing the Appropriate Classification System of Geographic Information Services

Geographic information service classification system is the cornerstone of geographic information center. Only geographic information service is appropriately classified and the service registration and discovery can be realized. Users of geographic information center may be senior professional users, general professional users and ordinary users, their demands and thickness for the service level are different. If the service level is thicker, the service abstraction is higher, the service interface is closer to users, and it is easier to understand and use for ordinary users. On the contrary, if the service level is thinner, the service interface is closer to system components, and it is easier to use for professional users. Therefore, following classification principles are adopted:

① The service classes should not be too more, so that geographic information service providers could go to their registration services types quickly.

② The service types should clearly partition the types of geographic information services, such as data service and function service. The types of data service should be able to indicate that whether the service is based on the OGC service criterions or is published by web services. The types of functional service should be able to recognize the service is encapsulated by the operation of spatial data display, conversion, integration and analysis.

(2) Creating Geographic Information Storage Strategy

Geographic information directory database is a core part of geographic information center. The registration and search of geographic information is actually operated the records in the catalog database. However, Geographic information directory database is mainly stored by geographic information service metadata. Geographic information service metadata is the structured description of service, and is the key point of service registration and discovery. Therefore, in the classification system of geographic information center, following three principles of service metadata storage strategy are adopted:

① For all types geographic information services, the storage and management of shared information are centralized, such as service provider information, service name, publication time, accessing entry and key words;

② For data services, the geographic range, coordinate system, sub-layer numbers of data service, and the name, title, summary, data ranges, such as coordinate system, style information of each sub-layer are stored to support the search and discovery of data services.

③ For function services, special tables of encapsulation methods and encapsulation parameters are built to store these detailed informations because of their differences. Encapsulation method numbers, each encapsulation method name, description, parameter numbers, each parameter name, description, type and order of the parameters and other informations are stored.

3.2 Geographic Information Encapsulation and Publication

Using OGSA-DAI grid middleware technology, geographic informations are encapsulated and published by extending the function of OGSA-DAI. OGSA-DAI is more advanced as data access and integration technology products at present, many domestic and foreign research institutions do scientific research based on OGSA-DAI. But OGSA-DAI's own capabilities are very limited, and developers need to extend the OGSA-DAI in order to meet their own requirements. Because the original functions of OGSA-DAI cannot meet the access and integration requirements of geographic informations, so it is need to extend OGSA-DAI, following two extension modes are mainly adopted:

① Extending OGSA-DAI data resources. Distributed relation databases and data files are encapsulated to OGSA-DAI data resources, and these data resources can be easily accessed by OGSA-DAI technology.

② Extending OGSA-DAI behaviors. By extending the OGSA-DAI behaviors, users can accurately access the contents of each data resource, including all tables of relation database and all data files of file system in each data resource server, and users can accurately get each table's content and each file's content.

3.3 Geographic Information Services Discovery and Evaluation

Numerous and enough geographic information are the foundation of geographic information services, but facing numerous and distributed geographic information services, how to find and discover services which meet specific user requirements and match the interfaces is a major challenge for users. Geographic information services discovery and evaluation aim to help users to efficiently even intelligently find the most optimized geographic information service, therefore the following research methods are used: Semantic idea is introduced in services description and matching to improve the problems of services model describing services no all-sidely and semantically lacking in services discovery, and the semantic description of services functions and behaviors is added. QoS restriction is introduced in services matching to improve QoS support lacking, and QoS evaluating model is established. Multi-level services matching is introduced to improve the services matching arithmetic flexibleness lacking. The research will be conducted from three aspects, which are semantic description of geographic information services based on ontology language (OWL-S), building geographic information services QoS evaluation model and multi-level geographic information services matching.

(1) Semantic Description of Geographic Information Services Based on OWL-S
Semantic web service is not a separated web service, and it is the extension of current service. Increased a semantic (knowledge) layer based on current service, semantic web services will be formed. In the semantic web services, information is given well-described meaning, which is conducive to promoting mutual cooperation of human and computer. Currently, researchers have proposed specific service semantic description language, which is OWL-S (Ontology Web Language for Service, OWL-S). OWL-S is an OWL ontology specification which is used to describe the web

service's properties and functions. OWL-S uses a series of basic classes and properties to describe the service, and provides a shared framework. Its goal is to make web services into computer-readable entities, so the automatic services discovery, selection, calling, interoperability, composition, performing monitor and other tasks could be achieved.

There are two optional ways of using OWL-S to describe geographic information services:

① Extending WSDL, adding semantic information using OWL-S;

② Extended OWL-S, adding the WSDL binding information as services Grounding content. We use the first way to do the semantic description of geographic information services, which can minimize the changes of existing services' WSDL.

(2) Building Geographic Information Services QoS Evaluation Model

Geographic information services QoS involves many factors, we will determine QoS evaluation indicators from the consumers view. They are: service availability (Availability), service response time (Time), service implementation cost (Cost), service reputation (Reputation), service reliability (Reliability). After evaluation indicators being determined, it is need to quantify these indicators. In the services discovery and matching process, there may be many candidate web services which meet the basic description matching and function matching. We build a QoS evaluation matrix for these candidate services, and the matrix line indicates all the QoS property values of certain service, the matrix column indicates certain QoS property value of all the services. Because the dimensions of QoS property values are different, dimensionless of QoS property values is needed. After normalization of QoS property values, we provide weights for each QoS property to calculate the QoS general values of web services. Then, we select the best by sorting the QoS general values.

(3) Multi-level Geographic Information Services Matching

Multi-level geographic information services matching algorithm is proposed for web services matching lack of flexibility. In this algorithm, the service matching should be a gradually accurate process. First, we get same services which meet the services catalog classification requirement through basic matching. Next, we do the services filter further through function matching. Finally, we consider the non-functional property matching including QoS matching. In multi-level service matching algorithm, each filter process can be used different services matching algorithm. The first level is basic description matching which is the match type of keywords, and is a syntax-level matching. The second level is services function matching, which is the most complex and the most crucial matching. For the function matching, the current typical algorithm is the "elastic matching algorithm", but this method classification is not specific enough and the semantic ability is not enough, so it is need to be extended. The specific idea is: Extending four classifications to infinite classifications, introducing binary relations between classes as an important reference in service matching, and proposing the concept and algorithm of semantic distance in the ontology concept and web services matching. The third level is QoS matching, which using geographic information services QoS evaluation model to qualify QoS.

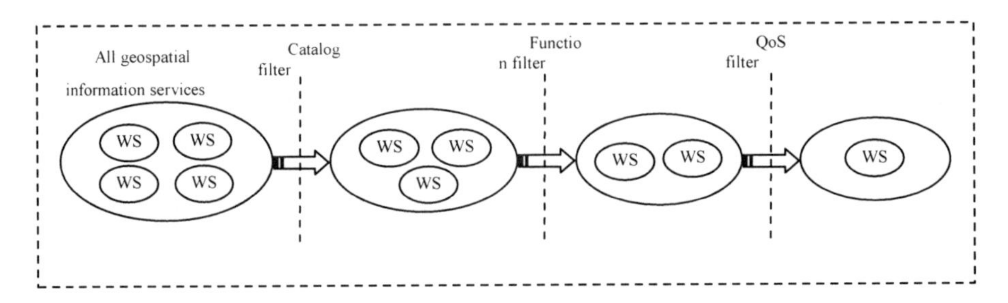

Fig. 2. Multi-level geographic information services matching

3.4 Geographic Information Calling and Overlaying

After finding and matching the geographic information services, a group of optimum geographic informations which meet user defined mapping are find out, sometimes these geographic informations had to been called and overlaid. Using geographic information service call of different types could realize geographic information service of different types overlaying to complete user defined mapping based on the geographic information sharing.

(1) Geographic Information Calling

According to the different standard interface protocol, either GIS data service or function service, geographic information could be divided into two types, which are standardized type and custom type. So the geographic information binding and calling can divide into two parts.

① Standardized geographic information binding and calling

Standardized geographic information used the standard protocol to provide a series of common public sealed interface service. Among these capabilities documents of services lots of service and sealed interface description information been included. Used these public interfaces could build the service calls. Several common popular standardized service and request style information were shown as followed table 1.

The geographic information of interfaces and result type defined clearly could be called according to the above uniform request format, and the result would be provided to users by pictures or documents. If the request was WFS service, the user had to call the GML codes to visualize.

② Custom geographic information binding and calling

For custom geographic information, the right data model could be concluded on the condition of the understanding the logical data structure, operation supported and application language. And the structure method was not current such as the SOAP protocol service. For realization of geographic information use, the common method was used for providing request strings or custom input of request documents, and providing SOAP request calls. The result was shown by string. And the user got the return structure by analytic strings.

Table 1. Popular standardized service and request style information

Service Types	Consistent HTTP Request Formats	Reference Introductions	Return Results
WMS	http://[IP Address]:[Port]/ [Content Provided by Provider]/[WMS Interface]? GetMap &Layers=[Level Name 1, Level Name 2, ……]&BBOX=[Left Bottom X Coordinate, Left Bottom Y Coordinate, Right Up X Coordinate, Right Up Y Coordinate]&WIDTH=[Width of Picture]&HEIGHT=[Height of Picture]&FORMAT=[Types of Picture]	"[]"is specific parameters. "&"is parameter separator. GetMap is WMS interface. BBOX is data range, which uses the longitude and latitude.	Picture
WFS	http://[IP Address]:[Port]/ [Content Provided by Provider]/[WFS Interface]? Query &Layer=[Level Name]&BBOX=[Left Bottom X Coordinate, Left Bottom Y Coordinate, Right Up X Coordinate, Right Up Y Coordinate]&FORMAT=[GML Version]&FILTER=[Condition 1,Condition 2,……]	"[]"is specific parameters. Query is WFS interface. BBOX is data range, which uses the longitude and latitude. FILTER is limited condition about attribution.	GML Coded Geo-feature Combin ation
WCS	http://[IP Address]:[Port]/ [Content Provided by Provider]/[WCS Interface]? GetCoverage&Layer=[Level Name]&BBOX=[Left Bottom X Coordinate, Left Bottom Y Coordinate, Right Up X Coordinate, Right Up Y Coordinate]	"[]"is specific parameters. GetCoverage is WCS interface. BBOX is data range,which uses the longitude and latitude.	Picture
WPS	http://[IP Address]:[Port]/ [Content provided by provider]/[WPS Inerface]? Execute&Process=[Method name]& DataInputs=[Input parameter]	"[]"is specific parameters. Execute is WPS interface. Process is method name.	GML Docum ent

(2) Geographic Information of Different Types Overlaying

① Data service overlaying

Both data service in the same zone were called by overlaying. If both of the results were pictures, pictured had been set transparency to meet several data service shown at the same time. If the result included the feature set, feature had been symbolized.

② Cooperation and combination between data service and function service

When the geographic information called by the user was accordant with the mapping need, the captured data had to be deal with such as projection, simplification, synthesize and spatial analysis etc. So cooperation and combination between data service and function service had to realize. While the user specified the order to the combining service, and set the input parameters between the current service and the next service accordantly to realize the cooperation and combination between data service and function service.

4 Conclusion

The user defined mapping system base on the geographic information center is a new type of GIS, that is self-service GIS. This technology refers to the user defined mapping technology to change the current situation that the isolation and independence lacking between geographic information management and service. And this technology might be used for digital earth, digital China, digital province, digital city, which is significant contribution to the resource share and cooperation among digital earth, digital province and digital city.

References

1. Huang, R.: Design and Implement on Geographical Information Services Registry Center. Information Engineering University Master Essay, Zhengzhou (2010)
2. Yue, P.: Research on the Key Technology of the Geographic Information Intalligent Service Supported by Semantic. Wuhan University doctoral dissertation, Wuhan (2007)
3. Gong, X.: Research on QoS-Aware Web Services Discovery and Composition Method. Chongqing University doctoral dissertation, Chongqing (2008)
4. Martin, D., Burstein, M., Hobbs, J., Lassila, O., McDermott, D., McIlraith, S., Narayanan, S., Paolucci, M., Parsia, B., Payne, T., Sirin, E., Srinivasan, N., Sycara, K.: OWL-based Web Service Ontology (OWL-S) (2004), http://www.daml.org/services/owl-s/1.1
5. Paolucci, M., Kawamura, T., Payne, T.R., Sycara, K.: Semantic Matching of Web Services Capabilities. In: Proeeedings of the AAAI Spring Symposium, Palo Alto, California, pp. 92–99 (2004)

Camera Calibration Based on the Template Paralleled to Image Plane

Enxiu Shi, Yumei Huang, Jiali Yang, and Jun Li

Institute of Machine and Automation, Xi'an University of Technology
Xi'an, 710048, Shaanxi, China
{shienxiu,kkarrie,jun1956}@163.com, hym_xaut@126.com

Abstract. It's an essential problem to look for a simple and effective method for camera calibration in the application of computer vision. A new simple and fast calibration method is proposed in this paper. It needs a calibration block and a plane template.Through the calibration block, the point of intersection which is the projection for the optic axis of CCD on the image plane or called image-plane's center coordinate is obtained directly. The focal distance is calculated by geometry method, too. Then applying the coplanar-feature point, other parameters of camera system could be calculated. The experimental results show that this method can do calibration for camera's extrinsic and intrinsic parameters fast and conveniently with a relatively high precision.

Keywords: Computer Vision, Camera Calibration, Image Plane, Camera's Parameters.

1 Introduction

The camera calibration is a process concerning the calibration of the camera's extrinsic and intrinsic parameters, which is to obtain geometry and optic parameters of the camera and three-dimensional position and posture (the camera coordinate system relative to the world coordinate system)[1-2]. The present calibration methods are mainly linear and non-linear. The former is simple and fast with low precision without taking the lens' aberrance into account; On the contrary, the aberrance is considered in the latter, it is complicated, slow and sensitive to the selection of initial value and noise. Among many methods reported, such as Zhang Plane Calibration[1], Double Plane[1,3] and Auto-Calibration[1-3], Tsai and so on. Tsai (or two-step calibration) in [3] is used frequently, but when the image plane parallels to the standard template, there would be morbidity equations.

In practical application, this situation exists extensively, such as the calibration on SCARA machinery arm and Wire Bonder packaged in the IC, etc. The calibration aimed at this situation is amended to Tsai in [4], but it needs to get some intrinsic parameters of the camera beforehand. This increases the complexity of calibration. In [5], it refers to the use of standard block, which could easily figure out the center of the image plane and the focus distance of CCD, but the method has less calibration

G. Zhiguo et al. (Eds.): WISM 2011, CCIS 238, pp. 67–73, 2011.

controlling points and the aberrance was not considered, the height of standard block was also a problem. The calibration precision was effected by the height of standard block[6].

In this paper, the calibration method is improved on the situation of image plane paralleled to the standard template. It could work out camera's extrinsic and intrinsic parameters without pre-calibration. The experiment confirms that the method has some advantages such as simple operation, little calculation and so on.

2 Camera Model

2.1 Camera Calibration Model

The ideal camera model is shaped like an aperture model (shown in Fig.1). This model does not consider the lens' aberrance. The system parameters of the camera would be got through linear transformation. World coordinate system ΣO_w is set for workspace of Automatic Guided Vehicle (AGV).ΣO is image coordinate system, which its origin is set the top and left point of the rectangle image plane and the units are pixel. Camera coordinate systemΣO_c is for CCD, which Z-axis parallels to that of ΣO_w and passes through the center point of the imaging plane, X-axis and Y-axis of it parallel to those of ΣO_c. The center point of ΣO is the projection point of optic axis on the imaging plane. The aberrance is not considered here.

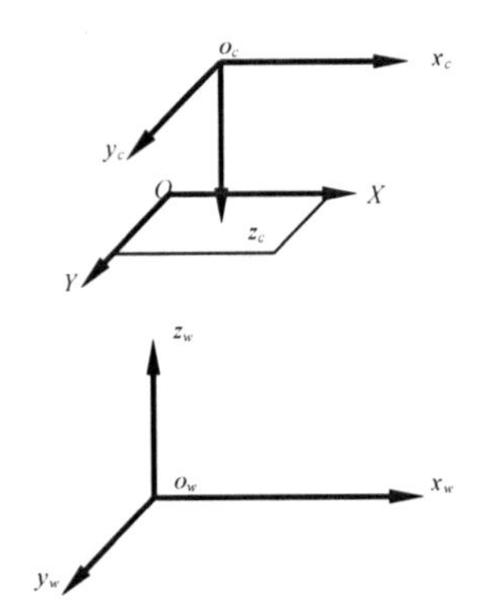

Fig. 1. The ideal camera model

The point P_c=$[x, y, z]^T$ in ΣO_c is the image for the point P_w=$[x_w, y_w, z_w]^T$ in ΣO_w. The relationship of them is as following:

$$\begin{bmatrix} x \\ y \\ z \end{bmatrix} = \boldsymbol{R} \begin{bmatrix} x_w \\ y_w \\ z_w \end{bmatrix} + \boldsymbol{T} = \begin{pmatrix} r_1 & r_2 & r_3 \\ r_4 & r_5 & r_6 \\ r_7 & r_8 & r_9 \end{pmatrix} \begin{pmatrix} x_w \\ y_w \\ z_w \end{pmatrix} + \begin{pmatrix} t_x \\ t_y \\ t_z \end{pmatrix} \tag{1}$$

Here, $\boldsymbol{R}$ is rotation matrix and T=$[t_x, t_y, t_z]^T$ is a vector for parallel translation between ΣO_c and ΣO_w.

$P_c=[x, y, z]^T$ is point in ΣO. The equation can be gotton:

$$\begin{cases} X = f \cdot x / z \\ Y = f \cdot y / z \\ Z = z \end{cases} \tag{2}$$

Here, f is the effective focal distance of camera. Considering the lens' aberrance, there is:

$$\begin{cases} X = x_d + \delta x = x_d + k_1 x_d (x_d^2 + y_d^2) \\ Y = y_d + \delta y = y_d + k_1 y_d (x_d^2 + y_d^2) \end{cases} \tag{3}$$

Here, δx, δy is the aberrance in the direction of X, Y in ΣO respectively; k_1 is radial aberrance factor. x_d and y_d is image plane coordinates. From (1) to (3), there are:

$$\begin{cases} x_d + k_1 x_d (x_d^2 + y_d^2) = f \dfrac{r_1 x_w + r_2 y_w + r_3 z_w + t_x}{r_7 x_w + r_8 y_w + r_9 z_w + t_z} \\ y_d + k_1 y_d (x_d^2 + y_d^2) = f \dfrac{r_4 x_w + r_5 y_w + r_6 z_w + t_y}{r_7 x_w + r_8 y_w + r_9 z_w + t_z} \end{cases} \tag{4}$$

Use Euler angle to replace rotation matrix, there is:

$$\mathrm{R} = \begin{pmatrix} C_\alpha C_\beta & C_\alpha S_\beta S_\gamma - S_\alpha C_\gamma & C_\alpha S_\beta C_\gamma + S_\alpha S_\gamma \\ S_\alpha C_\gamma & S_\alpha S_\beta S_\gamma + C_\alpha C_\gamma & S_\alpha S_\beta C_\gamma - C_\alpha S_\gamma \\ -S_\beta & C_\beta S_\gamma & C_\beta C_\gamma \end{pmatrix} \tag{5}$$

Where S_i and C_i ($i=\alpha, \beta, \gamma$) are the simplified forms for $\sin(i)$ and $\cos(i)$. α, β, γ are Euler angles.

For the points on the standard template is in the same plane, suppose that z_w is equal to 0 and the template plane parallels to the image plane, so $C_\beta \approx 1$, $C_\gamma \approx 1$, $S_\beta \approx \beta$ and $S_\gamma \approx \gamma$. The equation is obtained from (5):

$$\mathrm{R} = \begin{pmatrix} C_\alpha & -S_\alpha & C_\alpha \beta + S_\alpha \gamma \\ S_\alpha & C_\alpha & S_\alpha \beta - C_\alpha \gamma \\ -\beta & \gamma & 1 \end{pmatrix} \tag{6}$$

From (4) and (6), the following equation is gotten:

$$\begin{cases} x_d + k_1 x_d ({x_d}^2 + {y_d}^2) = f \dfrac{x_w C_\alpha - y_w S_\alpha + t_x}{-\beta x_w + \gamma y_w + t_z} \\ y_d + k_1 y_d ({x_d}^2 + {y_d}^2) = f \dfrac{x_w S_\alpha + y_w C_\alpha + t_y}{-\beta x_w + \gamma y_w + t_z} \end{cases} \tag{7}$$

Here, $x_d = (x_f - C_x)dx$, $y_d = (y_f - C_y)dy$. x_f and y_f are the pixel coordinate of the charactristic points. dx and dy are the distances in direction of X, Y between two CCDs.

Equation (7) is the model to be obtained. There are 6 extrinsic parameters and 4 intrinsic parameters needed to be calibrated, include f, (C_x,C_y), and aberrance factor k_1.

2.2 Calibration Block Imaging Model

In this paper, to pre-calibrate CCD, a cuboid is taken as the calibration standard block. The height of it is D. Point O_2 is the center point of the up plane and O_1 is that of the down plane(Shown as in Fig.2). Suppose the straight is vertical to the image plane P_I and is the center line of the standard block, which passes through the center point O'of the image plane P_I . The line is the optic axis of the CCD.

The center point coordinate (C_x,C_y) on the image plane is just the same image coordinate of O_1 and O_2. Through similar triangles principle, there are:

$$\begin{cases} \dfrac{oo_1}{f} = \dfrac{A_1C_1}{a_1c_1} \\ \dfrac{oo_2}{f} = \dfrac{A_2C_2}{a_2c_2} \end{cases} \tag{8}$$

Then, a equation is obtained as following:

$$\frac{D}{f} = \frac{A_1C_1}{a_1c_1} - \frac{A_2C_2}{a_2c_2} \tag{9}$$

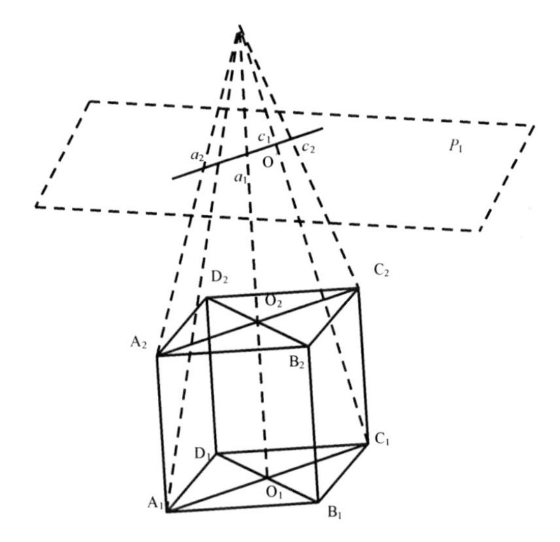

Fig. 2. The projection of the cuboid standard block on image plane

After adjustment, f is as follows:

$$f = D\frac{a_1c_1 * a_2c_2}{A_1C_1 * a_2c_2 - A_2C_2 * a_1c_1} \tag{10}$$

For $A_1C_1= A_2C_2$, equation (10) could be simplified as:

$$f = D\frac{a_1c_1 * a_2c_2}{A_1C_1 * (a_2c_2 - a_1c_1)} \quad (11)$$

3 Calibration

The process of calculation mainly includes three parts.

3.1 Parameters C_x, C_y and f

According to the analysis above, we know that the image coordinates for point O_1 and O_2 is the centre coordinate of the imaging plane (refers to Fig.2). There are 10 featured points on standard template, including 8 peaks and 2 center points O_1 and O_2. We could obtain 5 estimated values for focus distance f:

$$f_1 = D\frac{a_1a_2 \cdot b_1b_2}{A_1A_2 \cdot (a_1a_2 - b_1b_2)},\ f_2 = D\frac{d_1d_2 \cdot c_1c_2}{D_1D_2 \cdot (d_1d_2 - c_1c_2)},$$

$$f_3 = D\frac{d_1d_2 \cdot a_1a_2}{D_1D_2 \cdot (a_1a_2 - d_1d_2)},\ f_4 = D\frac{b_1b_2 \cdot c_1c_2}{B_1B_2 \cdot (b_1b_2 - c_1c_2)},$$

$$f_5 = D\frac{a_2d_2 \cdot b_2c_2}{A_2D_2 \cdot (a_2d_2 - b_2c_2)}$$

Here, $A_1A_2=D_1D_2=B_1B_2=A_2D_2=D$. Through experiment, many values of f would be gotten. The average of them is used as f.

$$\bar{f} = \frac{1}{n}\sum f_i \quad (12)$$

3.2 Parameters α, t_x and t_y

Take one point on the standard template as the origin of ΣO_w, so the coordinates of other points can be gotten (shown Fig.3).

If the radial aberrance is not considered, a equation can be gotten from (7):

$$\frac{x_d}{y_d} = \frac{x_w \cos\alpha - y_w \sin\alpha + t_x}{x_w \sin\alpha + y_w \cos\alpha + t_y} \quad (13)$$

Only t_x and t_y are not setten zero in the meanwhile, the following process is correct. In order to meet the above request, we just need to suppose the origin of ΣO_w far away from the center point of the image plane. If $t_y \neq 0$, then:

$$x_w \cdot y_d \cdot cos\alpha - y_w \cdot y_d \cdot sin\alpha + y_d \cdot t_x - x_w\, x_d\, sin\alpha - y_w\, x_d cos\alpha = x_d \cdot t_y \quad (14)$$

$$\begin{bmatrix} x_w \cdot y_d - y_w \cdot y_d & y_w \cdot y_d - x_w \cdot x_d & -y_d \end{bmatrix} \cdot C = x_d \quad (15)$$

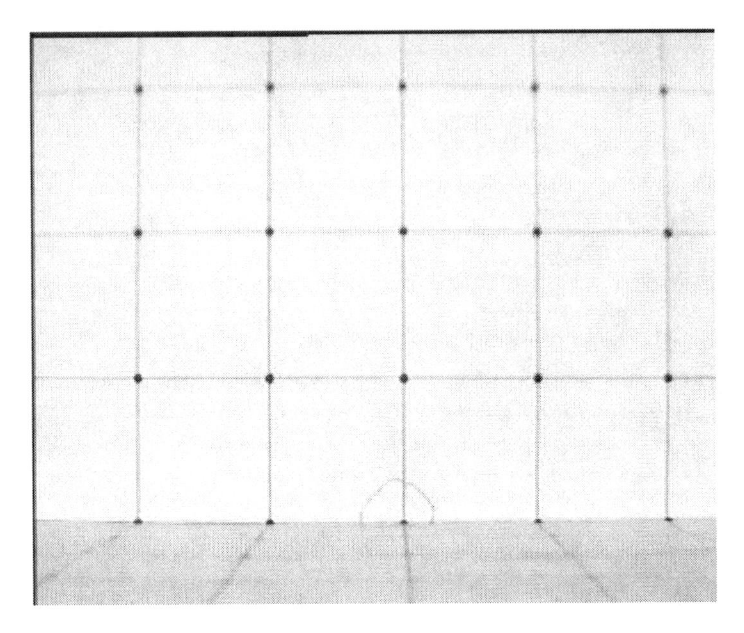

Fig. 3. Standard Template

Here, $C = [C_1 \quad C_2 \quad C_3]^T = [cos\alpha/t_y \quad sin\alpha/t_y \quad t_x/t_y]^T$.

Because $t_y \neq 0$, C_1 and C_2 could not be zero at the same time .If $C_1 \neq 0$, then:

$$\alpha = arctan\frac{C_2}{C_1},\ t_y = cos\frac{\alpha}{C_1},\ t_x = C_3 t_y \tag{16}$$

3.3 Parameters β, γ, tz and k_1

If the radial aberrance is considered, a equation can be gotten from (7):

$$\begin{cases} (-x_d \cdot x_w \cdot \beta + x_d \cdot y_w \cdot \gamma + x_d \cdot t_z)(1+k_1 \cdot r^2) = \\ \qquad f \cdot (x_w \cos\alpha - y_w \sin\alpha + t_x) \\ (-y_d \cdot x_w \cdot \beta + y_d \cdot y_w \cdot \gamma + y_d \cdot t_z)(1+k_1 \cdot r^2) = \\ \qquad f \cdot (x_w \sin\alpha - y_w \cos\alpha + t_y) \end{cases} \tag{17}$$

4 Experiment Results and Analysis

The camera type used in the experiment is SK-813 (Shi Jia CCD). Its effective area is 3.2mm×2.4mm, its resolution is 352*288 pixels. The circles with radius 5mm, which array by the distance 50mm, are printed on the standard template plane. The standard block is a cube with 70mm×70mm×70mm.

During the experiment, the standard block is adjusted to the proper position, we could get f and (C_x,C_y), then the extrinsic parameters and aberrance factor k_1 are obtained according to the the points on the standard template.

The experiment result :

f=6.7856mm, (C_x,C_y)=(183, 135), α=1.75,
β= -0.045rad, γ=-0.42rad; k_1= 0.0056.
$\boldsymbol{T}=[-85.4417\ \ -94.2545\ \ 495.5446]^T$

It is proved through the experiment that the camera's extrinsic and intrinsic parameters can be worked out conveniently, fastly and more precisely.

5 Conclusion

The calibration proposed in this paper needs that the image plane paralleles to the standard template. It could calibrate camera's extrinsic and intrinsic parameters without pre-calibration. Because it makes good use of the geometry information of the standard block, the center coordinate of the image plane and lens' effective f could be figured out fastly and precisely. Besides, we also considered the effect of lens' radial aberrance and improved the calibration precision on current conditions.

References

1. Ma, S., Zhang, Z.: Computer Vision: Calculation Theory and Algorithm Foundation. Science Press, Beijing (1998) (in Chinese)
2. Jain, R., Kasturi, R., Schunck, B.G.: Machine Vision. Mechanical Industry Press, Beijing (2003)
3. Shin, S.S., Hung, Y.P., Lin, W.S.: Accurate Linear Technique For camera Calibration Considering Lens Distortion by Solving an Eigenvalue Problem. Optical Engineer 32, 138–149 (1993)
4. Tsai, R.Y.: A Versatile Camera Calibration Technique for High-accuracy 3D Machine Vision Metrology Using off-the-shelf TV Cameras and Lenses. IEEE Journal of Robotics and Automation 3(4), 323–344 (1987)
5. Zhuang, H.Q., Wu, W.C.: Camera Calibration with-a-Near-Parallel (Ill-Conditioned) Calibration Board Configuration. IEEE Transactions on Robotics and Automation 12(6), 918–921 (1996)
6. Zheng, N.: Computer Vision and Pattern Recognition. National Defense Industry Press, Beijing (1998) (in Chinese)

Granularity Detection in Images of Feeding or Discharging in Jaw Crusher Based on Network Edition Software ImageJ

Zhiyu Qin[1,*], Shuo Rong[2], Rui Lv[1], and Xingfu Rong[1]

[1] College of Mechanical Engineering, Taiyuan University of Technology, 030024 Taiyuan, China
[2] Applied Biology Institute, Shanxi University, 030006 Taiyuan, China
rqs5619@163.com

Abstract. Granularity condition assessment is essential when developing an intelligent real-time monitoring system of feeding and discharging in jaw crushe, so this paper presents a method for such expectation. The images of crushing material to be broken in a jaw crusher are captured by the CCD camera. Analysis and calculation of boundaries and image segmentation of those images are implemented in a network edition image analysis software ImageJ based on its critical boundary points selecting and regional watershed algorithm computing. The results show that this method can detect granularities in feeding or discharging materials images with reasonable accuracy.

Keywords: Real-time Monitoring, Jaw Crusher, CCD Image, Granularity, Network Edition Software ImageJ.

1 Introduction

Jaw crushers have the advantages of simple structure, security and reliability, so it has become an essential equipment in mining production for a century. In recent years, based on the mining working environment caused extremely hurt to the health of workers, proposes of an intelligent real-time monitoring system of feeding and discharging in jaw crusher have caused widespread concern. Among them, monitoring of granularity of feeding and discharging needs to be improved [1],[2].

Computer vision-based image processing methods have been investigated as an alternative to provide solutions to practical measurement, identification, and size distribution analysis. Achieving similar outputs with manual means will be time consuming and painstaking.

Therefore, this article, on the basis of our existing studies [3], suggests a more perfect propose of real-time monitoring method for feeding and discharging granularity. Color images of feeding or dischargin particles are captured by an CCD camera in a fixed orientation and transfered to the computer-based processing platform, and then analysed and computed by the network edition image analysis

G. Zhiguo et al. (Eds.): WISM 2011, CCIS 238, pp. 74–80, 2011.

software ImageJ to deal with the granularity of objective images. First, approximate boundaries of objective particles images are analysed and calculated. Then some more complex adhesion particles images ranges are segmented using the watershed algorithm based on picking up critical boundary points method. Finally, particles granularity average is calculated by the results of the objective image segmentation. The results show that this method can be used to segment images that have particles adhesion components with concave, convex shape distribution.

2 Methods

The overall objective of this research is to test whether this process framework can detect, recognize and evalute the reasonable particles granularity of images recorded by CCD camera. The process is divided into three parts: objective image data collection, data transformation, boundaries determination and particles granularity average assessment. In current practice, the former is done by the CCD camera system automatically, while the latter two are performed in the computer-based processing plateform with a advanced interface technique and a specific image analysis software.

2.1 Objective Image Data Collection and Data Transformation

Objective is ore that either is conveyed to feeding equipment or discharged after being crushed in a jaw crusher. A ordinary CCD camera is installed above those particles and records their images data. Fig.1 shows such an color image which data are recorded by the CCD camera and transferred to the PC computer by its interface technique and opened by ordinary windows software.

Fig. 1. A objective color image by CCD camera

Fig. 2. A screenshot of objective image

2.2 Objective Image Procssing

2.2.1 Network Edition Image Analysis Software ImageJ

Specific computer vision applications of particle size and size distribution may require advanced programming using proprietary programming language environment such as Visual C, Visual Basic as well as MATLAB with specialized image processing toolboxes. Other methods can employ various morphological functions of commercial image processing software, such as Image Pro, National Instruments-

Vision Builder, and EPIX-XCAP for morphological features extraction and quantification. An attractive alternative is ImageJ plugin development for particle size and size distribution as a machine vision application. ImageJ is a Java-based, multithreaded, freely available, open source, platform independent, and public domain image processing and analysis program developed at the National Institutes of Health (NIH), USA [4,5]. ImageJ plugin is an executable Java code from ImageJ platform, tailor-made for a specific application including machine vision.

ImageJ can run on all the major computer-based plateforms, either as an online applet or as a downloadable application. Its image processing functions include a wide range such as calculation of image histograms, image morphology, thresholding, edge finding, smoothing, particle counting and computation of the dimensions and area of features. This software ImageJ is chosen to carry on our image processing.

For edge extract in image processing, ImageJ uses Sobel operator to extract edges based on taking a maximum value in the edge for first-order directional derivative variation. On the problem of segmenting adhesion images, the watershed algorithm is used to detect adhesion and overlapping edges of some particles. By this algorithm, the UEPs are computed in the Euclidian distance map (EDM), and then expansion operation are performed to each UEPs until to the end when an edge contactt occurs.

2.2.2 Image Preprocessing

Fig. 2 shows a computer screenshot of the objective image that is an operation interface the software ImageJ in which displays a picture that is an our objective image and waits to be processed further. This color picture contains much information.

In order not to lose image information, initial step on first is the background substraction preprocessing to the original color image, which is based on Stanley Sternberg's "rolling ball" algorithm. It first can form a 3D surface based on pixel values of the original image, and then perform a processing to the image through rolling the back side of rolling ball to create a background. This operation may to a large extent eliminate certain effects of noises and shadows and help a lot for confirming edges of images. The such processing result is shown in Fig.3. From Fig.3 it is not difficult to find that it still contains a rich information of ore textures after its being processed with reduced background algorithm, and appears fuzzy and complex adhesion among some boundary regions of particles.

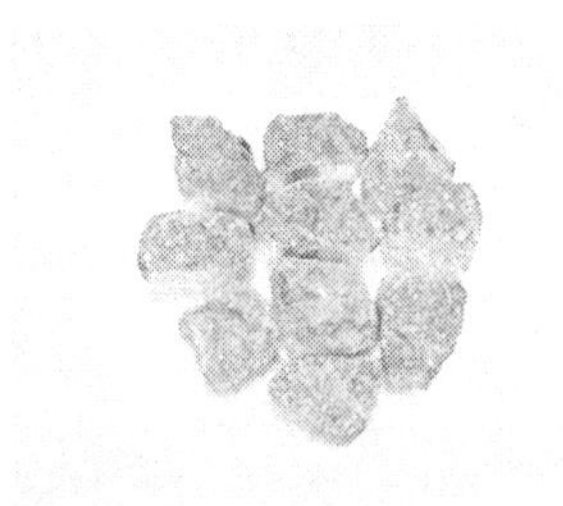

Fig. 3. An image by background substruction

Results

File Edit Font Results

	Area	Mean	Min	Max	X	Y	Slice
1	0	45	45	45	1244	788	1
2	0	19	19	19	1160	924	1
3	0	112	112	112	1208	1192	1
4	0	45	45	45	1316	1564	1
5	0	81	81	81	1756	1568	1
6	0	119	119	119	1752	1168	1
7	0	71	71	71	1756	928	1
8	0	123	123	123	1664	768	1

Fig. 4. A parameters table of picking up critical points

2.2.3 Set of Critical Poins Picking Up, Gaussian Blur and Grayscale Processing

After the background substraction operation, the Distance Map tool of ImageJ software is used to creates a set of critical points, which can be used to count and record pixel values and coordinates of the set and, of course, considered as the most important role for segmentation points of the complex adhesion image. The parameters table of the set of critical points is shown in Fig.4 with 30 as the pixel value of the domain and white as the color selected. As a result, this procedure can greatly improve the accuracy of segmenting area by the way of watershed processing. The image procssed by the method of set of critical point picking up is shown in Fig.5.

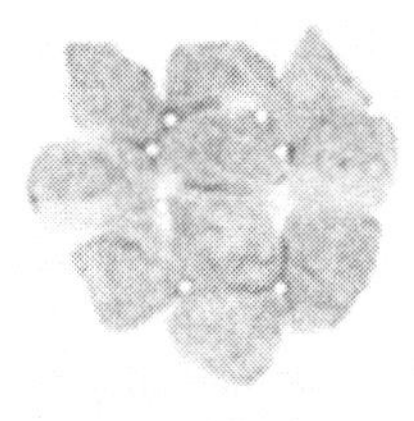

Fig. 5. An image by picking up critical points **Fig. 6.** An image procssed by Gaussian blur

The Gaussian blur operation in the ImageJ software is a smoothing filter utilizing the Gaussian function to make the convolution operation. Its algorithm treats external pixel values to be equal to ones of the most proximated edges and takes more attention to edge pixels instead of internal pixels, and such a relevant blur effect is determined by the edge pixels, especially, the corner pixels. This kind of blur visual effect is similar to browse images by a translucent screen, which is totally different from the shadowd of the lens stolen Coke image. The Gaussian blur function of ImageJ can reduce the effects of high-frequency component of the image and considered as a low-pass filter, so it is very effective to reduce image noise and detail of an image. The image procssed by Gaussian blur is shown as Fig.6.

A color of each pixel in a color image is determined by the *R, G, B* three components, and each component may take 255 color values, so that each pixel may have 16,581,375 color valums. But a grayscale image can be considered as a special color image which each pixel has the same value for the *R*, G, B three components and may just tack 255 color values. So in the digital color image processing it tends to turn a color image into a grayscale image at first, which can still reflect the entire image of global and local color and brightness levels of distribution, and also may makes subsequent image processing and calculation to be less. Grayscale processing has three main methods: maximum value, average, and weighted average. In our work we chose the thirdth one as our grayscale processing method. Experimental and theoretical derivations have proved that the value for *R, G, B* may be the value of brightness value when the $R=G=B=(0.299R+0.587G+0.114B)$ for the pixel. Fig.7 shows this kind of grayscale images after grayscale processing.

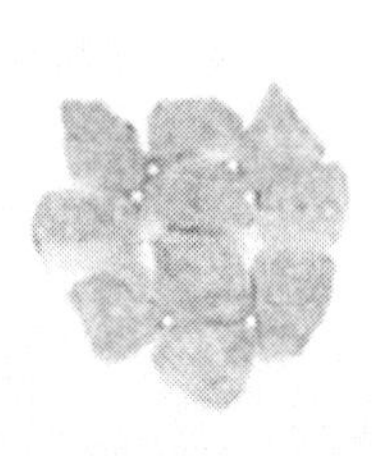

Fig. 7. An image after grascale processing

Fig. 8. An image segmented by minimum error threshold

2.2.4 Image Segmentation

First of all the image segmentation is carried out to the Fig.7 using a standard minimum error threshold method. Its principle is first to seek the image greyscale distribution and then use the statistical method to determine the optimum threshold. The processing result is shown in Fig.8, from which it is not difficult to know that there exsist complex adhesions among particles'edges. With its contours it is unable to determine a reasonable particle size.

To this end, the watershed segmentation algorithm in the JmageJ is chosen to handle the application shown in Fig. 8 which has a more complex adhesion problem. This algorithm is of a special feature of adaptive iteration. The algorithm needs to provide a target tag and is able to detect edges of adhesion and overlapping among objective particles. This segmentation is based on the calculation a corrosion set of points (UEPs) in the Euclidian distance map (EDM), and each UEP is treated by expansion operation until an adhesion between particles or an contact with a another UEP's edge that is being evaluated to happen. The processing result is shown in Fig. 9. Not hard to know from the Fig. 9, segmentations of adhesion regions between particles are well performed.

Fig. 9. An image segmented by wateshed

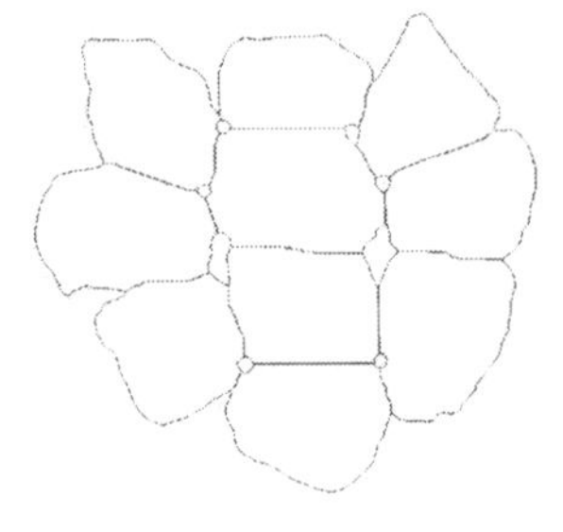

Fig. 10. An image by watershed

2.2.5 Determination of Edges or Profiles of Particles

It has to determine the particle's edges or profiles before performing a granularity operation. The ImageJ software has a Sobel operator to determine edges or profiles among particles in a target image. It uses a linear filter that contains three rows and three columns adjacents to offset the noise sensitivity with a simple gradient filter (by single row /single column). Sobel operator uses the following filters: $H_X=[-1,01;-1,0,2;-1,0,1]$ and $H_Y=[-1,-2,-1;0,0,0;1,2,1]$. It can, respectively, calculate an average gradient component through the adjacent three rows and three columns. The image procssed by the watershed method is shown in Fig.10.

3 Macro Command

Those procedures mentioned above may perform automatically by programing a macro command as a file named "startupMacros". After its being installed, next run may automatically load it, which exsists in the form of a command. Macro commands written as follows:

```
selectWindow("Broken Material.JPG");
run("Subtract Background...", "rolling=40 light");
//setTool("point");
makePoint(1216, 736);
makePoint(1160, 940);
makePoint(1664, 768);
makePoint(1768, 952);
makePoint(1764, 1188);
makePoint(1212, 1160);
makePoint(1308, 1572);
makePoint(1756, 1560);
run("Gaussian Blur...", "sigma=10");
run("8-bit");
setAutoThreshold("Default");
//run("Threshold...");
setThreshold(0, 235);
run("Convert to Mask");
run("Watershed");
run("Outline");
```

4 Conclusion

The method approached in this article focus on solving the problems of image segmentation for complex adhesions and edges extraction in the objective image. Main results are obtained as follows:

1. It gets initial results for resolving the problem of the image segmentation to the complex edge adhesions with a concave, convex shape distribution among particles.
2. Using its extensible programming advantages of the software ImageJ and Java language, some kinds of plugins are developed, which makes internal background substraction, Gaussian blur, grayscale, threshold selection, image segmentation algorithms being more flexible.

3. The automated operations are available through the macro programming, which reduces the man-machine conversations and accelerates the development forward to intelligent direction for resolving this kind of problem.

Acknowledgments. This material is based upon our work supported by the Returnee's Researching Project Fundation of Shanxi Province of China under Grants 2010-40 to the Taiyuan University of Technology.

References

1. Baoxian, L., Shiping, L.: Jaw Crusher. Metalligical Industry Press, Beijing (2008)
2. ASABE Standards.: Method of Determining and Expressing Particle Size of Chopped Forage Materials by Screening—ANSI/ASAE S424.1 DEC01. ASABE. St. Joseph, 619–621, Michigan (2006)
3. Song, Y., Qin, Z., Rong, X.: Approach to Detect Image Edge and Prospect by Imagej. Mechanical Management and Development 23, 180–181 (2008)
4. Rasband, W.S.: ImageJ. U.S. National Institutes of Health. Bethesda, MD, USA, http://rsb.info.nih.gov/ij/ (accessed, August 2010)
5. Writing ImageJ Plugins—A Tutorial. Version 1.71., http://rsb.info.nih.gov/ij (accessed, August 2010)

The Embedded Web Player Based on Davinci Structure*

Shi-qin Lv[1] and Xiao-jun Zhu[2]

[1] Department of Mathematics, Taiyuan University of Technology, 030024 Taiyuan, China
[2] College of Computer Science, Taiyuan University of Technology, 030024 Taiyuan, China
lvshiqin@163.com, zhuxiaojun@tyut.edu.cn

Abstract. Based on TI OMAP3530 processor development platform, Ångström desktop environment and an embedded web browser are ported, and an audio/video plug-in are designed, which can supports MP3, MPEG2/4, AVI, and H.264 data streams playback online. This player fully take advantage of the OMAP3530 dual-core features, which use DSP core to decode the audio/video data streams, ARM core to manage and control the system. Experiments show that player has low power consumption, high speed, and stable operation characteristics.

Keywords: Web Player, Plug-in, OMAP3530, Davinci Structure.

1 Introduction

The embedded player has a wide application prospect in industry control field, and has been widely used in all kinds of household appliances and portable net terminals, such as Internet-TV, iDVD, Web terminals, Digital STB and palm computer etc. For this, to design the high-performance Web players are always the goal of embedded system engineers.

In recent years open multimedia applications platform (OMAP) is on the market for wireless and mobile media by TI, which uses DAVINCI architecture integrates advanced ARM core of CortexA-8 structure and DSP64x+ core, and including the optimization of hardware 2D /3D accelerator[1]. Especially low-power consumption of the open multimedia applications platform devices offer distinct advantages in portable, handheld device, and thus to achieve higher performance and lower power consumption of the embedded player provides a reliable hardware basis.

This system is designed base on the Strong power of OMAP3530 in computing and Audio and video processing , in order to build a embedded Web player of Davinci Architecture, which can supports MP3, MPEG2/4, AVI, and H.264 data streams smooth playback online. This player should take advantage of the OMAP3530 dual-core features, which use DSP core to decode the audio/video data streams, ARM core to manage and control the system.

---

* This research was supported by the Soft Science Fund of Shanxi (2010041057-04) and Youth Fund of TUT (2011).

G. Zhiguo et al. (Eds.): WISM 2011, CCIS 238, pp. 81–86, 2011.

2 System Solutions

2.1 Overall System Architecture

The player consists of three layers. OMAP3530 processor and other peripheral devices (such as GPRS module, Wi-Fi module, etc.) constitute the underlying hardware platform layer. The middle layer, named system layer, is Composed of Linux kernel, drivers and graphics file system. The embedded browser and audio/video playback plug-in form the top-level application layer. The embedded browser via HTTP protocol to access streaming media server, when the page needs to streaming media support, audio/video playback plug-in is called as a plug-in. Overall system architecture as shown in Figure 1.

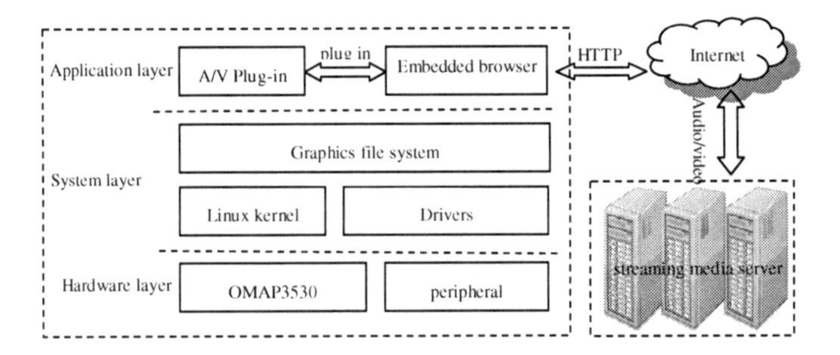

Fig. 1. Overall system architecture

2.2 Data Transfer Model

System's Data transmission based on HTTP request / response model[2], when man clicks on the hyperlinks of audio/video files through the embedded browser, a TCP link with the server will be established, then, the embedded browser will send the HTTP request to the server, the request contains the request method, URI, protocol version and associated MIME-style information. Server responds with a status line, contains the message of the protocol version, a success and failure codes, and the associated MIME-style message (contains server information, resources entity information and possible resources content). After the HTTP response message is received, the embedded browser to start checking the contents of the response message. When it comes to the resource name can not be parsed, the embedded browser will check the registration information of plug-in, when the MIME type and audio/video plug-in matching, Link library for this plug-in is loaded into memory, and the URL of media resources in the form of parameters passed to the player plug-in. DSP core to decode and playback the streaming media file which download from the server by audio and video plug-in. AC'97 standard audio data is output through the built-in or external speakers, as the same time, video data through the LCD display.

3 Hardware Design

The core of Web player hardware platform is an OMAP3530 processor, the net Interface, audio in/out interface, USB OTG, USB HOST, SD/MMC interface, serial interface, CAMERA interface and LCD Interface are extended as peripheral interfaces. The system hardware frame is presented in Figure 2.

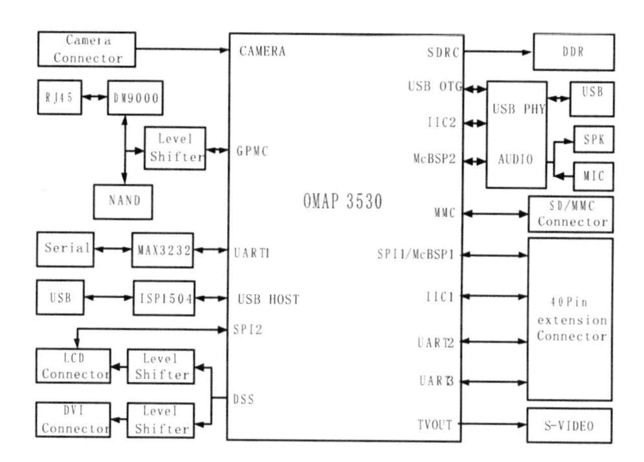

Fig. 2. The system hardware frame

Base on the idea of the core platform and scale board are separated in this system, the core platform adopts the 8F cables and mainly includes OMAP3530 processors, DDR2 memory, Ethernet and clock electric circuit; and the scale board adopts the 4F cables and mainly includes outer interface electric circuit. The detail context is as follows.

The OMAP3530 processor integrates the ARM Cortex™-A8 core of 600-MHZ and the 430-MHZ TMS320C64x+™ DSP core with advanced digital signal processing algorithms. This processor adopts 423 pins package for the pin pitch of 0.65mm.

The memory chip, MT29C1G24MADLATA-6IT, is produced by Micro, 137 pin BGA package, for the pin pitch of 0.8mm. This chip integrates a 1Gbit DDR and a 1Gbit FLASH.

TI audio codec chip, TPS65930, is produced by TI, which integrates power, full audio encoder/decoder and other features, you can achieve double-voice channel and downlink channel stereo, can play all standard audio.

DM9000 chip is selected for Ethernet port physical layer chip, and a 7-inch TFT LCD display is selected for this system, its outsider interface is parallel port, and with a transfer cable. In addition, the system also includes a high-speed USB 2.0 HOST port, which can be accessed WIFI module; a USB OTG interface, through the chip ISP1504 to control the receiving and sending data.

4 Software Design

4.1 Develop Environment

The audio/video files are decoded by DSP core, as the same time, the ARM core completes the management and control of system, so we need set up the cross compile platform for ARM side and the link compile platform for DSP side. The detail environment arguments are as follows.

1) Host machine operate system Ubuntu 8.10;
2) Cross compile Tools ARM GNU/Linux EABI 2008q1;
3) OMAP35x-PSP-SDK-02.01.03.11;
4) DVSDK_3_00_02_44.

From overall structure of the system, we can find software design includes the following aspects. Porting and building the systems software platform, including boot loader and embedded Linux, driver development and the construction of graphical file system. Port the embedded browser, and development audio/video plug-in.

4.2 Boot Loader Transplantation

The boot loader is composed of X-Loader and the U-Boot two levels, in this system the version of X-Loader and the U-Boot respectively is X-Loader-1.41 and U-Boot 1.3.3. After the system power, as the first level, X-Loader is automatically copied to internal RAM and executed by the CPU internal ROM. Its main role is to initialize the CPU, the X-Loader according to starting method copy U-Boot from NAND Flash or MMC/SD to the memory, and then transfers the domination to U-Boot. U-Boot, as the second level boot loader, is mainly used to user interaction, provide an image update, and boot the kernel etc.

4.3 Construction of Ångström Desktop Environment

Ångström was started by a small group of people who worked on the OpenEmbedded, OpenZaurus and OpenSimpad projects to unify their effort to make a stable and user friendly distribution for embedded devices like handhelds, set top boxes and network-attached storage devices and more[3]. Ångström system has provided a variety of software, including the document editor, Internet browser, audio/video player, graph editor, and we can control this system by USB keyboard and mouse. So we can construct the Ångström desktop environment on OMAP3530 to simplify our work, such as Linux migration, Driver development, Construction of graphical file system and transplantation embedded browser, and so on. We can construct the Ångström desktop environment by online tools. Configure parameters as shown in table 1.

Table 1. Configure parameters of Ångström

Machine	Beagle board	**Release**	Stable
Base System	Regular	**Device Manager**	udev
User Environment	X11	**Desktop Environments**	Matchbox、Gnome
Type of Image	tar.bz2	**Additional Packages**	Firefox, Gstreamer

The embedded browser, Minefield browser, is provided by Ångström desktop environment, which supports the NPAPI standard set of plug-in interface [4], by implementing this interface can be called audio/video player plug-in.

4.4 The Design of Audio/Video Plug-In

From the physical form, audio/video plug-in can be regarded as a dynamic link library, which independent of the browser. Minefield browser will get the description and the corresponding MIME types of the player plug-in from the specified folder while start, then carries on the registration in the browser according to the gain information. Minefield will inquire registration information of plug-in when the resources name unable to analyze. When the MIME type and registration information of plug-in is matched, Minefield will load dynamic link library to memory.

After the plug-in unit is loaded, the browser will assign memory for plug-in and initializes instance shared resource, then transmits media resources' URL by the parameter form for plug-in. Until browser to complete data transfer or data transfer is aborted, the related resources used by the data stream will be released, instance will be destroyed and the shared resources of plug-in will be released.

The player plug-in is composed of a main thread and a data receives sub-thread, a audio encode/decode, a video encode/decode sub-thread, a output sub-thread and the user interface sub-thread. First of all, the parameter which gains will be processed by main thread, the main purpose is removes redundant information in URL, transforms character set of parameters, and according to media data flow rate to determine the buffer size. Next, these child threads will be opened until Codec Engine Runtime and Davinci Multimedia Application Interface[5] initialized. When the received data reaches the minimum buffer data amount, the data receiving child thread to parse the media data obtained, by stripping header file to obtain the media data types, in order to achieve the separation of audio and video. Final, data is encapsulated into buffer handle format required by DMAI and is sent to the audio decoder FIFO and video decoder FIFO, respectively. According to different data types, audio/video decoding sub-thread to build the corresponding decoder, and then removed the decoded data from decoded FIFO and sent into audio/video display FIFO. The output sub-thread takes the audio as benchmark to complete the audio and video audio synchronization, then delivers the data to the audio/video output equipment realizes the broadcast output, respectively. Play, pause, stop and other functions are implemented in the User Interface sub-thread.

5 System Features

The hardware platform of Web Player is designed base on TI OMAP3530 processor. First, we transplanted X-Loader, U-Boot and the Ångström desktop environment, and then the plug-in of Minefield browser is designed for audio and video playback. Thus, an embedded Web video player is realized based on the Davinci structure. The function characteristic is as follows.

a) Hardware platform separation of the core board and backplane, the core platform adopts the 8F cables and the scale board adopts the 4F cables. This design not only benefit to the design depending on the need to re-floor and the development of new products, but also conducive to hardware debugging and cost savings.

b) Supports many kinds of audio and video formats, especially to support H.264 video standard. Take advantage of dual-core processor, ARM system-control tasks to complete, the video decode processing by the DSP.

c) Base on HTTP request / response model, in the embedded Linux platform, extending the browser functionality. The player plug-in is called by browser to realize the audio/ video online playback.

6 Test and Conclusion

Using the player, we visited the Web server and video server which in LAN, CERNET and Internet to test player performance respectively. Overall, the design has achieved the anticipated target. The main conclusions are as follows.

a) Player hardware performance is stable and reliable, boot loader is started normally, the operating system is stable, and the graphical user interface is friendly.

b) The player supports Flash, MMC/SD and so on many kinds of way to load the Linux kernel and the file system.

c) System support RJ45, WIFI input, and support for touch screen, external keyboard and mouse;

d) The embedded browser is stable, can playback audio/video smoothly, if the network band is enough. The audio/video plug-in can support many kinds of audio/video form, the audio form mainly has AAC, MP3 and so on, and the video form mainly has MPEG2/4, AVI as well as H.264.

References

1. Texas Instruments. OMAP35x DVEVM Getting Started Guide, http://processors.wiki.ti.com/index.php/OMAP35x_DVEVM_Getting_Started_Guide
2. Wang, C., Hu, C., Liu, x.-n.: Implementation of HTTP Protocol in the Embedded System. Journal Of Electron Devices 25(1), 93–96 (2002)
3. LinuxToGo. Ångström Manual, http://www.linuxtogo.org/gowiki/AngstromManual
4. Liu, W., Chen, S.-y., Wu, X.-s.: Media Plug-in Based on Embedded Linux. Computer Systems & Applications 18(4), 127–130 (2009)
5. Texas Instruments. Davinci Multimedia Application Interface, http://processors.wiki.ti.com/index.php/Davinci_Multimedia_Application_Interface

The Design of Violating-Traffic Statistical Analysis Module

Liping Chen

Department of Electronics and Information Engineering,
Shanxi polytechnic college, Taiyuan, 030006, China
ty-clp@163.com

Abstract. Violating-traffic statistical analysis is an important part of railway cargo traffic information statistics. This paper uses "rail freight traffic information statistical system" which is one of three main modules in internet application. Violating--traffic statistic methods monitor the implementation of full-flow and improve the level of information.

Keywords: Violating-traffic Sstatistics, Extreme Table, JSP.

1 Introduction

Railway cargo traffic information statistics is one of the main rail systems. The product of railway cargo transport is cargo ton-km , which is expressed in cargo turnover. Cargo turnover means the sum of the product of tonnage of each shipment and transport distance during the reporting period, which is the basis of calculating the transportation costs, labor productivity, the liquidation distribution, the average fortune of goods, the average transport density of goods and so on. It is the key indicator to reflect the transport efficiency [1].

The path vehicle transported from the departure station to the final station is named traffic running path, referred to as the traffic flow path. In addition to a specific traffic path, we usually use the shortest path. Traffic roundabout path is the temporary specified traffic path when the railway line occur some temporary changes. Bypass traffic roundabout also known as transport, has a certain proportion in practice. According to monitoring data traffic on the roundabout project by relevant statistics is the roundabout traffic flow statistics, also known as violating-traffic statistic.

2 Violating-Traffic Statistics in the Rail Freight Traffic Violation Information Statistical System Paper Preparation

Railway cargo traffic information system is designed to statistic and analysis the traffic flow in rail spacing and monitoring point, to accurately reflect the transport through the adoption of a railway administration traffic, traffic arrival and departure traffic statistics of the actual situation and provide a reliable actual basis, to provide the theoretical basis for accurately preparing statistical reports, to achieve integrated management such as traffic testing, inquiry, traffic monitoring

G. Zhiguo et al. (Eds.): WISM 2011, CCIS 238, pp. 87–92, 2011.

and adjusting flow (violating-traffic adjustment) and so on. Violating-traffic statistical analysis is one of the main modules of this system.

The system including the system log, checking the data correction, the dictionary via the maintenance, traffic information query, generate violating-traffic calculation, data network transmission, violating-traffic adjustment and statistical analysis. Upon completion of the database structure design, and using SQL Server 2000 database management system created on the basis of the database system, the system consists of three main parts: the login screen, traffic information query module and the violating-traffic statistical analysis module.

3 Design of Violating-Traffic Statistic Analysis Module

Violating-traffic statistic analysis module is responsible for checking the information from grass-roots unit, update station name, name and other basic dictionary, write and maintain traffic-specific path files, aggregate traffic classification, calculation and analysis of violating-traffic information, Accordingly carry through telegraph keying of foreign bureau violating-traffic ,violating-traffic electronic text production, upload, complete violating-traffic adjustment, and the traffic analysis. Violating-traffic statistical analysis module system design is one of the main design modules. The design of this module is for the purpose of violating-traffic statistical analysis, the JSP page of query results designs by Extreme Table, and implements paging display.

3.1 Introduction of Extreme Table

Extreme Table [2] is one of the best paging components which is very popular currently. Compared to other similar controls, it is beautiful and practical. It is one of Extreme Components which is the open source JSP custom tag library, Extreme Table is powerful and easy to configure, extend, and customize. The purpose to use this component is to display the data page according to people's habits, and sort data, summary and export to Excel format. It also can filtrate, export, and use the limit function to extract database data, custom column styles, etc.

3.2 Relational Markov Networks

Paging is the most common web application functionality, It is not need to write the next page from scratch in the current pace of IT development, but should integrate components of a fully functional page [3].

Extreme components default search all the records from the database, if the result set is large, ec will use the Limit object passed to the Controller PageNo, PageSize, OrderBy and other paging information. The servicer will return the total number of records and the contents of the current page to Ec. Specific implementation steps:

1) Obtain data from the database and save to the Collection;

2) Store the Collection in the session or the variable of request;

3) Remove the variable with taglib from the page, and traverse the Collection to display each record.

3.3 The Sort and Export Capabilities of Extreme Table

Extreme Table is embedded sorting function. By default, the sortable in all columns are true in the property. Click the name of any column will see a triangle symbol: the ascending triangle means upward, and the adown triangle means descending [3].

Extreme Table default support for the file of xls, pdf, csv format, it can also be custom export format. If you want to use the export feature of Extreme Table to export excel file, the operation requires the following steps:

1)Take the jar package under the lib/xls directory in the compressed package into the WEB-INF/lib directory package

2) Register a filter in the web.xml, the code is as follows:

```
<filter>
   <filter-name>eXtremeExport</filter-name>
   <filter-class>
       org.extremecomponents.table.filter.ExportFilter
   </filter-class>
</filter>
<filter-mapping>
    <filter-name>eXtremeExport</filter-name>
    <url-pattern>/*</url-pattern>
</filter-mapping>
```

3.4 Concrete Realization of the Application

3.4.1 Design of Parameters Table

Establish the tables of violating-traffic adjustment of the outside and inside council in Sybase database. Input the relevant entry code, exit code, distance and other parameters to the outside council, and input the ID number and the appropriate path parameter to the inside Council for calculations.

3.4.2 Design of Query Parameter Input Interface

Create inputparam.jsp documents to provide input interface including council not, query start and end dates and scheduling orders flags, and specify the action to VehicleStream.action.

3.4.3 Design of Action Class and Struts.xml Skip Control Configuration

1) Establish VehicleStream class, and define the four properties such as bureau, qsyf, zzyf and order, generate their getter, setter methods.

2) Define execute method in the VehicleStream class, and pass the query parameters to the DAO interface layer, call the query methods defined in the DAO implementation class , and introduce the query results to session for the calls of query JSP page.

3) Create struts.xml configured profile in the src directory and configure the VehicleStream, when the result is "success", jump to the test.jsp to display query results.

3.4.4 Design of DAO Interface

1) Define the DAO class interface SwapDAO, and define the implementation method gethj () of query operations in the interface.

2) Define a concrete implementation class SwapHj1Imp of SwapDAO class, and implement gethj () query in such class. Query operation is connected to java: comp / env / jdbc / jmtj, using JdbcTemplate class of Spring's framework to implement. The main Code:

```
            Context initCtx=null;
            List list=new ArrayList();
            try {
                  initCtx=new InitialContext();
                  DataSource ds=(DataSource)initCtx
                  .lookup("java:comp/env/jdbc/jmtj");
      JdbcTemplate jdbctemplate=new JdbcTemplate(ds);
                  String sql ="select
a.adjid,b.notes,b.tn_bz,sum(a.toned),c.qjzjyj,c.qjzjyj*su
m(a.toned) from adj_ton a,adj_def b,zq_bjtzds c where
a.adjid=b.adjid and a.adjid=c.adjid and (a.ny between ?
and ?) and b.tn_bz=? group by
b.tn_bz,a.adjid,b.notes,c.qjzjyj order by
b.tn_bz,a.adjid";
                  Object[] params = new Object[3];
                  params[0]=(String)qsyf;
                  params[1]=(String)zzyf;
                  params[2]="B";
      list=jdbctemplate.query(sql,params,new RowMapper(){
   public Object mapRow(ResultSet rs,int rnum) throws
SQLException{
                  if (rnum<0)
                     return null;
                  Hj1Result hj1result=new Hj1Result();

                    hj1result.setJjm(rs.getString(1));
                    hj1result.setJm(rs.getString(2));
                    hj1result.setRkdm(rs.getString(3));
                    hj1result.setHzrk(rs.getString(4));
                    hj1result.setCkdm(rs.getString(5));
                    hj1result.setHzck(rs.getString(6));
                    return hj1result;
                    }
                  });
            }catch (Exception e){
                  e.printStackTrace();}
            return list;
```

3.4.5 Design of Servlet in the Service Layer

The method of calling the SwapHj1Imp in DAO layer is gethj (). Pass the query results into the hj1results object, and pass the query results into the results screen by the setAttribute () method.

3.4.6 Design of the JSP Query Results Show Pages

This page is designed by Extreme Table, which receive query results h1results from the service layer. Take all attribute values of h1results to generate the columns of query results table, specify rowsDisplayed value setting the number of lines per page, and export Excel report format by setting exportXls. The main code is as follows:

```
<%@taglib uri="http://www.extremecomponents.org"

prefix="ec" %>
<link rel="stylesheet" type="text/css"
href="<%=request.getContextPath()%>/styles/extremecompone
nts.css">
<ec:table
   items="hj1results"
   var="employee"
action="${pageContext.request.contextPath}/VehicleStream.
action"
imagePath="${pageContext.request.contextPath}/images/*.gi
f"
   rowsDisplayed="25"
   width="800"
   cellpadding="0"
   title="调流信息">
   <ec:exportXls                    fileName="presidents.xls"
tooltip="Export Excel"/>
   <ec:row highlightRow="true">
    <ec:column property="jm" alias="局名"/>
    <ec:column property="hzrk" alias="入口"/>
    <ec:column property="hzck" alias="出口"/>
   <ec:column property="zjds" alias="增加吨数"/>
   <ec:column property="qjzjyj" alias="全局增加运距"/>
   <ec:column property="qjzzl" alias="全局增加周转量"/>
    </ec:row>
 </ec:table>
```

3.4.7 The Result of the Program

Enter the query address In the IE browser, you can query the detailed information of the adjust flow, grasp the dynamic information of the adjustment of roundabout traffic flow, and provide the basis for the development of violating-traffic adjustment programs, to achieve a win-win between volume and efficiency of Railway Administration. Result of the program shown in Figure1:

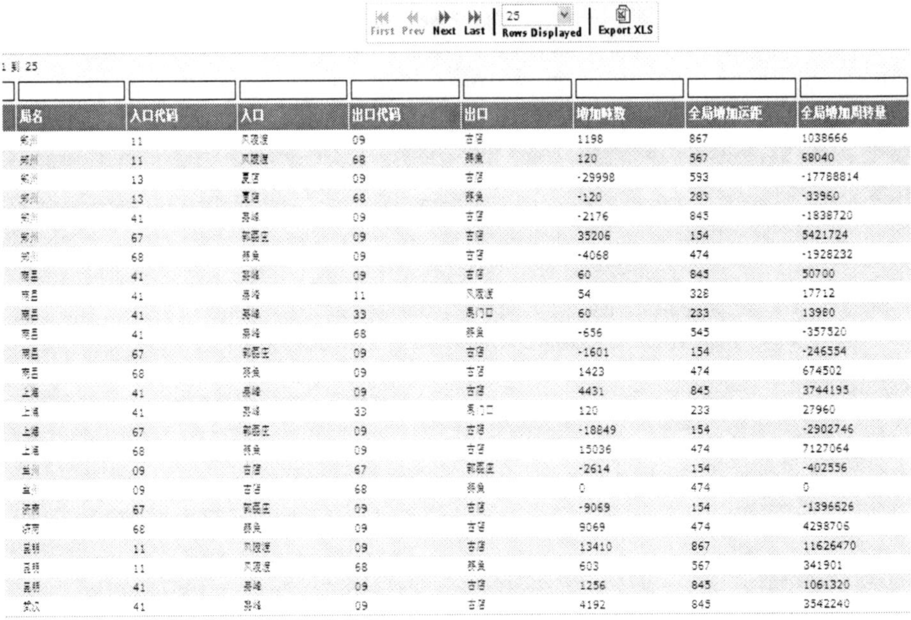

First Prev Next Last | 25 Rows Displayed | Export XLS

1 到 25

局名	入口代码	入口	出口代码	出口	增加吨数	全局增加运距	全局增加周转量
郑州	11	风陵渡	09	古营	1198	867	1038666
郑州	11	风陵渡	68	[illegible]	120	567	68040
郑州	13	[illegible]	09	古营	-29998	593	-17788814
郑州	13	[illegible]	68	[illegible]	-120	283	-33960
郑州	41	[illegible]	09	古营	-2176	845	-1838720
郑州	67	郭磊庄	09	古营	35206	154	5421724
郑州	68	[illegible]	09	古营	-4068	474	-1928232
南昌	41	[illegible]	09	古营	60	845	50700
南昌	41	[illegible]	11	风陵渡	54	328	17712
南昌	41	[illegible]	33	[illegible]	60	233	13980
南昌	41	[illegible]	68	[illegible]	-656	545	-357520
南昌	67	郭磊庄	09	古营	-1601	154	-246554
南昌	68	[illegible]	09	古营	1423	474	674502
上海	41	[illegible]	09	古营	4431	845	3744195
上海	41	[illegible]	33	[illegible]	120	233	27960
上海	67	郭磊庄	09	古营	-18849	154	-2902746
上海	68	[illegible]	09	古营	15036	474	7127064
兰州	09	古营	67	郭磊庄	-2614	154	-402556
兰州	09	古营	68	[illegible]	0	474	0
济南	67	郭磊庄	09	古营	-9069	154	-1396626
济南	68	[illegible]	09	古营	9069	474	4298706
昆明	11	风陵渡	09	古营	13410	867	11626470
昆明	11	风陵渡	68	[illegible]	603	567	341901
昆明	41	[illegible]	09	古营	1256	845	1061320
武汉	41	[illegible]	09	古营	4192	845	3542240

Fig. 1. Queries shown figure about violating-traffic statistical analysis module

4 Conclusions

Taiyuan Railway Bureau assumes the task of transporting the coal to allover the country, so its cargo transport is the largest in the eighteen railway bureaus in our country (18% or so). The proportion of violating-traffic transport was high (12%). Rail freight traffic information statistics system in the Council's trial achieves good results.

References

1. Zhang, Q.: Strengthen the Inspection for Bypass Train Flow and Realize statistic information within Whole Administration. Railway Transport and Economy 29(6), 3–5 (2007)
2. Wright, B.: JSP Database Programming Guide, vol. 6, pp. 1–263. Beijing Hope Electronic Press, Beijing (2001)
3. Perry, B.W.: Java Servlet & JSP Cookbook. O'Reilly Media Inc., U.S.A (2004)

The Visible Research of 6DOF Ballistic Trajectory Base on VRML Technology

Chanyuan Liu

School of Mechatronic Engineering, North University of China, Taiyuan, Shanxi 030051
lcy@nuc.edu.cn

Abstract. The numerical results of external ballistic equations are lists of data tables or data curves, which can't display the motion of shell vividly. SIMULINK is a new software to solve differential equations by numerical methods, VRML offers a simple way for data visualization. SIMULINK software is used to solve the equations and VRML for data visualization. The trajectory can be displayed exactly and vividly by the two methods above.

Keywords: VRML 6DOF Ballistic Trajectory Visualization.

1 Preface

The exterior ballistic was described by a rank differential equation, and only few can use elementary method to get the answers, and most of the questions' research must draw support from the numerical solution. SIMULINK technology is based on the modular modeling of a new type of numerical simulation method which have been applied in ballistic field literature [1,2], the document 1 has simulated the particle the literature of the rotating 2 missile trajectory simulation six degrees of freedom, but they have not study about visualization ballistic.

In the process of research of ammunition dexterity, agrument stage always needs cartoon demonstrates for work principle and the animation process of external trajectory. 3DMAX software were used to make cartoon in some years ago, but because of the moves of 3DMAX was controlled beforehand, and only can get through the exterior ballistic equation. But after the initial ballistic's condition change it must be anew to advance the ballistic' s count and animation rendering of a complicated process .

Visualization simulation programming has OpenGL, Direct X SDK and so on, but the require of the programming ability is too high, and the development cycle is also relatively long. VRML provides an easy ways to the scene visual design personnel which can get rid of weapons designer heavy programming, and achieve the ammunition simulation visualization quickly. It has been used in rocket weapon teaching and training system [3]. It is convenient to use in search of ballistic visualization.

G. Zhiguo et al. (Eds.): WISM 2011, CCIS 238, pp. 93–98, 2011.

This text is combined with the two aspects, which use simulink to build the 6 dof ballistic and use VRML to build the visualization scene and combined them with simulink application luterface, to simulate the exterior trajectory visualization simulation rockets.

2 The Mathematical Model of 6DOF Ballistic Trajectory

For the different ballistics, according to the different emphasis may establish different ballistic equation. For the sake of not lossing the general in this text, there are some 6 dof trajectory equations which according the external ballistic's theory:

(a)Dynamics equation

$$m\begin{bmatrix}\frac{dv_x}{dt}\\ \frac{dv_y}{dt}\\ \frac{dv_z}{dt}\end{bmatrix}=\begin{bmatrix}\cos\varphi\cos\psi & -\sin\varphi & \cos\varphi\sin\psi\\ \sin\varphi\cos\psi & \cos\varphi & \sin\varphi\sin\psi\\ -\sin\psi & 0 & \cos\psi\end{bmatrix}\begin{bmatrix}P\\ 0\\ 0\end{bmatrix}+$$

$$\begin{bmatrix}\cos\theta\cos\sigma & -\sin\theta & \cos\theta\sin\sigma\\ \sin\theta\cos\sigma & \cos\theta & \sin\theta\sin\sigma\\ -\sin\sigma & 0 & \cos\sigma\end{bmatrix}\begin{bmatrix}Q\\ Y\\ Z\end{bmatrix}+\begin{bmatrix}g_x\\ g_y\\ g_z\end{bmatrix}$$

$$J_x\frac{d\omega_x}{dt}=M_x$$

$$J_y\frac{d\omega_y}{dt}+(J_x-J_Z)\omega_x\omega_Z=M_y$$

$$J_x\frac{d\omega_z}{dt}+(J_y-J_x)\omega_x\omega_y=M_z$$

In the formula, m,θ,δ are stands for rocket instantaneous respectively, quality, elevation Angle, ballistic PianJiao; Ψ, ϕ respectively is yaw Angle, the pitch; P, Q, Y, Z respectively, for the engine thrust, lift, and the aerodynamic drag; lateral force. And Jx, Jy, Jz are respectively , projectile coordinate system for the three axis of inertia; For the moment, Mx, My, Mz in the body coordinate axis component 3; And $\omega_x,\omega_y,\omega_z$ for the angular velocity in the body; the projection coordinate, v_x,v_y,v_z for the projectile, the speed of the launch coordinate system; At launch g_x,g_y,g_z for gravity, the weight of the frame.

(b)Kinematics equations

$$\frac{d\varphi}{dt} = (\omega_y \sin\gamma + \omega_z \cos\gamma) / \cos\psi$$

$$\frac{d\psi}{dt} = \omega_y \sin\gamma - \omega_z \cos\gamma$$

$$\frac{d\gamma}{dt} = \omega_x + \tan\psi (\omega_y \sin\gamma + \omega_z \cos\gamma)$$

$$\frac{dx}{dt} = v_x$$

$$\frac{dy}{dt} = v_y$$

$$\frac{dz}{dt} = v_z$$

In the formula, x, y, z stands for rocket launch coordinate system, in the coordinates.
（c） Geometric relation formulae equation

$$\sin\beta = \left[\cos\gamma \sin\psi \cos(\theta-\varphi) + \sin\gamma \sin(\theta-\varphi)\right]\cos\sigma - \cos\gamma \cos\psi \sin\sigma$$

$$\sin\alpha = -[\cos\sigma \sin\gamma \sin\psi \cos(\theta-\varphi) + \cos\sigma \cos\gamma \sin(\theta-\varphi) - \sin\sigma \cos\psi \sin\gamma] / \cos\beta$$

$$\cos v = (\cos\alpha \sin\gamma \cos\psi - \sin\psi \sin\alpha) / \cos\sigma$$

Type,α,β,v stands for attack, Angle and roll Angle; γ stands for rocket roll corner.

3 Ballistic Simulation Model Based on SIMULINK

Visualization six degrees of freedom trajectory simulation models seperate into three most, external force and torque computation module, motion equation module and visualization module, the general structure of the model is shown in figure 1 below.

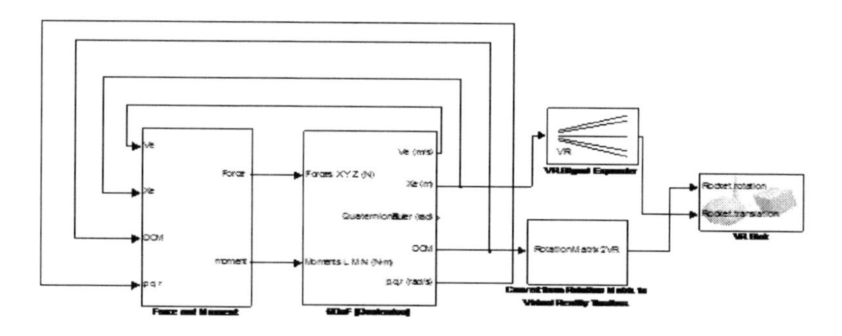

Fig. 1. Model of Visualized 6 DOF trajectory simulation

For any ballistic equations the core is the kinematics and dynamics equation. in the will all the external force by rockets have split into the ground and coordinate: All outer torque decomposition to coordinate the projectile, the model shown in figure 2. The output of model for three velocity component, position, the rotation rate, orientation cosine matrix and said the body posture of the euler Angel.

The simulation with 6 dof ballistic model needed basic data included: shoot horn, velocity, rocket engine thrust cures; Aerodynamic data, including resistance coefficient, pitch moment coefficient, yaw moment coefficient ,damping moment coefficient and pressure heart coefficient, etc; quality and rotational inertia data, centroid data; Geophysical data and local wind field data and other environmental parameters. When the above data, can be determined, using Simulink to simulate calculation results, the calculation results obtained can be used directly visualization simulation .

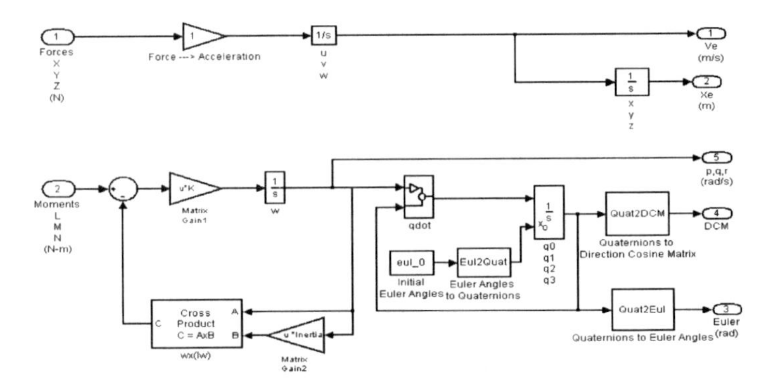

Fig. 2. Model of equations

4 Establishing Virtual Reality Scene

6 dof ballistic by virtual reality scene has 3major components: fire positions, launchers and rockets and background. VRML has strict syntax definition and powerful functions, but to construct a more complex three-dimensional object, a lot of VRML codes are needed .therefore, this paper use of MatLab bringing VRML constructor V-out to build a rocket launch Builder 2.0 virtual scene MatLab bringing VRML constructor V-an is a important part in the Builder out more emphases (Object Library is Object) ,it contains which has designed good ready-made virtual Object .It can directly object library to join the specimen object of virtual world, also can bring the object added to establish targets. This simulation import the background to the target stock, and then export to the unreal scene from the target stock. According to the need again from object repository in rockets, launch vehicles derived with the object of scenes for the launch. Finally establishing virtual scene fig. 03 shows, the literature four method, can build a more real 3d terrain.

VRML's standard is the child node under the father node, and under the child nodes there still have child nodes, and child nodes under the child node. Each node is composed by domain, each domain has threshold, these threshold appointed a scene features. According to the actual situation of branches launch can find corresponding object type attribute box, and make appropriate modification, until each part of the position relations are more fit. Rocket position and posture need to update according to calculation results of SIMULINK

Fig. 3. Virtual reality scene

5 Visualization Simulation

Simulink about virtual reality in the module has three categories: the core module, special input device module, signal module. One core modules for VR Sink and VR returned module, VR Sink module provides to the virtual world from Simulink output signal user interface, VR returned from module provides a virtual world input user interface to Simulink interface. Virtual reality toolbox can move in the MATLAB interface and Simulink interface, this paper chose Simulink interface with the virtual reality toolbox, and to realize the establishment linkage virtual reality simulations. On the need to adopt RotationMatrix2VR module will orientation cosine matrix transformation for virtual reality models need rotation vector. Operating model, can be in different perspective ballistic simulation observed under, as shown in figure 4 above for scene showed ballistic. The typical view include: launcher rear, perpendicular to the ballistic landing nearby, the plane far above the position, ballistic etc.

Fig. 4. Trajectory simulations scene

6 Conclusion

Ammunition engineering visualization simulation as weapons design field is a very important research methods for weapons design which provides a powerful help, including warheads power evaluation visualization, ammunition fired ballistic visualization, ammunition and mutilate and goal, vulnerability assessment visualization and so on, it can do: simulation objects trajectory, direction, azimuth of the change; Simulation ammunition effects, such as fire, explosion and sound effects; Realizing complex models such as aircraft, tanks, etc, the geometrical model of texture map, etc; Realize more complex calculation and analysis.

Based on SIMULINK produces, six degree-of-freedom missile trajectory model of VRML and outside and visualization simulation intuitivey reflect the rocket external trajectory characteristics. The results are accurate and intuitivey reflect the rocket external trajectory process. Basic accord with the actual circumstances of the flight. Available for design parameter estimation, reporting demo and teaching use.

References

1. in hua: based on Matlab/Simulink lowa, the rotation of the trajectory simulation six degree-of-freedom missile. Journal of the Arrows and Guidance 27(2), 222 (2007)
2. Yang, l.: Matlab in the application of trajectory simulation. The Computer Simulation 25(1), 58–61 (2008)
3. Zheng, Y.: The VRML rocket weapon teaching and training system. Computer Engineering and Design 29(15), 4037–4039 (2008)
4. Ping Z.H.: 3d, terrain simulation of VRML. Naval Ship, 27 Work Process, Electric Son (6), 128–131 (2007)
5. Peng, W.H.: The aircraft is virtual reality simulation study. The Modern Electronic Technology 276(13), 46–48 (2008)

Research on Modeling Microblog Posts Scale Based on Nonhomogeneous Poisson Process

Hongcheng Zou, Gang Zhou, and Yaoyi Xi

National Digital Switching System Engineering & Technological Research Center, ZhengZhou 450002, HeNan, China
{zouhch123,gzhougzhou}@gmail.com, Brain3333@163.com

Abstract. Research on modeling microblog posts scale was the foundation of predicting the number of microblog posts. To predict the microblog posts was beneficial to reasonably schedule load and estimate flow. Firstly, this paper explained the mathematical definition of microblog posts scale problem, compared the three microblogging characteristics random,independent and orderly with the four conditions of poisson process and established the nonhomogeneous poisson process model for microblog posts scale problem. Then, took sina microblog as experimental platform, established the model and calculated the arrival intensity function and mean function based on the measured datas of sina microblog posts. Finally,partial mathematical test was done to the model, then conducted predicting the number of sina microblog posts, which showed quite good predicting performance by comparing with the measured datas. Finally summarized the whole paper and pointed out the future work.

Keywords: Microblog, Network Behaviour, Nonhomogeneous Poisson Process.

1 Introduction

Microblog was a new media form emerging from Internet recently. It applied network technique, wireless communication technique and so on to conveniently satisfy the users' demand to communicate in anytime and anywhere. Microblog was a kind of blog form that allow users renewing short text(usually less than 140 words) in time and posting publicly. It allowed everyone or only group designated by the author to read. After years of development, microblog posts could be published in many ways, including short message,MMS, instant message, email and web. Microblog message not only supported text type, even multimedia type, such as pictures or vedio clips[1]. Compared with the traditional blog, microblog had following four characteristics:

First was brief of content. Microblog required the posts in no more than 140 words length. Brief was the most important characteristic different from traditional blog. In the meantime, as the length of microblog posts was limited, the microblog users were casual in using words. Irregular abbreviations,network language and expression symbols were being widely used. Irregularity was quite obvious for microblog language.

Second was realtime of post. Because of the short characteristic of microblog posts and the multiple terminals to post microblog message, especially the support of the

G. Zhiguo et al. (Eds.): WISM 2011, CCIS 238, pp. 99–112, 2011.

mobile terminal, users could publish his seen, heard and felt on their microblog in the first time. Microblog played an important role in the Japan big earthquake which showed that microblog become a real new media form.

Third was extensive of spread. As a new thing, Microblog reached hundreds of millions users in the course of just a few years' development. Microblog users could learn the dynamic news of whom they were interested of by adding concerns and spread their own points of view and messages by being added to be the concerned. For example, the number of fans who added Kai-Fu Lee as their concern reached more than four millions which fully showed that the microblog platform spread extensively.

Fourth was the long-term of influence. Microblog usually expressed spontaneously, released at any time, so its language was quite oral, but easily created some new words. These words were quit refining and meaningful also with humor that could easily spread, went deeply into people's hearts, then influenced the society for a long time. Although microblog emerged for quite a short time, it already formed a kind of microblog culture. For example, "microblogging control" was a new word in the microblogging time,which meaned people who were extremly mad of microblog.

Microblog developed with rapid speed because of the above microblog characteristics. Facing the rapid development of microblog, it become an outstanding problem for web site management that how to better estimate microblog site traffic in order to better upgrade and optimize the hardware and software configuration of the site and reasonably schedule the load. Based on the above, this paper underwent researching on microblog posts scale.

In fact, there was already related research on microblog. Paper[2] worked on the implementation of the microblog search engine based on nutch. Paper[3] studied microblog sentiment detection and paper[4] researched on the microblog feature selection. Research on microblog posting behaviour characteristics was the foundation of modeling the microblog posts scale. Microblog posting behaviour would lead to change of web directly. There was connection between posting behaviour and change of web. Actually, there was related research about change of web both at home and abroad. For example, in the study of a single web page, the paper[5] considerd life cycle of a single web page as poisson process and estimated the time of next change according to its history locus. The paper[6][8] researched into the ways to estimate and traced the change of web and proposed ways to keep the freshness of web. The process of web change was also considered as poisson process in the research of the paper[7][10][11]][12]. For research on a large number of web pages, the paper[6] got their changing rule by sampling,monitoring and statisticsing WWW pages. The paper[14] researched into the time law of web change in the overall, described the law of change by introducing the heap and stack distance model and did experiment to test the time locality law of web page change. The paper[15] studied the change of vedio web site review and underwent prediction. The result achieved good result in the application of improve searching algorithm. Whether the above fruit could apply in predicting microblog posting behaviour directly was still a problem because of microblog's own characteristics.

Poisson process was one of the basic mathematic model for descripting accumulated number of random events. Generally,the poisson process was random,indepent and ordered. Intuitively, if random events happened indepently in disjoint time interval,and happened no more than once in sufficiently small time interval, then their accumulated number was a poisson process[16]. Poisson process model could model many real problem,in addition to applying in discribing the law of web pag change. For example, the paper[17] used nonhomogeneous poisson process to monitor transactions on a customer's account for deviations from the customer's established behaviour pattern. The paper[18] used nonhomogeneous poisson process model in research on software reliability. The paper[19] estimated species populations by using nonhomogeneous poisson process.

The paper was organized as follow: part2 analyzed three characteristics of microblog posting behaviour,explained the formal definition of microblog posts scale and compared with the four conditions of poisson process. The comparing result showed that the formal definition of microblog posts scale satisfied the conditions of poisson process. Part3 detailed the process of experiment on the sina microblog platform. After analyzing the experiment datas,part4 further simpified the model as nonhomogeneous with the same daily arrival intensity function,then established the microblog posts model. Part5 partly tested the model. Then based on the model, the paper did prediction of total number of microblog posts in several different time interval,and compared the result with the experiment data. Finally, part6 summarized the whole paper and pointed out the future work.

2 Microblog Posts Scale Modeling

2.1 Research on Microblog Posting Characteristics

Because research on microblog posting behaviour was the foundation of microblog posts scale modeling,the paper firstly needed to analyze the microblog posting behaviour characteriscs. To sum up, microblog behaviour had three characteristics as follow:

1. Orderly. Although the number of microblog users was quite large, it was possible that different users posted posts at the same time,the behaviour of updating the user's mainpage just happened on the site's data server. The updating could be considered as premitive opration. It meaned that no more than one microblog user's data updated in a sufficiently short time,which meaned that the posts updated on the server in order.
2. Independent. That was to say that all the users' posting behaviour were independent in any two non-overlapping time intervals.
3. Random. Random meaned the uncertainty of posting time. It wasn't sure that any user woule post in some time. That is,there wasn't any time interval could ensure that there must be posts posting in the time interval.

Formal description of microblog posts scale detailed as follow:

Definiton 1

$$\begin{cases} \text{random count process } \{X_t\ ;\ t \geq t_0\} \\ \text{inside :} \\ (1):\ X_t = \sum_i X_{ti};(2):\ X_{ti} = Y_{ti} - Y_{t_0 i};(3):\ X_{t_0 i} = 0 \end{cases}$$

Inside:

t_0: starting time

Y_{ti} : absolute posts' number of microblog user i in time t

X_{ti}: posts' number of microblog user i in time t related to time t_0.it was a random variable. Especially, $X_{t_0 i}$:meaned posts' number of microblog user i in starting time t_0 related to itself which equaled to 0.

X_t : posts' accumulated number of all microblog users in time t related to starting time t_0. It was also a random variable. When t changed, the random variable X_t followed changing. Then got the random process $\{X_t\ ;\ t \geq t_0\}$. That was the mathmatic formal description of microblog posts scale.

For the microblog posts scale problem, the paper chose the nonhomogeneous poisson process as the model. This was because: Firstly,the above random process had characteristics called random, orderly and independent. As mentioned above, a random process which intuitively had the three characteristics probably was consistent with poisson process. Secondly,more formal analyzation were explained as follow:

Firstly, the four conditions of poisson process were showed up. Assuming: $\{N_t\ ;\ t \geq t_0\}$ was an accompany counting process of point process in interval $[t_0, \infty)$. If point process had the following conditions[20]:

1. For $t \geq t_0$, point process was consistent orderly in interval $[t_0, \infty)$.
2. The point process was of no aftereffects evolution; it meaned that for any $t \geq t_0$, the points happening in $[t, \infty)$ had nothing to do with the points happening in $[t_0, t)$.
3. For interval $[t_0, \infty)$, There was no any certain limited interval in which there must be points happening.
4. $P_r\left[N_{t_0} = 0\right] = 1$. it meaned that the number of points in the starting moment t_0 was constant 0.

Then the counting process $\{N_t\ ;\ t \geq t_0\}$ was poisson counting process with continuous parameter function.

Inside,Orderly defined as follow:

There was a counting process $\{N_t\ ;\ t \geq t_0\}$, if for any given number $\varepsilon > 0$,there existed a number $\delta \equiv \delta(t, \varepsilon) > 0$ which made the formular (1) established for all the $\delta' \in (0\ ,\ \delta)$:

$$P_r\left[N_{t,t+\delta'} > 1\right] \leq \varepsilon P_r\left[N_{t,t+\delta'} = 1\right] \tag{1}$$

Then the counting process was orderly in moment t. If the point process was orderly in interval $t \geq t_0$, then the point process was orderly in every moment of interval $t \geq t_0$.

If there existed a number $\delta = \delta(\varepsilon)$ which was inrelevant to t ,and made fornular(1) established in every moment of interval $t \geq t_0$. Then the point process was consistent orderly in interval $t \geq t_0$.

If $P_r\left[N_{t,t+\delta'} = 0\right]$ didn't equal to 0,then formlar(1) was equivalent to formular(2) :

$$\lim_{\delta \to 0} \frac{P_r\left[N_{t,t+\delta'} > 1\right]}{P_r\left[N_{t,t+\delta'} = 1\right]} = 0 \quad (2)$$

Formular(2) meaned that:for an orderly point process,as long as the given interval was short enough,ratio of probability of which the number of points happened was more than 1 in the interval and probability of which the number of points happened was equal to 1 in the interval was equal to 0. Intuitively explanation was that the point couldn't happen in a same moment.

From the above random process definition of microblog posting scale, X_{t_0} is constant 0,conditon 4 obviously established. The ordered characteristc of microblog posting behaviour satisfied the condition 1. The independent characteristic of microblog posting hehaviour corresponded to conditon2. And condition 3 could be acquired according to the random characteristic of microblog posting behaviour.

From the above analysis, the paper considered using poisson process as the mathematical model of the microblog posts scale. Taking into acount the following common sense,because microblog user posted in different frequency at different time,so posts' number was different in different moment,it was more reasonable to use nonhomogeneous poisson process as mathematic model. Following showed the definition of the nonhomogeneous poisson process[21]:

Definition 2. if one counting process $\{X(t), t > 0\}$ satisfied three conditions as:

(1) : $X(0) = 0$

(2) : $X(t)$ was an independent incremental process

(3) : $P\{X(t+h) - X(t) = 1\} = \lambda(t)h + o(h)$

$P\{X(t+h) - X(t) \geq 2\} = o(h)$

inside : $\lambda(t)$ indicates the arrival intensity of nonhomogeneous poisson process

$o(h)$ indicates infinitesimal.

Then the counting process was nonhomogeneous poisson process. Its arrival intensity function was $\lambda(t)$.

2.2 Establish Microblog Posts Scale Model

To establish microblog posts scale model using nonhomogeneous poisson process, the first step needed to do was ensuring the arrival intensity function of nonhomogeneous poisson process $\lambda(t)$.

Datas form in which Random point process gathered was usually as follow:devided the observation interval $[t_0,T)$ into several non-overlapping sub-interval $[t_0,t_1)$, $[t_1,t_2)$, $\cdots$, $[t_{j-1},t_j)$, $\cdots$, $[t_{k-1},t_k=T)$ following a certain rule,then gathered points' number n_1, n_2, $\cdots$, n_k happened in these interval,which kind of datas were called histogram datas. Statistic theory generally used count rate histogram to estimate the arrival intensity function[20]. The data points in the count rate histogram could be abtained by the following formular:

$$\left(\frac{n_j}{t_j-t_{j-1}},\frac{t_j+t_{j-1}}{2};j=1,2,...,k\right) \tag{3}$$

Especially,if devided $[t_0,T)$ every a same interval,and took the interval as basic time unit, formular (3) could be simplified as follow:

$$\left(n_j,\frac{t_j+t_{j-1}}{2};j=1,2,...,k\right) \tag{4}$$

Once got the count rate histogram, as long as the expression of the figure was obtained then obtained the arrival intensity function $\lambda(t)$.Known by characteristc of nonhomogeneous poisson process[21], the mean function of nonhomogeneous poisson process with arrival intensity function $\lambda(t)$ could be figured out by expression (5):

$$m_x(t)=\int_0^t \lambda(s)ds \tag{5}$$

After the mean function was figured out, by theorem 1[21],the distribution function of count process in interval $[t,t+s)$could be acquired.

Theorem 1. if there was a nonhomogeneous poisson with mean function,then probablity of which the number of count process points happening in interval $[t,t+s)$equaled to *n* could be calculated by formular(6):

$$P\{X(t+s)-X(t)=n\}=\frac{\left[m_x(t+s)-m_x(t)\right]^n}{n!}\exp\{-\left[m_x(t+s)-m_x(t)\right]\} \tag{6}$$

The theorem showed that distribution function of point process points' number was consistent with poisson distribution in the interval $[t,t+s)$. Its mean value was $[m_x(t+s)-m_x(t)]$ which was the difference of the mean function value of the two interval endpoints. Predictive value of the points' number in any interval could be estimated by the mean value of distribution function in the corresponding interval. That is $[m_x(t+s)-m_x(t)]$.

Generally speaking,it was possible to abtain absolute posts' number Y_{ti} of any microblog user i in any time t. Further ,posts' number X_{ti} corresponding to start time t_0 could be acquired. Then got the posts' number of any user between any adjacent time t_{j-1} and t_j ,denoted by n_{ji}. So it was easy to get n_j which is the accumulated number of posts' number of all the users between any adjacent time t_{j-1} and t_j. Inside, $n_j = \sum_i n_{ji} = \sum_i X_{t_j i} - X_{t_{j-1} i} = \sum_i Y_{t_{j-1},t_j} = \sum_i Y_{t_j i} - Y_{t_{j-1} i}$. If the collection interval in an experiment was the same,by the above theory,estimated number of posts' in any interval could be acquired,which was $\left[m_x(t+s) - m_x(t)\right]$. The actual measured number could be figured out by formular $X_{(t+s)} - X_t$. So it was possible to compare the predictive value and actual measured value of microblog posts' number for any time interval.

3 Experimental Process

As mentioned earlier, for the sake of predicting the microblog site posting scale, first needed to collect necessary datas in order to estimate the arrival intensity function and solve the distribution function. The paper took sina microblog as research object. Sina microblog was the first microblog in china launched by Sina. It began testing in a August 2009. Now it owned more than 100 million users which was the largest in china. The following experiment studied the changes in posts' number of 2000 users. Experiment process detailed as follow:

Experimental enviroment:

Table 1. Enviromental parameters table

Machine model	HP ProLiant DL380 G5 Rack Mount Chassis
Operating system	Windows 2003 Enterprise Edition (32 bit/ SP2 /DirectX 9.0c)
CPU model	Intes Xeon E5440 @ 2.83GHz
Memory size	4 GB (DDR2 FB-DIMM 667MHz)

Experimental steps:

1. Acquired users' ids. Specific process was:JAVA crawler using multithread technique crawled the first level users by width first algorithm from a seed user,got the users' ids and wrote into file and searched the second level users' ids by taking the first level ids as a clue. Repeated the algorithm until the ids' number reached 2000.
2. Collected the data of the users' posts number in one time. Once got the users's ids,we could gather the data of the amount of users' posts. The experiment got the users' posts directly by obtaining the posts' number on the users' homepage(as shown by fig 1).

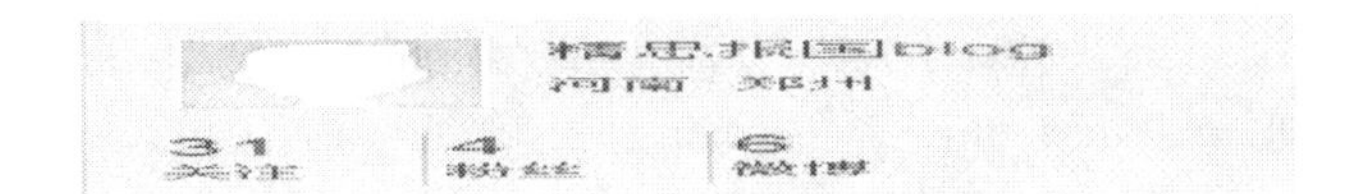

Fig. 1. Screenshot of sina microblog homepage

3. Collected data periodically. Programme set a timer,gathered the 2000 users' posts by step2 algorithm every one hour from a set time and save the datas into file. Programme gathered from October the 29th to December the 31st in total of 32 days. Partial days didn't gather 24 hours because of some Irresistible power, such as sina microblog stopping service on December 1st and power failure.

4 Model Solution

4.1 Model Simplification

In the light of the datas gathered from the above experiment,we could get the total posts of the 2000 users in every hour related to the start time point. The count rate histogram could be easily got according to formular (4). Plotted curves of the image shown in Figure2:

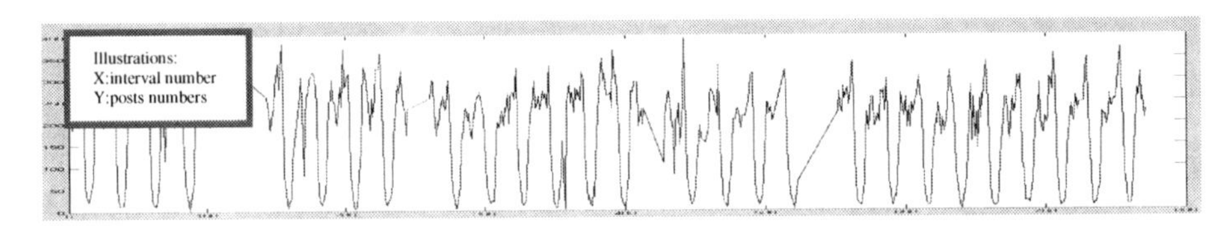

Fig. 2. Curve image of the posts' number of 2000 users every one hour

If these points are represented in the form of histogram,then got the following image:

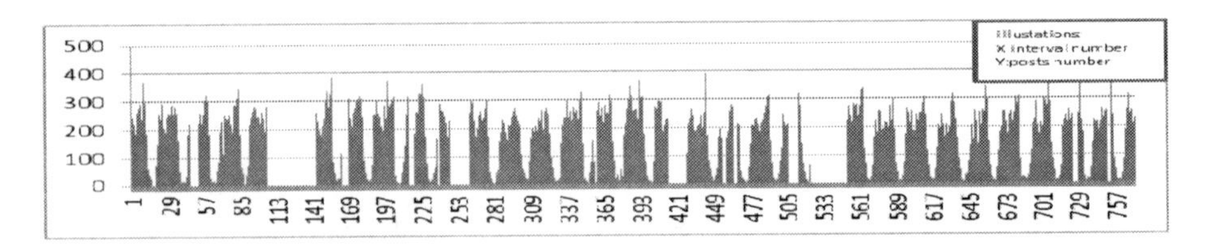

Fig. 3. Histogram of the posts' number of 2000 users every one hour

We could see apparent periodicity characteristic from the shape of the above curve. In order to observe the picture clearly,we chose datas in five days from December 23th to 27th, and dipicted as follows:

We could see shaking periodic in everyday's picture. Further more, Eliminating the error and random vibration factor,the daily model could be simplified as independent nonhomogenerous poisson process with the same daily arrival intensity.

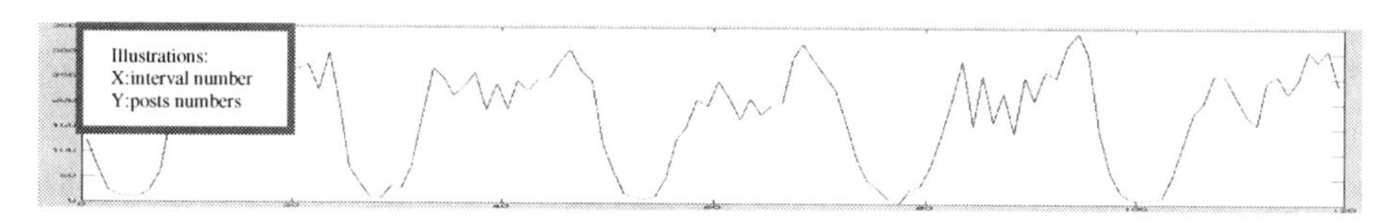

Fig. 4. Curve image of the posts' number of 2000 users every one hour from 23rd to 27th

4.2 Solution of Arrival Intensity Function

The folloing solved the function expression of arrival intensity function $\lambda(t)$ using polynomial fitting.

In order to reduce the error's impact on the model,we overlaid datas in the days which had complete cycle of data,and figured out the arithmetic mean value every moment. The overlaid image was as follow:

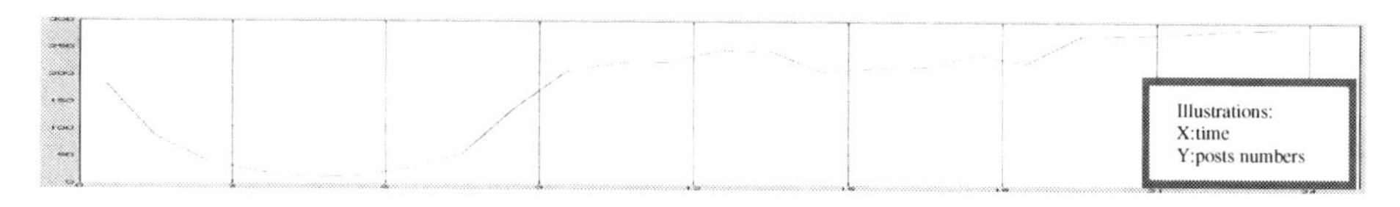

Fig. 5. Posts' curve in a cycle after overlapping and avaraging

In order to get a better expression, it needed to fit the curve. First of all, a natual choice was polynomial fitting. It was needed to determine the fitting order first for polynomial fitting. In order to determine fitting order,we tried to fitting and compare the experiment datas which had all the 24 hours datas in a day to determine the suitable order using the tool of ORIGIN statistic software. The indicators being used were Adj.R-square(A.R.S) and Residual Sum of Squares(R.S.S). For the sake of convenience of explanation and illustration, the following table showed datas which were the fitting results from third-order to ninth-order of the three days' including December 2nd,7th and 11th. The table was shown as follow:

Table 2. Fitting results from third-order to ninth-order on December 2nd, 7th and 11th

A.R.S / R.S.S	December 2nd	December 7th	December 11th
Third-order	87199/0.636	89245/0.637	47909/0.699
Fourth-order	48411/0.787	47042/0.798	27542/0.820
Fifth-order	37833/0.824	30764/0.861	14775/0.897
Sixth-order	37573/0.815	28997/0.816	9585/0.929
Seventh-order	35573/0.814	22187/0.887	5451/0.950
Eighth-order	35562/0.802	16636/0.909	3919/0.960
Ninth-order	35458/0.789	11792/0.930	2785/0.970

The goal of fitting was to make A.R.S greater and R.S.S smaller. Determining the order should consider the changes in two indicators synthetically in order to achieve the best fitting result. From the above datas,it could easily see: Firstly, A.R.S ascended and R.S.S descended when the order was below 5; however A.R.S ascended sharply when the order was above 5. So the fitting order shouldn't be below 5. Secondly,when the fitting order changed from 6 to 7, A.R.S just descended slightly for December 2nd,but ascended sharply for December 7th. At the same time,A.R.S ascended but R.S.S descended sharply for December 11th . Summarized the above two points, the fitting order was best identified as 7. After the analysis by the ORIGIN software, the coefficients of the fitting polynomial were as follow.

Table 3. Fitting result table

	Value	Standard Error
Intercept	192.50045	34.73802
B1	-192264	52.75515
B2	-39.27537	25.12517
B3	12.71322	5.34681
B4	-1.49543	0.58558
B5	0.0842	0.03431
B6	-0.0023	0.00102
B7	2.43639E-5	1.21338E-5

The result expression of seventh-order fitting polynomial was as follow:

$$\lambda(t) = 2.43639E^{-5}t^7 - 0.0023t^6 + 0.0842t^5 - 1.49543t^4 + 12.71322t^3 - 39.27537t^2 - 19.2264t + 192.50045 \quad (7)$$

The corresponding values of two indicators A.R.S and R.S.S were shown as following table:

Table 4. Values of 2 indicators

Residual Sum of Squares	5239.97431
Adj. R-Square	0.96594

It could be seen:The A.R.S indicator reached above 0.96,which illustrated good fitting result. The fitting graph and the origin graph drawn by MATLAB[22] were shown as follow:

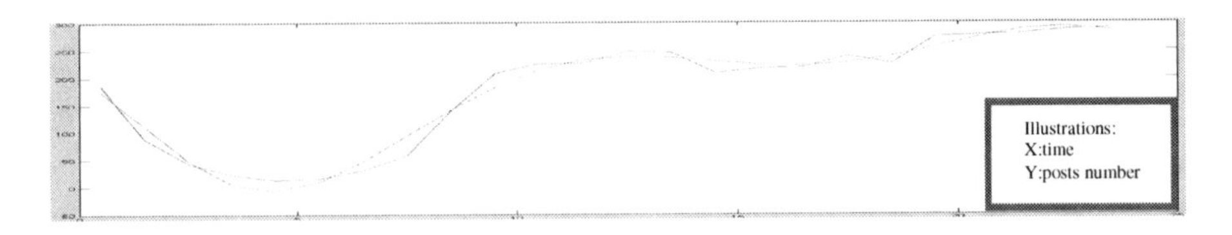

Fig. 6. Superimposed curves of the fitting curve and the origin curve

It could be found:The fitting curve quite coincided with the origin curve from the above graph.

The fitting expressing was the daily arrival intensity function $\lambda(t)$. We could easily get the mean function $m_x(t)$.Once got arrival intensity function. Integral to the formular (7), then got mean function as follow:

$$m_x(t)=2.43639E^{-5}\frac{t^8}{8}-0.0023\frac{t^7}{7}+0.0842\frac{t^6}{6}-1.49543\frac{t^5}{5}+12.71322\frac{t^4}{4}-39.27537\frac{t^3}{3}-19.2264\frac{t^2}{2}+192.50045t \quad (8)$$

5 Model Validation

5.1 Model Test

In order to validate the model,according to the testing method of nonhomogeneous poisson process[23], it could be devided into two steps: firstly,the random variable of daily total posts should obey poisson distribution with a same coefficient for nonhomogeneous poisson process with the same daily arrival intensity function, so,the first step was to conducting Statistical Inference to the samples which were the total posts' number of days which had the complete 24 hours' datas in order to make sure these samples obeyed a same poisson process distribution. Secondly,for nonhomogeneous poisson process with a same daily arrival intensity, in theory, instantaneous arrival intensity value of the same moment everyday should be the same. So,the next step was to conducting Statistical Inference to the samples which were the posts' number of intervals which were the same intervals of a day in order to make sure these samples obeyed a same poisson process distribution.

The following introduced ways to test whether discrete random variables obeyed a same poisson process distribution. Known by 《Random Point Process and Its Application》 [24],Statistics

$$d=\sum_{i=1}^{n}\left(n_i-\bar{n}\right)^2/\bar{n}$$

was similar to χ^2 distribution with $n-1$ degree of freedom, in the condition of random variables n_1 , n_2 , ... n_k obeying the same poisson distribution with coefficient μ . So, Statistics d could be used to test whether random variables obeyed the same poisson distribution. Detailed as follow:

Assuming: $H_0: n_1, n_2 ... n_k$ were all samples of poisson distribution with coeffecient μ.
H_1 : not all $n_1, n_2 ... n_k$ were samples of poisson distribution with coeffecient μ

As mentioned earlier,when assumption H_0 was true,had:

$$d=\sum_{i=1}^{k}\left(n_i-\bar{n}\right)^2/\bar{n}\propto\chi^2(k-1) \quad (9)$$

Then got the rejection region as[25]:

$$\sum_{i=1}^{k}\left(n_i-\bar{n}\right)^2/\bar{n}\le\chi^2_{1-\alpha/2}(k-1)\ or\ \sum_{i=1}^{k}\left(n_i-\bar{n}\right)^2/\bar{n}\ge\chi^2_{\alpha/2}(k-1) \quad (10)$$

First test:chose eight samples of daily total posts' number of days which had complete 24 hours datas,samples were:3926,4060, 3962,3992,3979,3960,3979,4094. Their mean value was 3994, variance was 3087.7. The variance to mean ratio was close to 1. Further more,after computing formular (9), got the value 5.41. Let significant level α equaled to 0.5, then got: $\chi^2_{1-\alpha/2}(k-1)=\chi^2_{0.975}(7)=1.69$, $\chi^2_{\alpha/2}(k-1)=\chi^2_{0.025}(7)=16.013$. Because 1.69<5.41<16.013, the value of d was in the acceptance region,so accepted the assumption H_0 , which meaned that the assumption that daily total posts obeyed a same poisson distribution was accepted.

Second test:chose twelve samples of posts' number of time intervals from 21:00 to 22:00 everyday.They were:268,288,289,270,266,233,306,275,276,284,286,316. Their mean value was 279.75, variance was 441.66. The variance to mean ratio also was close to 1. Further more, after computing formular (9), got the value 17.37. Let significant level α equaled to 0.5, then got: $\chi^2_{1-\alpha/2}(k-1)=\chi^2_{0.975}(11)=3.816$, $\chi^2_{\alpha/2}(k-1)=\chi^2_{0.025}(11)=21.920$. Because 17.37 were between 3.816 and 21.920, so the assumption H_0 was accepted. So the assumption that total posts' numbers in interval from 21:00 to 22:00 everyday obeyed a same poisson distribution was accepted.

After the above two steps of test,the assumption that the microblog posts scale model was nonhomogeneous poisson process with the same daily arrival intensity function was partial evidenced.

5.2 Model Prediction

Using the obtained model, it could do prediction to total posts of 2000 users for a period of time in the future and do comparison with the result gathered from experiment in order to verify validity of the model. Taking predicting the daily total posts on December 23rd as an example,known by the model, the estimate value of the daily total posts on December 23rd could be figured out by the following formular:

$$m_x(t+24)-m_x(t)=m_x(24)-m_x(0)=4257$$

According to the the experiment,The actual daily total posts gathered on December 23rd was 3992. Error of the predicted value was 6.2%. Similarly,the paper did ten groups prediction, the acquired result was as follow:

Table 5. Predicting result table

Interval / Value	1/2 day	5/8 day	3/4 day	7/8 day	1 day
Experiment value	1230	1827	2452	3192	3992
Prediction value	1124	1953	2625	3393	4257
Error (%)	8.6	6.8	7.0	6.3	6.2
Interval / Value	**2 days**	**3 days**	**4 days**	**5 days**	**6 days**
Experiment value	8460	12439	16399	20378	24472
Prediction value	8514	12771	17028	21285	25542
error (%)	0.6	2.6	3.8	4.5	4.3

It could be seen from the above result that it could obtain relative satisfactory predicting result of posts' number by establishing nonhomogeneous poisson process with the same daily arrival intensity.

6 Summaries and Future Work

The paper showed the formal mathmatical definition of the microblog posting scale,and established the nonhomogeneous poisson process model by comparing with and analysing the four conditions of poisson process. Based the model,the paper solved the arrival intensity function and mean function of the model according to the measured data of sina microblog,and conducted the test and prediction,achieved relative satisfactory predicted result.

Because it needed some time to gather datas of microblog posts in a time,there was a certain error for the raw datas. For the reason,when predicting for a short interval,there might be bigger error. It is needed to improve the efficiency of the crawl in order to reduce the error of the raw datas. In addition, The lasting time of the experiment was nearly a month,and some datas didn't be gathered which resulted lack of the sample datas. So the effect of the test and prediction was influenced to some extent. It was worthy going a step further. Finally,in order to serve better for the site's scheduling load and estimating flow work,further study could analyse the ratio of different information form such as audio,vedio,picture and text also the avarage information size,which probably achieved more accurate predicted effect.

References

[1] Wikipedia, http://zh.wikipedia.org/zh/%E5%BE%AE%E5%8D%9A%E5%AE%A2
[2] Zhang, K., et al.: The Study and Implementation of Micro-blog Search Engine Based on Nutch. In: 2010 2nd International Conference on Future Computer Communication, WuHan, China (2010)
[3] Li, G., et al.: Micro-blogging Sentiment Detection by Collaborative Online Learning. In: 2010 IEEE International Conference on Data Mining (2010)
[4] Liu, Z., et al.: Short Text Feature Selection for Micro-blog Mining. In: 2010 International Conference on Computational Intelligence and Software Engineering (2010)
[5] Cho, J., et al.: The Evolution of the Web and Implications for an Incremental Crawler. In: Proc of 26th International Conference on Very Large Database, pp. 527–534 (2000)
[6] Cho, J., Garcia-Molina, H.: Synchronizing a Database to Improve Freshness (2000)
[7] Douglis, F., Feldmann, A., Krishnamurthy, B.: Rate of Change and other Metrics: A live Study of the World Wide Web. In: Proc. of the USENIX Symp. on Internet Technologies and Systems, pp. 147–158 (1997)
[8] Cho, J., Garcia-Molina, H.: Estimating Frequency of Change. ACM Transactions on Internet Technology 3(3), 256–290 (2003)
[9] Fetterly, D., Manasse, M., Najork, M.: A Large-scale Study of the Evolution of Web Pages. In: Proc.12th International World Wide Web Conference, New York, pp. 669–678 (2003)

[10] Brewington, B.E., Cybenko, G.: How dynamic Is the Web? In: Proc. of the 9th Int'l. World Wide Web Conf., pp. 257–276. Elsevier Science Publishers, North-Holland (2000)
[11] Li, X.M.: An Estimation of the Quantity of Web Pages ever in China. Journal of Peking University (Science and Technology) (2003)
[12] Bar-Ilan, J., Peritz, B.D.: Evolution, Continuity, and Disappearance of Documents on a Specific Topic on the Web: A Longitudinal Study of "informetrics". Journal of the American Society for Information Science and Technology 55(11), 980–990 (2004)
[13] Pandey, S., Olston, C.: User-Centric Web Crawling. In: Proc. of the 14th Int'l. Conf. on World Wide Web, pp. 401–411. ACM Press, New York (2005)
[14] Meng, T., Yan, H., Wang, J.: Time Local Law of Changes of Web Information and Validation. Intelligence Journal (2005)
[15] Li, X.: The Research of Crawling Policies Based on Web Page Freshness Analysis. China Mining University (2010)
[16] http://baike.baidu.com/view/1278827.htm
[17] Scot, S.L.: Detecting Network Intrusion Using a Markov Modulated Nonhomogeneous Poisson Process. Journal of the American Statistical Association (2000)
[18] Zhao, J., et al.: A Software Reliability Growth Model Considering Differences Between Testing and Operation. Computer Research and Development 43(3) (2006)
[19] Leite, J.G., Rodrigues, J.: A Bayesian Analysis for Estimating the Number of Species in a Population using Nonhomogeneous. In: Statistics & Probability Letters. Elsevier, Amsterdam (2000)
[20] Snyder, D.L. (ed.), Liang Z., Deng Y.: Translation. Random point process. People's Education Press (1982)
[21] Liu, C.: Stochastic Process, 4th edn. Huazhong University of Science and Technology Press (2008)
[22] Jiang, J., Hu, L., Tang, J. (eds.): Numerical Analysis and MATLAB Experiment. Science Press, Beijing
[23] Ross, S.M. (ed) Wang, Z., et al. : Translation. Simulation. People's wiley&sons Press (2007)
[24] Liang, Z., Deng, Y.: Random Point Process and its Application. Science Press, Beijing (1992)
[25] Sheng, Z. et al.: Probability and Mathematical Statistics. Higher Education Press (2008)

A Density Granularity Grid Clustering Algorithm Based on Data Stream

Li-fang Wang and Xie Han

School of Electronics and Computer Science Technology,
North University of China, Taiyuan, 030051, China
{Wsm2004,Hanxie}@nuc.edu.cn

Abstract. In the field of data mining, conventional algorithms aren't very suitable for data stream analysis mainly because these algorithms can not adapt to the dynamic environment of data stream mining process, and mining model and mining results can not meet the users' actual application. To this problem, this paper presents a density granularity grid clustering algorithm to effectively accomplish the analysis task of data stream. The algorithm breaks the shackles of the traditional clustering algorithms, divides the entire mining process into off-line and on-line, and finally realizes the data stream clustering.

Keywords: data stream, density, granularity, grid clustering.

1 Introduction

With the rapid development and wide application of information technology, data collection, transmission and sharing technologies have reached a very high level, and all kinds of businesses are generating and accumulating large amount of data stream in an alarming rate. To effectively analyze these data and find the hidden knowledge are particularly important. However, the data sets to be mined are considered to be fixed in many conventional clustering algorithms and focused on how to minimize processing time and memory. And new requirements of traditional clustering algorithms are proposed because of the dynamic nature of data stream.

1) In the data stream analysis, the data can not be repeatedly processed unless specifically preserved. That Large-scale data stream completely stored needs the high time and space costs. Historical dynamic data can not be freely searched and visited.

2) With data points constantly inflowing and the data amount gradually increasing, space and time required for receiving data points should not rapidly expand.

3) Mining results can reflect the variation rules of data stream clustering with time. Data points on the timeline are generated sequentially in a certain rate to form data stream. And analyzing the continuous variation of data stream is valuable. Users not only want to know the clustering results of a time, but also want them of a certain period and how the clustering changes with time to obtain the generating rules of data stream.

G. Zhiguo et al. (Eds.): WISM 2011, CCIS 238, pp. 113–120, 2011.

To those problems, this paper presents a density granularity grid clustering algorithm (A density granularity grid clustering algorithm based on data stream, DGGDS) to effectively complete the analysis task of data stream.

2 Related Concepts

The grid clustering first divides the multi-dimensional data space into a certain number of units and then clusters on this structure. Its characteristic is that the processing speed only depends on the number of grid. The density clustering regards the clustering as high-density object area in data space which is separated by the low-density area and its characteristic is that it can find the arbitrary shape clustering. The clustering based on density granularity grid combines the two kinds of clustering ideas and it can well deal with high-dimensional data and large data sets and find arbitrary shape clustering. DGGDS divides data space into grid cells and then clusters on the basis of the dense grid cells. Grid structure is shown in Table 1.

Table 1. Grid Structure

0.0	0.2	0.4	0.6
1.0	1.2	1.4	1.6
2.0	2.2	2.4	2.6
3.0	3.2	3.4	3.6

First, it divides each dimension of data space into m small equal intervals and obtains data space of multi-dimensional granularity grid. In the n-dimensional data space it uses a multidimensional array A [a1] [a2]... [an] to express the grid logical structure in which each dimension is arranged in order a1, a2 ..., an and small intervals on each dimension are ordered by the boundary value. Each element in the array is mapped to a particular grid of data space. Each grid is identifies with a string label.

Shown in Figure 1, two-dimensional data objects is as an example, LB(A[j][k])=j.k Similarly, high-dimensional data objects can also get such a mapping and each data point can be located in the corresponding grid. The i-arrival data point, the value of data point on k-dimensional and arrival time are labeled by si, si.[j], ti. The storage structure of each grid is defined as follows:

$$grid :< \mu, \partial, sum, lb > \tag{1}$$

In them: μ, ∂ : d-dimensional vector, $\mu[k], \partial[k]$:component in k-dimension. μ, ∂ as statistical information, data points mean, variance ,arrival time mean in grid can be obtained by simple calculation ($\mu^{'}, \partial^{'}$:mean and variance); sum: number of data points in the grid; lb: grid label.

$$\mu' = \frac{\sum_{s_j \in grid} S_j[k]}{sum} = \frac{\mu[k]}{sum} \tag{2}$$

$$\begin{aligned} \partial' &= \sqrt{\frac{\sum_{s_j \in grid} (S_j[k]-\mu)^2}{sum}} \\ &= \sqrt{\frac{\partial[k] - 2\mu\mu'[k]}{sum} + \mu'^2} \end{aligned} \tag{3}$$

Time mean and variance can be similarly derived. With the arrival of new data points, new information can be easily added to the original information. $s_{new}, \mu_{new}, \partial_{new}$ express new arrival data point and updated statistics. Update process is as follows:

$$\mu_{new}[k] = \mu[k] + s_{new}[k] \tag{4}$$

$$\partial_{new}[k] = x_{new}[k] + \partial[k] \tag{5}$$

3 Analysis of Dggds Algorithm

3.1 Mining Process

This paper mainly analyzes dynamic data stream to meet the needs of users. The mining process is divided into two parts: on-line and off-line. Data stream quickly received in on-line process is as mined intermediate results which are timely updated with the influx of new data points. Some particular time intermediate results periodically selected according to a certain time frame are saved to external memory and as an off-line process input. Users call the off-line process and query the final mining results. Through two processes we can dynamically and quickly deal with data stream and satisfy the user demands with the properly designed statistical information.

3.1.1 On-line Process

On-line part is a never-ending data processing engine which timely receives and adds the updated results to the original intermediate results to form the latest model of intermediate knowledge. And according to a certain time frame it can write the knowledge of a certain moment back to intermediate knowledge base. So the intermediate knowledge base is formed by periodic write-back.

With WithWiwthe influx of data points, each data point can be navigated to the corresponding space grid, and at the same time the statistical information in grids is updated. In the algorithm, we only save those grids with data points, makes hash tree and visits the corresponding grid by grid label. So for the high-dimensional data a large amount of storage space can be saved. The grids with data points are gradually added to the list of Table_Grid. Table_Grid is the intermediate knowledge of on-line. Pseudo codes are as follows:

```
On-line(a(S1,…,Sn),b_time)
{
lb=get(a);
// detect the grid with label id whether added to Table_Grid
 if(exist(Table_Grid,lb))
  {
  b_grid=get(Table_Grid,lb);
//detect the current grid whether contains the isolated
points which need be deleted
   if(dense(b_grid))
//delete the isolated points in the current grid
      {
   add(Table_Grid,lb);
          update(lb,a,b_time);
      }
   }
else
          delete(Table_Grid,lb);
}
```

Input parameters a(s1, ..., sn), b_time respectively express the new data point and its arrival time. Get(a) expresses the corresponding grid label of data point p. exist (Table_Grid, lb) is used to determine the grid with label id whether exists in the list. dense(b_grid) is used to determine the grid whether dense. First we calculate the value δ of corresponding density of this grid. When $\delta > \gamma$ (γ is a user-defined threshold), the grid is dense.

$$\delta = \frac{data - po\text{int}\, s}{total - number - data - po\text{int}\, s} \tag{6}$$

New(b_grid) is used to check the data points in grid whether be with the great time span. update(lb,a,b_time) adds statistical information in new data points to the corresponding grid.

3.1.2 Off-line Process

Off-line part is a analysis environmental and is used to get users mining results. Its working basis is the accumulated intermediate knowledge in on-line process. Users can set the parameters and the mining time window according to their needs, get mining results of corresponding time or clustered cases of data points in a given period of time and further study the change process of data stream.

a). Results Cluster According to Intermediate Knowledge in a time

From any grid it finds all dense grids connecting with current grid by depth-first traversal and forms a cluster. Init: the initial grid label, curr: the current cluster label, E: the dimension of data space. The Codes are as follows:

```
Dcs(init,curr)
{
num=n;i=1;
while (j <=E)
 {
  //examine the left neighbor of init in dimension i
  cl=lb(i,init,left);
  if (cl is dense) and (cl is undefined)
  Dcs(cl,curr);
//examine the right neighbor of init in dimension i
  cn=lb(i,init, right);
if ( (cn is dense)&& (cn is undefined))
Dcs(cn,curr);
i++
 }
}
```

b). Getting the cluster results of data stream in a certain historical time period and completing the analysis of data stream.

The problem is described as follows:

Input: $\Delta t = t_2 - t_1, t_2 > t_1$,output: the cluster case of the arrival data points in Δt

Solution: getting the clustering case in the period by the intermediate knowledge of the two moments. F1: intermediate knowledge grid at time t_1. F2: intermediate knowledge grid at time t_2. F2-1: intermediate knowledge grid of the arrival data points among the time (t_1, t_2).

Treating the grids among F1 and F2 one by one, and then adding them to F2-1. The specific process is as follows: To the grids both in F1 and F2: statistics in F1- statistics in F1, then added to F2-1.To the grids in F1 and not in F2:the grids are formed by noise points and deleted in the period. So they are not necessary to be added. To the grids in F2 and not in F1: obviously the points in the grid appearing in this period are directly added to F2-1.Finally it clusters in F2-1 by the Dcs() and gets results.

3.2 Analysis of Isolated Points

The definition of isolated points and the appropriate approach in the dynamic data stream environment are different from the cases in the static environment. To this problem, the algorithm proposes the following solution: exclude the isolated points in intermediate knowledge base. The isolated points are very strong noises in data and can't reflect the general characteristics of data sets and on the contrary can bring big error in the statistics of data sets. The definition of isolated point is as follows: a) The space isolation of isolated point: Isolated point appears in remote areas of space and reflects the contingency of emergence in space. This is the same as the case in a static environment. b) The time contingency of isolated point: In the dynamic data stream environment, a current isolated point in space can not conclude that it must be isolated forever and never form concentrated regions. If it remains isolated for a very long time, its appearance at that time is considered to be very accidental.

The point satisfying the above two conditions is considered to be isolated point and deleted. The point appearance probability is very small in time and space by theory and its location information and time information can bring error statistics to the entire clustering. As time continuously goes on, Table_ Grid constantly receives new data points in accordance with the mining model. Meanwhile, according to the pyramid time frame mentioned in [1], it periodically writes intermediate knowledge of corresponding time to external memory. So the intermediate knowledge base of the external memory is truly formed, which is calculated for off-line process to meet users different analytical requirements.

4 Experimental Results

In the actual data and experiment, we use the KDDCup 1999 data in UCI data sets which include 4898431 records and 23 different categories in which the number of the three kinds: smurf, normal and nep-tune is 98% of the total data. We check out the three kinds and 48984 records as experimental data. In the experiment, we divide the data sets into three equal parts and save once intermediate knowledge to disk every 16 320 data records. The correctness of the final mining results is shown in Table 2.

Table 2. Mining Results

Type Name	Data Distribution	Algorithm Name	Mining Results
Neptune	10874	DGGDS	11133
		K-Means	12572
Smurf	28280	DGGDS	27420
		K-Means	23540
Normal	9830	DGGDS	10431
		K-Means	12872

Fig.1 shows the data stream processing time in the three added process of off-line and on-line. At the same time it compares the time efficiency with the classical algorithm K-Means of static data sets. We can see the time which DGGDS algorithm uses in absorption data point on-line is basically positive proportional to the number of data points and the time used in the off-line clustering does not sharply rise with the increase of data points. This is because the time complexity of off-line clustering is a linear with the number of grid. The time used in K-Means grows with the increase of data points. With the increasing of data point number and the gradually concentrating, DGGDS can show better time characteristics.

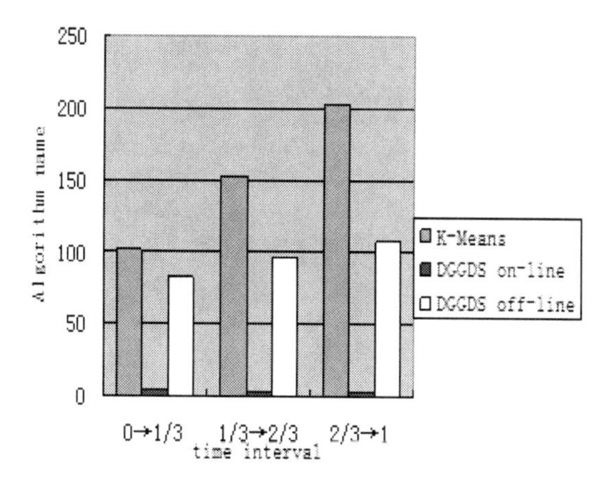

Fig. 1. Comparison of different algorithms

DGGDS as an inheritable and supporting incremental clustering method has a strong ability to process data streams. At the same time, DGGDS is a real cluster based on density granularity and grid. This makes the mining model very suitable for data stream analysis.

5 Conclusion

To the data stream mining, this paper presents a density granularity grid clustering algorithm which breaks the traditional mining model, sets focus on intermediate knowledge base and mines in two steps. It not only can efficiently handle dynamic data, but also well meet the needs of users. Through analysis we can see that DGGDS ensures the efficiency of mining and improves the system capacity. It readily provides the knowledge show of the latest data to easily make decision by periodically combining with new data to the original knowledge base.

References

1. Barbara, D.: Requirements for Clustering Data Streams. SIGKDD Explorations 1, 23–27 (2008)
2. Aggarwal, C.C., Han, J.w., Wang, J., et al.: A Framework for Clustering Evolving Data Streams. In: Proceedings of the 29th VLDB Conference, Berlin, Germany, vol. 1, pp. 87–90 (2010)
3. Zhang, T., Ramakrishnan, R., Livny, M.: BIRCH: An Efficient Data Clustering Method for Very Large Databases. In: ACM SIGMOD Conference, vol. 3, pp. 200–201 (2010)
4. Agrawal, R., Gehrke, J., Gunopulos, D., et al.: Subspace Clustering of High Dimensional Data for Data Mining Applications. In: ACM SIGMOD Conference, vol. 1, pp. 115–116 (2010)
5. Guha, S., Mishra, N., Motwani, R.: Clustering data streams. In: Proc.41st IEEE Symp. on Foundations of Computer Science, vol. 6, pp. 359–366 (2010)

An Automated Grading Model Integrated LSA and Text Clustering Together for English Text*

Huang Guimin, Cao Guoyuan, Zhou Ya, Qin Kuangyu, and Zhang Yan

Guilin University of Electronic Technology
Guilin 541004, China
{gmhuang,ccyzhou,qky}@guet.edu.cn, caoguoyuanwo@163.com, cathyzhangyan@yahoo.com.cn

Abstract. This paper proposes an automated grading model (AGM) for English text. The AGM combines the Latent Semantic Analysis (LSA) with Text Clustering (TC) to measure the content quality of English text. In the AGM, LSA represents the meaning of words as vectors in semantic space using statistical analysis technique applied to large corpus, and TC is applied to classify the texts in the corpus into different categories. By comparing the similarity of a text with reference texts which are belong to the same cluster on basis of semantic content, our model can grade the text in the same cluster. In addition, the AGM can judge the to-be-graded text whether related to the given subject. Our experiments show that the AGM is competent in grading the content of texts.

Keywords: Automated Grading Model; Latent Semantic Analysis; Text Clustering.

1 Introduction

Writing is an important part of the English study, many teachers find it difficult to give large numbers of writing assignments due to the effort required to evaluate them. Therefore, AGM is needed to grade the text written by students. In recent years, automated grading in natural language processing study has been paid more attention. Actually, automated grading is a method that to evaluate and grade a text through computer technology. It is a challenge task of how to grade the text which written by students. While with the development of compute and natural language processing technology, several foreign grading models have been put into practice. Such as Intelligent Essay Assessor (IEA)[5], Apex[6], and so on. These automated grading models have their own advantages and disadvantages in respect of technology and application, but it greatly improve grading efficiency while reducing the amount of time devoted to grading. The related research in China has also obtained the certain achievements.

* This work is supported by the Educational Science & Research Foundation of Guangxi (No.200911MS83).

G. Zhiguo et al. (Eds.): WISM 2011, CCIS 238, pp. 121–129, 2011.

Various AGM have different grading criteria. While lots of studies suggested that texts analysis should move to unearthing even more semantic features in texts. According to LSA, an innovative AGM is proposed in this paper. Our research is mainly based on LSA to evaluate the text on its content. But, it is a waste of time to compare the similarity of all the texts in the corpus. Then, TC is applied in AGM to classify the texts in the corpus into different clusters. So, the text can be graded according to reference texts which belong to the same cluster. At the same time, the text can be judged whether it is related to the given subject.

The paper is organized as follows: in section 2, we introduce previous work on the AGM. Then we will describe the LSA and TC. Next, it presents how to integrated LSA and text clustering together to AGM in section 3. In section 4, it describes experiments and analysis of its experimental results. Finally, in section 5, this paper concludes research work in the paper and discusses future works.

2 Related Work

LSA is both a computational model of human knowledge representation and a method for extracting semantic similarity of words and passages from text [1]. While, TC is studies for improving the precision and recall in information retrieval systems and as an efficient way of finding the nearest neighbors of a text. We will introduce the LSA and the text clustering.

2.1 An Introduction to LSA

In order to compute the semantic similarity of pieces of texts, LSA[2] according to large corpus of texts to build a semantic high-dimensional space of all words and texts, by means of a statistical analysis technique. Actually, the semantics of a word is determined by all of the paragraphs in which the word occurs [4].

The first step is to represent texts as a matrix A in which each row stands for a unique word and each column stands for a text passage. This matrix is built by considering the number of occurrences of each word in each piece of text. So far, it is just straightforward occurrence processing.

Next, singular value decomposition (SVD) is applied to the matrix A. According to the equation (1), SVD decomposes a matrix A into three new matrices.

$$A_{m\times n} = U_{m\times r} S_{r\times r} V^T_{r\times n} \tag{1}$$

Where $U_{m\times r}$ is an matrix whose columns are the eigenvectors of the AA^T matrix. These are termed the left eigenvectors. $S_{r\times r}$ is a matrix whose diagonal cell are the singular values of A. Then the singular values were ranked in descending order. $V_{n\times r}$ is a matrix whose columns are the eigenvectors of the A^TA matrix. $V^T_{r\times n}$ is the transpose of $V_{n\times r}$. Here the U and V column vector are orthogonal vectors. Assume for the rank of A is r, then:

$$u^T u = v^T v = I_r \tag{2}$$

The row vector of U and V respectively as word vector and text vector.

The essence of LSA is dimension reduction. There is a mathematical proof that any matrix can be so decomposed perfectly, using no more factors than the smallest dimension of the original matrix. [3] One can reduce the dimensionality of the semantic space simply by deleting the coefficients which have a comparatively small impact on the results in the diagonal matrix, ordinarily starting with the smallest.

The choice of dimensionality has a significant impact on the results of LSA. However, there is currently no efficient method for automatically selecting the optimum dimensionality. So, the selection of number of dimensions must be empirically determined.

2.2 An Introduction to TC

Given a text sample set $X=\{X_1, X_2......X_n\}$ according to the similarity of texts between the sample set into k cluster: $C=\{C_1, C_2......C_k\}$ called TC. A good clustering method generated the cluster has high similarity within cluster and low similarity between clusters.

In fact, TC is just the clustering algorithm applied to the texts. Clustering [10] is a technique for grouping objects based on similarity. According to the concrete realization method of TC, the cluster algorithm is divided into hierarchical clustering and non-hierarchical clustering.

1) Hierarchical Clustering

Hierarchical clustering result can be expressed as a tree graphics. According to the way category tree generated, hierarchy clustering method can be divided into Agglomerative and Divisive.

a) Agglomerative: Start with the points as individual clusters, then merge the most similar or closest pair of clusters.

b) Divisive: Start with one, all-inclusive cluster, then split a cluster until clusters of individual points remain.

Summarize the agglomerative hierarchical clustering algorithm procedure as follows:

a) Compute the similarity between all clusters.
b) Merge the most closest two clusters.
c) Update the similarity matrix.
d) Repeat steps 2 and 3 until only one cluster remains.

2) Non-hierarchical Clustering

The non-hierarchical clustering algorithm is an iterative process. Usually, there are a number of non-hierarchical clustering techniques, now the K-means algorithm which is widely used in text clustering will be described.

Basic K-means Algorithm

a) Randomly select K points as the initial centroids.
b) Assign all points to the closet centroid.
c) Recompute the centroid of each cluster.
d) Repeat steps 2 & 3 until the centroids don't change.

This bisecting algorithm [7][9] has been recently discussed and emphasized. Bisecting k-means [8] tends to fast produce balanced clusters of similar sizes in the Semantic feature space. In the study, the bisecting K-means algorithm is adopted. Now summarize the bisecting K-means algorithm procedure as follows:

a) Pick a cluster to split (split the largest).
b) Find 2 sub-clusters using the basic K-means algorithm.
c) Repeat step 2, the bisecting step, for ITER times and take the split that produces the clustering with the highest overall similarity.
d) Repeat steps 1, 2 and 3 until the desired number of clusters is reached.

3 Automated Essay Grading Model

To apply AGM, the text must be expressed in the form which the computer can understand. So, texts are represented as a matrix in which each row stands for a unique word and each column stands for a text.

In order to obtain the feature list, the unique words which occur in at least two texts can be selected. Then delete the stop words such as the, is, not etc. Each cell of the matrix represents the frequency of the word which occurs in the text. But the eigenvector simply using frequency as vector of the weights method has its great limitation. It ignores much semantic information of the text words. In order to reflect the effect of word in the text, so the equation (3) is adopted:

$$W_{ij} = tf_{ij} * idf_i = \frac{n_{ij}}{\sum_k n_{kj}} * (\log(\frac{N}{df_i}) + 1) \tag{3}$$

Where tf_{ij} is local weight and idf_i is global weight. Local weight is function of how many times each word occurs in a text, global weight is function of how many times texts containing each word appears in the corpus. n_{ij} is the count of ith word in jth text; $\Sigma_k n_{kj}$ is the total number of words in the jth text; N is the total number of text in the corpus; df_i is text frequency which is the count of text which contains term i. The equation (3) shown that the importance of the words in texts as it is proportional to the number appeared with it, but at the same time in corpus is inversely proportional to appear in frequency.

Next, SVD is applied to the matrix, so the text vectors can be obtained. The similarity of texts is computed through the equation (4).

$$\cos(V_i, V_j) = \frac{V_i \bullet V_j}{|V_i||V_j|} \tag{4}$$

Where V_i and V_j respectively stand for a text vector of a matrix. So, TC can be applied to the matrix. Then the texts in the corpus can be classified into different categories according to the similarity. Now we need present the text which is for evaluation as a text vector through the equation (5):

$$d = d^T us^{-1} \tag{5}$$

Where u and s is the result of SVD, d is the text vector of the text in semantic space.

While, the text is judged which cluster does it belong to and compute the result of the similarity between the text and the other reference texts in the same cluster. So we compute the score with the equation (6) showed as the follows:

$$score(p;\theta) = \frac{\theta_1 * p_1 + \theta_2 * p_2 + ... + \theta_n * p_n}{\theta_1 + \theta_2 + ... + \theta_n} \tag{6}$$

Where θ is the result of the similarity between the text for evaluation and the reference text in the corpus; p is the pre-score of the reference text in the corpus.

Through this method the computation with the text which does not belong to the same cluster can be avoided, so it can improve efficiency and accuracy of the automated grading model. At the same time, the text can be judged whether it is related to the given subject.

The pseudo code is shown as follows:

Input: The corpus M which prepared; the text for evaluating; the stop words list N.

Output: The feature terms list F and the matrix we have trained.

```
/*Training corpus process .*/
Begin
  Import M and N
  Begin
   Extract unique words w which occur in at least two texts
    If (The word w_i exists in N)
      Then Delete w_i
    Else Add the word to the feature terms list F
      Begin
          Output F
      End
   End
   Begin
    Create a matrix A according to F
    Begin
      Weight every elements of A
    End
```

```
  Begin
    Then SVD was applied to A
      Begin
        Output the trained matrices S, U and V
      End
   End
      Begin
        Classify the texts in the corpus into different clusters
      End
      Begin
        Output the result of TC
      End
   End
 End
/* The part two of algorithm is to score the text*/
Begin
  Import the text for evaluating
  Begin
    Represent the text as a text vector d according to F, U and S.
  End
  Begin
    Judge which cluster is closet and compute the similarity between the other reference
    texts in the same cluster.
    Begin
      Compute the score of the text
      Then Output the score of the text
    End
  End
End
```

4 Experiment and Analysis

First, the ST4 Chinese learner corpus is prepared as the training corpus and the other as test texts. The distribution of the ST4 Chinese learner corpus which is as training texts is shown as Table 1.

Table 1. The Distribution of Titles and the Number of Texts

Title/Score	6 7 8 9 10 11 12	Sum
My View on Fake Commodities	64 61 58 45 30 10 10	278
Haste Makes Waste	42 52 60 55 33 16 10	268
My View on Job-Hopping	25 30 35 22 17 10 /	139

4.1 Grading Ability Evaluation

The experiment is designed to assess the Pearson correlation between the automated grading model (AGM) and the pre-score of test texts. Three titles in the test texts are respectively graded by AGM. Then, the Pearson correlation between the pre-score and the score which evaluated by AGM is computed. Finally, the result of the Pearson correlation is shown in table 2. Here the correlation is the Spearman Rank Correlation, the equation (7) shown as follows:

$$R = 1 - \frac{6\sum D^2}{n(n^2 - 1)} \tag{7}$$

Where n is the number of the test texts of the titles, D is the variance of the n texts.

Table 2. Correlation between AGM and Human rater

Correlation	H_t1	H_t2	H_t3	H_aver
AGM	0.69	0.70	0.73	0.71

From the Table 2, the Spearman Rank Correlation between pre-score and the score graded by AGM in the range of 0.69 to 0.73, the average is close to 0.71 in this experiment. The result is close to the pre-score. So the AGM can be competent in grading the text absolutely.

4.2 Text Clustering Ability Evaluation

The experiment is designed to show the selection of number of dimensions is very important. The final score of the text is affected by the text clustering result. After standard processing of SVD to the texts mentioned above, a 1896 by 686 word-by-text matrix was created and a set of K parameters are selected. Then the result of the text clustering is shown as the follow Fig3:

Fig 1 shows that the accuracy rate of TC varies with the different dimension. As shown in the chart, the accuracy of TC can obtain a better result when the K in the range of 350 to 500. When K is 472, the result of TC is optimal. Then the final score is affected by the result of TC. The final score of the text is the most precise when the K is 472. So, the selection of dimension is very important to the accuracy rate of AGM.

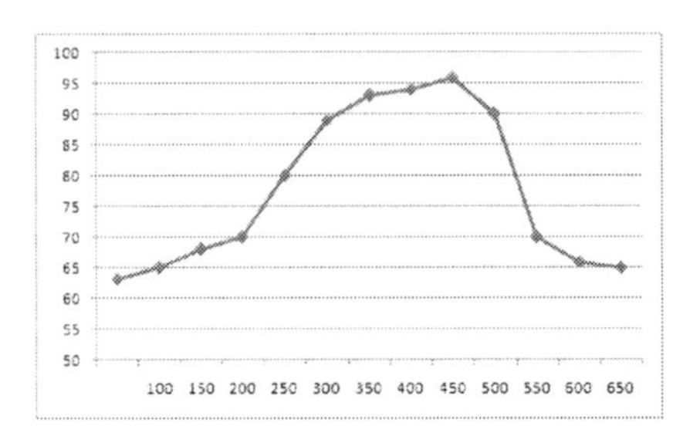

Fig. 1. The accuracy rate of text clustering

5 Conclusion

In this work, we discussed how to integrated LSA and the text clustering together in the automated grading model. Through several experiments we have showed that the automated grading model can be competent in grading the text. In the other automated model, it is a waste of time to compare the similarity of all the texts in the corpus. While, this automated grading model applies text clustering to classify the texts in the corpus into different clusters. So, we can grade the text according to reference texts which belong to the same cluster. So the efficiency of the automated model can be improved. In addition, the text can be judged whether it is related to the given subject.

In this paper we pay attention to score the text just on the content of the text. While grading and commenting on written texts is important not only as an assessment method, but also as a feedback device to help students better learn both content and the skills of thinking and writing. Therefore, we should join misspelling intelligent analysis approach for checking misspelled words and grammatical analysis approach for checking grammar error. Then the part-of-speech tagging and named entity content in order to improve the accuracy of the final score.

References

1. Foltz, P.W., Laham, D., Landauer, T.K.: The Intelligent Essay Assessor: Applications to Educational Technology, `http://imej.wfu.edu/articles/1999/2/04.htm` (1999-02-04)
2. Landauer, T., Foltz, P.W., Laham, D.: Introduction to Latent Semantic Analysis. Discourse Processes 25, 259–284 (1998)
3. Lemaire, B., Dessus, P.: A system to assess the semantic content of students essays. Journal of Educational Computing Research 24(3), 305–320 (2001)
4. Deerwester, S., Dumais, S.T., Furnas, G.W., Landauer, T.K., Harshman, R.: Indexing by Latent Semantic Analysis. Journal of the American Society for Information Science 41(6), 391–407 (1990)
5. Debra, T.H., Pete, T., Anne, D.R., Marian, P.: Seeing the Whole Picture Evaluating Automated Assessment Systems. Journal of Innovation in Teaching and Learning in Information and Computer Sciences 6(4), 203–224 (2007)

6. Fridolin, W., Christina, S., Gerald, S., Yoseba, P., Gustaf, N.: Factors Influencing Effectiveness in Automated Essay Scoring with LSA. Frontiers in Artificial Intelligence and Applications 125, 947–949 (2005)
7. Steinbach, M., Karypis, G., Kumar, V.: A Comparison of Document Clustering Techniques, Technical Report, 00-034 2000
8. Gautam, B.P., Shrestha, D., Members Iaeng: Document Clustering Through Non-Negative Matrix Factorization: A Case Study of Hadoop for Computational Time Reduction of Large Scale Documents. In: Proceedings of the International MultiConference of Engineers and Computer Scientists, vol. I, pp. 570-575 (2010)
9. Savaresi, S.M., Boley, D.L.: A comparative analysis on the bisecting K-means and the PDDP clustering algorithms. Intelligent Data Analysis 6, 1–18 (2002)
10. Ward Jr., J.H.: Hierarchical grouping to optimize an objective function. Journal of the American Statistical Association 58(301), 244–263 (1963)

Design of Learning Ontology Framework Based Text*

Wu Qin-xia and Liu Yong-ge

School of Computer and Information Engineering
Anyang Normal University,
Ayyang Henan China
wqx0218@163.com

Abstract. In this text give a learning ontology framework. The framework includes term extraction 、 ontology building and ontology pruning three function modules. Term extraction can identify words in the web text that are relevant for ontology building; Ontology building can automatism build ontology bases extract terms, In the build ontology process we give a automatism build ontology method base ontel, this method can solve some ontology learning methods can't completely learning relations between different concepts; Ontology pruning filters out some irrelevant concepts and redundancy concepts, so we can capture a prefect domain ontology.

Keywords: ontology; ontology learning; ontel; ontology pruning.

1 Introduction

As the development of network technology, World Wide Web entered the revolutionary transformation time. Future Web should be take the machine understandable information and the service as characteristic semantic Web, realizes the free space which truly between the machine and the human cooperates. The Ontology is an explicit description of the form of shared conceptualization, Ontology gradually attracted people's attention as a hot field of computer research, mainly due to Ontology can be reached between people and application systems to share the d meaning of terminology and common understanding, which is precisely the key to building the Semantic Web [1].

Today, World Wide Web is it in high speed development period produces, A lot of HTML pages are isomerism, chaotic. To realize the Semantic Web crucial work is how extract the semantic information from these HTML pages, Construction of high-quality domain ontology. To simplify these concepts and the terminology confusion using the domain Ontology, so that the user can be shared electronic documents and each type information resource. But domain ontology building is a complicated project, which needs a large amount of manpower and material resources. This is bottleneck to obtain domain ontology. Its effective solution is studies the omain ontology from the existing resources, Ontology learning target is technology, These technologies can help and guide users to semi-automatically build the required domain ontology effectively.

---

* This paper supported by NSFC(60875081).

G. Zhiguo et al. (Eds.): WISM 2011, CCIS 238, pp. 130–136, 2011.

Because much of the information resources available to Web text format exists, this paper presents a text message from the ontology learning methods, first by taking out the word to learn in areas relevant to knowledge; followed by the construction of terms extracted from the body; Finally, Through high-quality ontology domain ontology pruning.

2 Ontology and Ontology Learning

2.1 Ontology

At the earliest period, Ontology is a philosophical category, and later with the development of artificial intelligence, it has been given a new definition by artificial intelligence field. There are many definitions of ontology and the frequently-used one is that ontology provides the shared, clear, formal, standardized conceptual model for a certain domain knowledge [2].

The goal of Ontology is to capture relevant domain knowledge; to provide a common understanding of knowledge in this area; to confirm the common recognition of the vocabulary in this area; and to give the clear definition of the mutual relationship between these vocabulary (terms) and vocabulary from different levels of formal models. Therefore, ontology can be drawn is the key technology of Semantic Web, and has a direct impact on whether the Semantic Web can be successful implementation or not.

2.2 Ontology Learning and Existing Defects

Ontology construction is a very complex process that requires the participation of experts from various fields. Although the tools of Ontology are more mature, but the manual ontology construction is still a cumbersome structure and hard task, and eventually which can result in a so called knowledge acquisition bottleneck. Therefore, the techniques of ontology learning have emerged as the times require. Ontology Learning (Ontology Learning) technology that is currently a hot topic, whose aim is to develop automatic construction of ontology which can machine learning techniques to help to build the ontology of knowledge engineering, ontology learning tasks include: (1) Ontology Acquisition: including the creation of ontology, model (schema) extraction, and ontology instance extraction. (2) Ontology Maintenance: including ontology integration and navigation, ontology update and expansion of ontology [3].

The main objective of ontology learning: a document from the Web automatically for the field of terminology and their mutual relations; using the information extraction technology to determine the semantic relationships between concepts and building the Ontology in the obtained concepts and the base of their relationship. Existing ontology learning technologies are not only from free text (free text), dictionary and existed ontologies to learn ontology, but also from the database (schemata) and from XML documents to learn ontology [4]. There are many shortcomings of theses existed ontology learning techniques, which seriously limit the ability of ontology learning. These drawbacks are as follows:

1) Most of the ontology learning technology is based on grammatical analysis and statistical methods, so that the semantic information in the sentence can not be applied in the process of ontology learning.

2) Existed ontology learning method is extremely dependent on the WordNet system, and it will be restricted by the WorldNet system in a special domain knowledge field when learning classifier relations, inheriting the concept

3) The information of corpus can not be used fully; most of the ontology learning methods only use the syntactical information in the corpus, rather than get the detailed semantic information from the Collected Works.

4) Some methods of ontology learning is attributes to the learning of concepts and examples and its property, through the linkages between different concepts of rules and matching method to find the relationship between the concepts, but this method only can learn few relationships of the concepts.

3 Ontology Learning Frame

Ontology learning frame can be divided into three parts, terminology extraction, ontology creation and ontology pruning. see Figure 1.

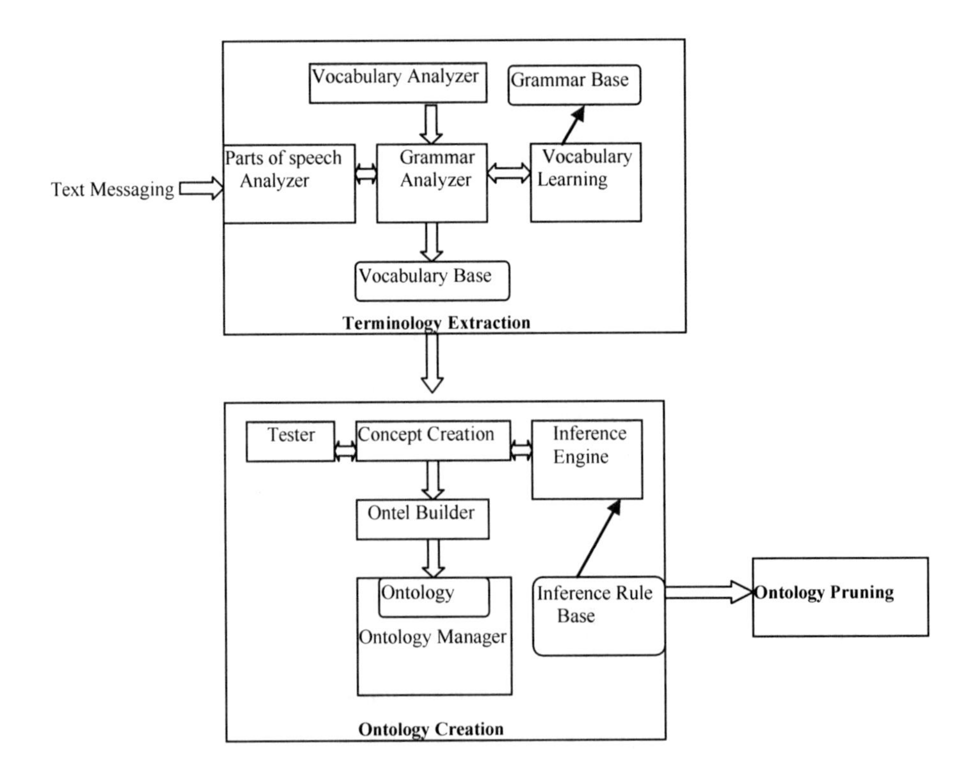

Fig. 1. Ontology learing frame

3.1 Terminology Extraction

Terminology extraction [5] is the first step of ontology learning, it is about indenting the vocabularies that relate to ontology creation from the corpus. There are two steps to extract terminology, the first one is language analysis phase, in which standard language information in the corpus is analyzed; the second one is extract terminology that the ontology interests in from the corpus by means of a series of extraction rules of the language information. Ontology extraction also contains the text of natural language processing features to analyze the text that entered. It includes five major functional components: lexical analyzer, parser, word form analyzer, vocabulary and vocabulary learner.

The general introduction of the five major functional components are as follows: lexical analyzer inputs the text which will be divided into words and sentences, and then transported to the parser; parser uses a top-down chart method transform the analysis of from the sentence into the corresponding tree, This process will also use vocabulary and vocabulary learner, word form analyzer will be put into use if there is any new word; word form analyzer then will compare the new word with the stored ones individually, the vocabulary learning device will be activated if the corresponding word cannot be found; though learning the structure and characteristics of the vocabulary, the vocabulary learning device can know the word and put it into the vocabulary.

There are not only the known vocabulary but also the grammatical information and the vocabulary information before the learning of the vocabulary. Vocabulary learning has two missions to study: the characteristics of words and the meaning of the words. Methods of learning vocabulary are generally based on template driven and heuristic language learning. The vocabulary learning device can study the origin of the vocabulary via the analyzing of form, meaning and grammar. Vocabulary learning can be divided into the following categories :

- Learning vocabulary via grammar rules

1) Using the parser constructor from the form analyzer to learn the affixes, in order to learn new words.

2) According to grammar rules, learning the form of the new words to learn vocabularies.

- Heuristic-based vocabulary learning

This method can easily extract the vocabularies that the ontology creation interests in, depend on grammatical relations to learn vocabulary.

3.2 Ontology Creation

Ontology creation includes knowledge extractor, Ontel builder, ontology manager, inference engine and testers. The role of each feature are as follows:

Knowledge Extractor: its main task is to extract knowledge from the concept of sentence structure, the knowledge extractor can extract knowledge from both the levels of the sentence and the documents by means of semantic template and inference. After dealing with the sentence, the knowledge extraction which includes extracting field concept and the relations among concepts, then put these concepts and relations into the element body builder to generate body elements.

Element body builder: it is mainly responsible for the conceptual information, and includes transforming the verb and noun phrase into some relating conception and determine their position in the ontology. The other relations among the conceptions will be determined according to their position.

Tester: it means testing knowledge (including concepts that cannot be recognized by the system) that is to be extracted.

Inference engine: it means inferring new knowledge from the existing one. Inference rule in the rule base is base of reasoning, in is inferring new ontology elements from the existing knowledge elements.

Tester: it means testing knowledge (including concepts that cannot be recognized by the system) that is to be extracted.

3.3 Ontel

Ontel means build out element ontology though the learning of the initial concept first, and then build the ontology that base on element ontology, it includes simple concept, relations and operators[6]. When learning a new vocabulary, one should define its conception and the relations with other words, in order to build the element body. Here we look at the steps to build element ontology:

(1) Creating an initial concept for each unknown word, at this stage the system will find or create a right concept for the words, and will give examples for them according to the sentences.

(2) Though the method of semantic analysis tree, putting the new learned concept under the parent node, and parent nodes can be obtained through the following two ways: (a) drawling from the current sentence, according to the relationships between the upper and lower meaning. For example: Tom is a human, here 'Tom' is a new concept that learned, while ‘human’ is a new concept of the parent node. (b) inferring the parent node according to the vocabulary form. When one cannot drawn parent node from the relationships between the upper and lower meaning, this method is available. For example, the noun’s parent node is object, the adjective’s is property and the verb’s is action.

(3) Extracting relationships between the initial concepts according to the sentence structure and semantics of the template. For example: one can extract ‘Tom’ ‘drank’ ‘water’ from the sentence "Tom drank water", then find parent nodes for each one, ‘Tom’belongs to drinkers class, water’belongs to drinkables class, and one infer the relationship between ‘Tom’and‘water’is‘drank’.

(4) Learning axioms. The structure and quantifier phrases of conditional or compound sentences can be transformed according to the following axioms: (a) extracting precondition and results from the sentence; (b) creating element ontology for each separate part, and naming them as parent element body and results element body; (c)the relationship between parent element body and results element body can be connect by (=>); (d) transforming the unbound instance into the variables into the axioms; (e) representing the implicit relationship between condition axioms and result axioms by extracting and adding arguments for the axioms.

3.4 Ontology Manager

The main task of ontology manager is updating the existing ontology according to the new element ontology. It also includes guiding position of the inheritance hierarchy; eliminating redundancy and concept conflict; updating concepts and relations; merging or separating the existing concepts in order to gain new concept[6].

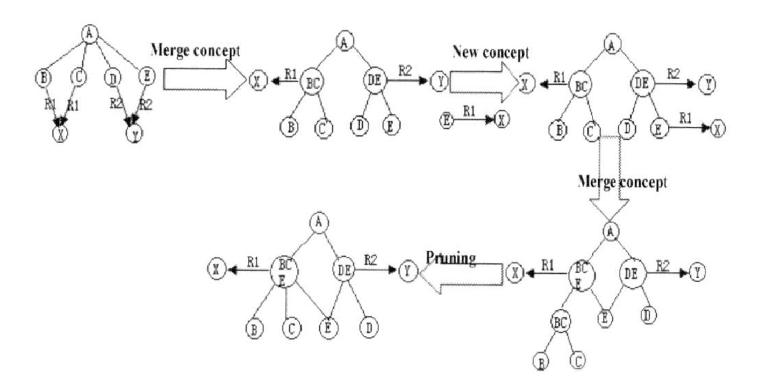

Fig. 2. A Simple Example of Ontology Manager

Here we can use a simple chart to illustrate how the ontology manager merge the existing concepts, the process of learning and trimming new concept. The concept can be represented by a circle, the relationship between parent node and subordinate node can be represented by an arrow which from the former to the latter, and point out type of the relationship on the arrow.

4 Ontology Pruning

In the framework of term extraction and ontology creation are the two main parts, but the term was created by extracting the body is often not perfect, each of the concept of ontology building and the area is irrelevant. So that when the ontology must be trimmed after the success of the module to filter out the irrelevant concepts.

There are two trim body pruning strategies[7]: (1) Based on the frequency. high frequency concepts using remain in the domain ontology, some low frequency will be trimmed from the body. (2) Based on the relationship pruning. This pruning strategy not only consider the concept of the frequency used in the corpus, but also consider the relationship with other concepts. If the concept high frequency –using, but also the concept keep relationship with other concepts ,As a result , trimming body will be preserved.

Currently widely used is based on the frequency of pruning strategy, this method can define threshold frequency of concept of t, For frequency is less than the threshold of all the concepts will be trimmed. Defined as follows:

Definition 1. Using $\overline{Freq}$ to exprees the concept of learning in the corpus which n times the average probability , $\overline{Freq} = \frac{\sum_{i=1}^{n} freq(concept_i)}{n}$ express the pruning threshold, $freq(concept_i)$ express concept t frequency using, For all the words Less than the average probability are cut off from the domain ontology.

5 Conclusions

Ontology engineering has become hotspot for research in many fields, but the body building quite time consuming, which serious obstacles development and application of ontology. Automatically create ontology by learning ontology is a very effective solution. We build a framework for ontology learning, through the term extraction, ontology creation, ontology pruning three functional modules to automatically create domain ontology.

For existing shortcomings Ontology Learning Method, the paper generated from the learning concept ontology, ontology is proposed based element to automatically create a domain ontology approach. This would solve the relationship concepts which can not effectively find when ontology learning by learning concepts instance and attributes .Finally, filter out irrelevant and redundant concepts from the study of the body by body trim module, resulting in robust ontology.

References

1. Hendler, J.: Ontologies on the Semantic Web. IEEE Intelligent Systems 17(2), 73–74 (2002)
2. Sump up the content of Ontology concept, descriptive language, methodology, `http://bbs.xml.org.cn/viewfile.asp?ID=142`
3. Staabs, Maedche, A.: Ontology learning for the Semantic Web. IEEE Intelligent Systems 16(2), 72–79 (2001)
4. Li, M., Du, X., Wang, S.: Learning ontology from relational database. In: Proceedings of 2005 International Conference on Machine Learning and Cybernetics, August 18-21, vol. 6, pp. 3410–3415 (2005)
5. Apted, T., Kay, J.: Automatic construction of learning ontologies. In: 2002 Proceedings.International Conference on Computer in Education, December 3-6, vol. 2, pp. 1563–1564 (2002)
6. Shamsfard, M., Barforoush, A.A.: Learning ontologies from natural language texts. International Journal of Human-Computer Studies 60(1), 17–63 (2004)
7. Sabou, M., Wroe, C., Goble, C.: Learning domain ontologies for semantic Web service descriptions. Web Semantics: Science, Services and Agents on the World Wide Web 3(4), 340–365 (2005)

Energy Management Contract Period Design

Herui Cui, Ruyu Zhang, and Zhaowei Xuan

Department of Economy Management,
North China Electric Power University,
Baoding 071003, Hebei, China

Abstract. According to the actual situation in China, aiming at the design of energy management contract period, the article determines the time parameters and the economic parameters which affect the period of the contract, and makes a design-model of energy management contract period described through a case.

Keywords: Existing public buildings; energy-saving; contract energy management; Period Design.

1 Introduction

Contract energy management is a new way of market-based energy-saving investment. "Energy Service Companies" (called ESCO) sign "Energy Management Contract"(called EMC) with the users of energy-saving projects, and are responsible for financing and undertaking technical and financial risks, reduce energy costs and share the energy efficiency of running projects with the customers. During the contract period the customers get a set of energy services: retrofitting energy conservation, investment or financing, energy efficiency audit, design on energy-saving project, raw material; when the contract expires, all the energy-saving equipment and benefits belong to the users of energy-saving projects.

The EMC also has the corresponding period of the contract like the "Build Operate Transfer " (called BOT) projects Franchise period, contract energy management, which largely affected the risk and the rights of enjoy on both sides etc. If the term of the contract has been set too long, the interests of customers will be affected, at the same time, EMC will also assume many great operational risks; similarly, if the period is too short, the EMC will be likely to not complete the contract or to recover the project costs caused losses. Therefore, the determination of the term of the energy management contract is in a very important position throughout the entire project, and usually is one of the key points to the contract negotiations. In China, however, the phenomenon of specifying the contract period optionally is very common, such as "take-tuning","5-year package ","10-year package" etc. To solve this problem, here will be more in-depth discussion. [1-4].

G. Zhiguo et al. (Eds.): WISM 2011, CCIS 238, pp. 137–143, 2011.

2 Analysis of Energy Management Contract Period

2.1 The Time-Parameters which Affect the Duration of EMC

The whole process of EMC project include project preparation, project construction, project operation, payback period, and several other stages, which the time parameters of the project affect the term of EMC. Every time parameter's changes will directly affect the project to determine the duration of contracts. The relationship between time parameters is shown below [6]:

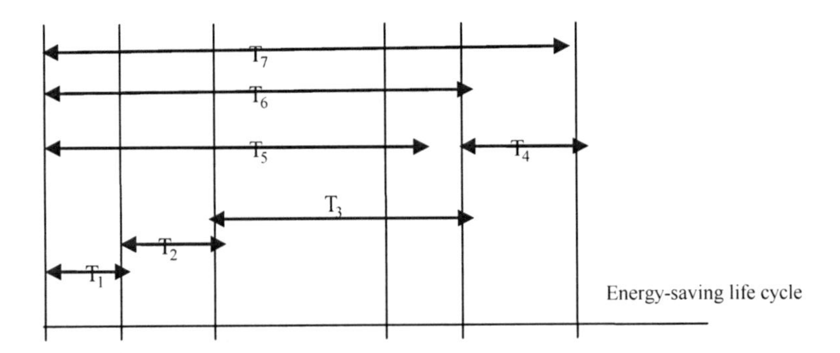

Fig. 1. Energy-saving Lifetime: T_1 : Project preparation, T_2 : Project construction, T_3 : Project operation (Sharing of energy-saving effect), T_4 : Energy-saving effect only for customers, T_5 : Payback period, T_6 : The contract period, T_7 : Economical life of the project

The figure shows that the project contracts include three sections: project preparation, project construction and the project operation; the economic life of the project involves project preparation, project construction, project operation and the transfer period of the project.

As:

$$T_6 = T_1 + T_2 + T_3 \tag{1}$$

$$T_7 = T_1 + T_2 + T_3 + T_4 \tag{2}$$

To determine a reasonable period of EMC, as is mentioned earlier, it must be took the interests of both ESCO and the owners into account. Therefore, when it is in determining the length of EMC, from the perspective of ESCO, the contract period should be greater than payback period; from the owner's point of view, the project should meet the conditions that the economic life of the project is greater than the contract period. These two conditions are expressed as a mathematical formula below:

$$T_6 > T_5 \tag{3}$$

$$T_7 > T_6 \tag{4}$$

2.2 The Economic Parameters Which Affect the Duration of EMC

In the EMC project, which affect the term of the contract has two main economic parameters: Rate of return on investment (ROI) and net present value (NPV). These two important economic parameters are expressed as:

$$\text{RIO} = \frac{\text{Ending assets} - \text{Investment}}{\text{Investment}} \times 100\% \tag{5}$$

$$\text{NPV} = \sum_{t=1}^{n} \frac{I_t}{(1+r)^t} \tag{6}$$

n: the economic life of the project; t: year; I_t : the cost of energy-saving difference after the first t years; r: predetermined discount rate.

3 Design of Energy Management Contract Period

Calculation, analysis and application of the net present value and Investment return determine the duration of contracts.

3.1 Calculation of the NPV and ROI from the Perspective of ESCO

$$\text{ESCO's NPV} = \sum_{t=1}^{T_c} \frac{I_t}{(1+r)^t} \tag{7}$$

NPV: the NPV which ESCO obtains during the contract period; T_c : the term of the contract; t: year; I_t : the cost of energy-saving difference after the first t years; r: predetermined discount rate.

Equation (7) shows that, the higher NPV ESCO have the greater benefits investors will obtain. However, what kind of NPV is considered large or NPV compared with what data is displayed as large, are fundamental issues. The key to solve the problems is, when the investors are in deciding whether to invest in EMC projects they have a basic criterion, which is the NPV must be greater than the amount of investment returns a certain ROI measured. The relationship between the two sides is formulated as:

$$\text{ESCO's NPV} = \sum_{t=1}^{T_c} \frac{I_t}{(1+r)^t} \geq I_c R \tag{8}$$

$I_c R$: The amount of investment returns; I_c : the total investment; R :The expected ROI. According to the formula (8), what can be determined is the minimum number of years the EMC period can not be less than.

3.2 Calculation of the NPV from the Perspective of the Customers

$$\text{Custmers' } NPV = \sum_{t=T_c+1}^{n} \frac{I_t}{(1+r)^t} \tag{9}$$

NPV: NPV the customers received after the expiration of the contract; n: the economic life of the project; the other symbols are same as before.

By equation (9), what can be determined is the economic life of the project. What meets the conditions required that the client determines the economic life of the EMC project is: The NPV of energy conservation obtained by customers in the follow-up operation after the expiration of the contract must be greater than or Equal to 0. This condition is expressed as a mathematical relationship below:

$$\text{Custmers' } NPV = \sum_{t=T_c+1}^{n} \frac{I_t}{(1+r)^t} \geq 0 \tag{10}$$

According to the formula (10), EMC can determine the maximum of the economic life of the project.

3.3 The Model of the EMC Project Contract Period

From the analysis above, in order to determine the EMC period, the first two conditions must be satisfied that not only to make the ESCO's NPV greater than or equal to the amount of investment returns, but also to make the customer's NPV greater than or equal to 0. Therefore, the first EMC period model is a combination of formulas (8) and (10). Advantage of this not yet complete model, the EMC period ultimately can be determined on the basis of determining the minimum term of contract and the longest age of the economic life and coupling with another condition. This third condition is: to make ESCO's NPV greater than or equal to the amount of investment returns and less than or equal to the NPV of the economic life of the project. This condition is expressed as a mathematical relationship:

$$I_c R \leq ESCO's\ NPV \leq \text{the NPV of the economic life} \tag{11}$$

In summary, the model of the EMC project contract period is composed of the following three equations:

$$\begin{cases} \text{ESCO's NPV} = \sum_{t=1}^{T_c} \frac{I_t}{(1+r)^t} \geq I_c R \\ \text{Custmers' } NPV = \sum_{t=T_c+1}^{n} \frac{I_t}{(1+r)^t} \geq 0 \\ I_c R \leq ESCO's\ NPV \leq \text{the NPV of the economic life} \end{cases}$$

4 Application and Analysis of a Case

A following case is taken to illustrate the use of the EMC period mode. Because so far, all of EMC projects are at the implementation stage, that have not yet completed the process of energy management projects throughout the contract, so practice is still unable to provide real data, the data in the following cases are simulated.

An ESCO for energy efficiency carried out in a campus building. Project uses finance model of guaranteed savings structure. The main project is an energy management data control systems, the annual expected energy 556,128kwh. This item is a composite type. In addition to the control system, it also includes lighting improvement, engine efficiency, heating system upgrades, gas water heaters, heat recovery devices, new doors and double glazing and so on. For the repayment of the project, the Campus loans $ 1,156,000 to the local bank for 10 years (4% per annum) to meet the labor, the cost of raw materials and ESCO (total 135,217 $). Owners also invested $ 350,000 of its own funds. ESCO guarantees to the owners about $ 1,500,000 energy savings benefits.

ESCO guarantees to save at least 118,041$ in the first year, and increasing year by year in the after 10 years, the two sides compare the energy-saved bills with the energy bill when energy-saving equipment is not installed before. If the amount of energy is below the guaranteed amount of energy, ESCO will provide the compensation for the difference. If the actual energy savings exceed the minimum amount, the campus will give 15% of the excess return to ESCO as a reward. If the expected benefits can be achieved, it will be paid to the ESCO 50,000 $ for saving results. It assumes that ROI of the ESCO is 15%. After the implementation of this project, the results of energy-saving benefits after 10 years are showed as follows: [7]

Unit: $

Year	Energy saving benefits	Total energy saving benefits	The cumulative energy savings after deducting the amount of guarantee
1	119135	119135	-1380865
2	133250	252385	-1247615
3	153273	405658	-1094342
4	204356	610014	-889986
5	255634	865648	-634352
6	266752	1132400	-367600
7	277328	1409728	-90272
8	295842	1705570	205570
9	317233	2022803	522803
10	324349	2347152	847152

According to the case's determined conditions, what determine the course of the project contract is:

(1) According to the conditions given by case the amount of investment returns can be calculated.

$I_c R = 135217 \times 15\% = 20282.55\$$

(2) According to the formula (8), the conditions must be met:

$$\text{ESCO NPV} = \sum_{t=1}^{T_c} \frac{I_t}{(1+r)^t} \geq I_c R$$

Therefore, due to the data from the table it can be determined the project contract should last at least 7 years. Because in the 7th year, the cumulative energy savings after deducting the amount of guarantee is only \$ -90,272 below the minimum standard of \$ 20,282.55, for that it has not been completed to ensure energy savings. To determine specifically which year after the 7th years is the termination of the contract, this issue is still being unresolved.

(3) According to the formula (10), the conditions must be met:

$$\text{Custmers'}\ NPV = \sum_{t=T_c+1}^{n} \frac{I_t}{(1+r)^t} \geq 0$$

$$\text{Custmers'}\ Investment = 350000 + 1156000 \times (1+4\%)^{10} = 2061162.4\$$$

From what the table above shows, the 10th year shall be to recover the investment, and then calculate the 10th year's remaining balance that deducts both customers' investment and ESCO's energy savings.

ESCO's energy savings=50000+847152×15%=177072.8\$

Remaining balance=2347152-2061162.4-177072.8=108916.8\$>0

Now temporarily assume that the end of the contract is in the 10th year, it is showed from the type and table data that in the 10th year, the remaining balance that deducts both customers' investment and ESCO's energy savings is 108,916.8 \$. So it is certain that, every given year after the 10th year, may act as the project's ultimate economic life. To determine the contract period, a condition must be satisfied: on the one hand to make the ESCO has a certain profit on the basis of cost recovery; on the other hand to allow customers to recover the investment and get some return. Therefore, due to the calculation above, customers have recovered its investment and had a certain income after paying the ESCO at the 10th year. However, if the term of the contract is set at 11th year or later, the customer will need to from share a part of energy efficiency with ESCO, which resulted in customers' interests of the transfer for no reason. Therefore, in this case, it is more appropriate to take the contract period for ten years than others.

5 Conclusion

Energy Management Contract in China has great potential for development. But there is the problem that determining the of EMC at discretion, to solve that, this issue determines the time parameters and economic parameters which greatly affect the EMC period, and makes a model of the EMC project contract period and instructs through a case.

References

1. Lei, B.: Recommendations and Development of EMC in China. China Venture Capital (08) (2010)
2. Kang, Y., Zhang, Y., Yin, Z.: Developing Status and Prospect of China s Energy Service Industry. Energy of China (08) (2010)
3. Zhao, L.: The Mode of EMC operation. China Venture Capita (08) (2010)
4. Wang, Z., Liu, F.: EMC: a new mechanism for energy efficiency investments. Innovation & Technology (01) (2010)
5. Liu, Y., Liu, C.: Economic Benefit Evaluation about Energy Efficiency Reconstruction of Existing Buildings Based on Whole Life-cycle Cost Theory. Construction Economy (3), 58–61 (2009)
6. Hu, L.: Study of BOT Concession Period. Enterprise Economy (6), 89–93 (2008)
7. Sri-Lanka, Model ESCO Performance Contracts. USAID-SARI Energy Program (November 2002)

Empirical Analysis on the Economy - Energy - Environment System of Hebei Province

Herui Cui and Libing Fan

Department of Economy Management, North China Electric Power University,
Baoding 071003, Hebei, China

Abstract. The Economy-energy-environment system is an organic entirety, each subsystem affects and constraints mutually one another, can not be viewed in isolation. We must have a thinking on coordinate and sustainable development, whose core is the development, whose goal is social development, whose basis is economic development, whose necessary condition is environmental protection. The resources are the material basis on achieving sustainable development and energy is an important part of resources which is closely related to the environment and economic. In this paper, first we introduce the general conditions of the economic, energy, environment in Hebei Province; next we have correlative analysis to the economy-energy-environment system, use the Energy Intensity to analyze the relationship between the economy and energy, use the Environment Kuznets Curve to analyze the relationship between the economy and the environment, use figure to have comparative analysis to the relationship between energy and the environment; finally we obtain conclusion.

Keywords: the Economy-energy-environment system, the Energy Intensity, the Environment Kuznets Curve, sustainable development.

Energy is the engine of the economic growth, is the strategic resource of human survival, is also driving force of generalized social and economic activities. Economic, energy, and environment are closely related and affected one another. In the contradictory relationship on the Economy -energy - environment system, the environment is the basis of economic development, the economy is the dominant of environment, the energy is essential production factor and input factor of economic growth. Economic growth can not be separated from energy, the energy development is the premise of economic growth. The gradual depletion of energy and the ecological and environmental issues which is caused by energy, will be a serious impediment to economic development. We must ensure coordinal development among economic, energy, environment, which ensures sustainable development of society.

1 The Condition of Economy-Energy-Environment System of Hebei Province

1.1 The Economic Condition of Hebei Province

Into the twenty-first century, economic development in Hebei Province is rapid, is currently in the stage of rapid development. From 2004 to 2008, the pace of

G. Zhiguo et al. (Eds.): WISM 2011, CCIS 238, pp. 144–149, 2011.

development of GDP in Hebei Province are higher than 18% in addition to 2006, is significantly higher than the national average level. The pace of development in 2004 is 22%, GDP is over one trillion yuan in 2005 and over 2 trillion yuan in 2010. In Hebei Province, the per capita GDP is over one million yuan and is 10251 yuan in 2003, the per capita GDP is over 2 million and is 23,239 yuan in 2008, which is more than 2003 by 127%.

1.2 The Energy Consumption of Hebei Province

With the pace of rapid economic development in our province, energy consumption also increases year by year, and the major consumption of energy is coal-based. Because our province is not the major coal-produce area, coal production is relatively small, is mainly purchased from Shanxi Province and other places, we can see the supply of coal resources in our province is very low. At present rapid economic development of our province will inevitably require substantial coal resources. In 1990 coal consumption in our province is 61,242,200 tons standard coal, and in 2008 the coal consumption is 242,256,800 tons standard coal, in less than 20 years in our province coal consumption, increased by nearly 3 times. Since the reform and opening up economic development of our province is the cost that is consume a large number of coal. We should improve technology of coal use, it is necessary to pay attention to the amount of energy use and also focus on the quality of energy use, effectively make good use of each ton coal. Since 1987, in our province, the proportion that coal consumption accounted for the total resources consumption is more than 90% all other years in addition to1998 for 89.68%, coal occupies a pivotal position in the economic development of our province.

1.3 The Environmental Condition of Hebei Province

At present the pace of economic development in Hebei Province is gratifying, but the environment has serious environmental pollution, which becomes increasingly prominent. Since 1990, a large number of industrial, enterprises continue to rise, which drive economic the rapid development, the same time problem to environmental pollution continues to emerge that SO2 emissions from industrial enterprises and sewage is not compliance, resulting in acid rain and other disasters.

The above figure shows that the concentration of SO2 in 2009 in Hebei Province reduces by 62% compared to 2003, and reduces by 33% compared to 2005; NO2 concentration has decreased, but it is not obvious

Hebei Province, the frequency of acid rain in 2008 was 2.4%, compared with 2007, the number of cities with acid rain fairs last year, frequency of acid rain decreased by 3.3 percentage points, but the strength of acid rain increases. Hebei Meteorological Research Institute shows that acid rain frequently appear in various regions across our province, appears 2 times in Qinhuangdao Station and 3 times in Palace Station in April 2009.

In recent years Hebei Province does a lot of work on pollutant emissions, invests heavily capital in pollution treatment, strengthens the pollution exclusion and treatment measures, but no city in Hebei Province meets the national emission standards, has to be further efforts.

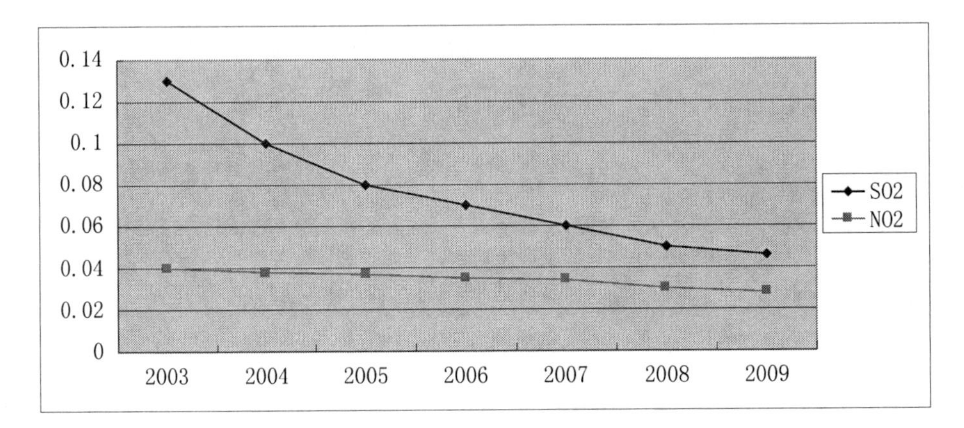

Fig. 1. Average concentration of major pollutants (SO2, NO2) in Hebei Province from 2003 to 2009(Data from the Hebei Environment Bulletin)

2 The Correlative Analysis on Economy-Energy-Environment System of Hebei Province

Economic, energy, environment are closely related and interrelated, is interdependent and an interactive system, is called the Economy - energy - environment system. We must take it as a system to study, not be seen in isolation.

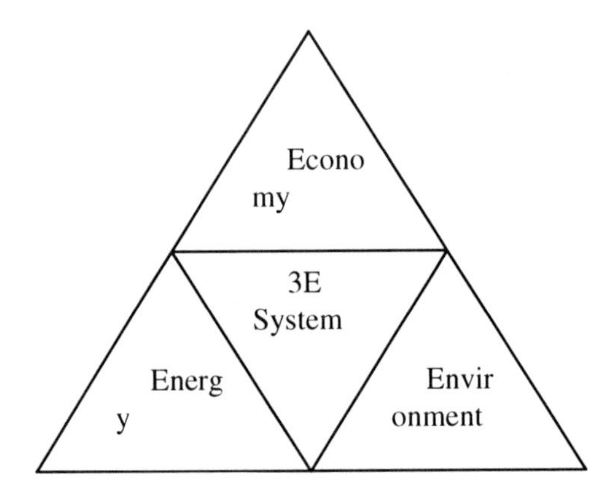

Fig. 2. The Economy-energy-environment system

2.1 The Relationship between Energy and Economic Growth

According to research, in different historical periods and under different condition energy consumption has a very difference and has its regularity. We use the Energy Intensity to illustrate the relationship each other. The formula of the Energy Intensity is expressed as: $e = E / G$, where e says the Energy Intensity, E says the energy consumption, G says GDP. the Energy Intensity is the final energy demand of per unit GDP, is an important indicator of the energy efficiency. The smaller the Energy Intensity, the greater the output of per unit energy.

Table 1. Total energy consumption, GDP and the Energy Intensity of Hebei Province each year from 2003 to 2008(Data from Hebei Statistical Yearbook in 2009)

Years	2003	2004	2005	2006	2007	2008
Total energy consumption	15298	17348	19745	21690	23490	24226
GDP	6921	8478	10096	11516	13710	16189
Energy intensity	2.210	2.046	1.956	1.884	1.713	1.496

The above table shows that the Energy Intensity of Hebei province decreases year by year in addition to 2002, which shows that our province has been expanding output of per unit energy, energy use efficiency has been improving. Rapid economic development in Hebei Province is the cost that it consumes a large amount of energy.

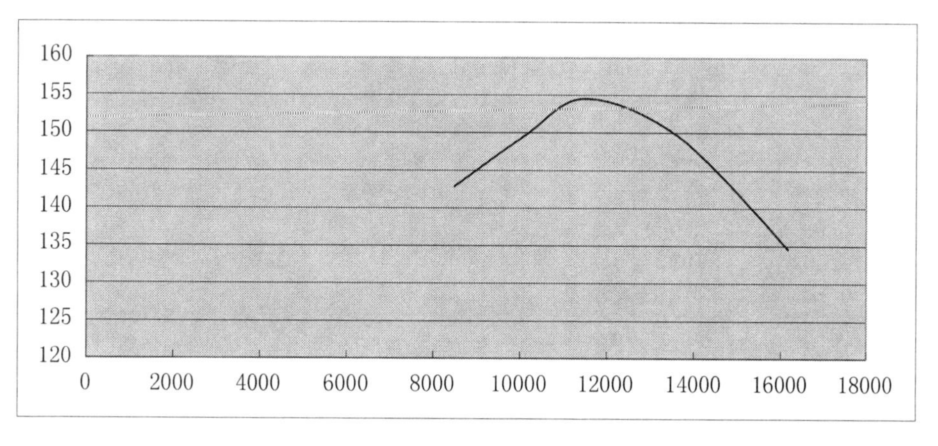

Fig. 3. The Environment Kuznets Curve for GDP and SO2 emissions of Hebei Province from 2004 to 2008 (Data from Hebei Statistical Yearbook and Hebei Environment Bulletin in 2009)

2.2 The Relationship between Economic Growth and Environment

In this paper, we use the Environment Kuznets Curve to study the relationship between economy and environment of Hebei Province, which is a doctrine that is used to analyze, the relationship between the fair distribution and per capita income level and is proposed by Kuznets. In 1991 Grossman and Krueger for the first time made the empirical study to the correlation between economic growth and environmental quality and found that they present a "inverted U " shaped relationship between economic growth and environmental quality, that is, in the early stages of economic development environmental quality declines with economic growth and when economic growth overgoes a specific "turning point", environmental quality will improves with economic growth. This is the famous the Environment Kuznets Curve, referred to as the EKC.

In Figure 3 the horizontal axis represents GDP and its unit is 100 million yuan; vertical axis represents the SO2 emission amounts and its unit is 10 thousand tons.

From Figure 3 it can be seen that the "turning point" appears in 2006, which showes that since 2006 with increasing GDP of Hebei Province, SO2 emissions gradually declines and environmental quality has been greatly improved. GDP of Hebei Province increases year by year and the economic development brought about the deterioration of the environment, which will necessarily obstacle and constraint economic development. Environmental pollution is serious in "the Ninth Five-Year" period and is improved in "the Tenth Five-Year" period.

2.3 The Relationship between Energy and the Environment

Energy consumption of per unit GDP is as ton standard coal per 10 thousand yuan to unit, SO2 emissions is as million tons to unit.

From Figure 4 it can be seen that in Hebei Province from 2004 to 2009 energy consumption of per unit GDP other years declines in addition to 2007; SO2 emissions the remaining years was also in decline in addition to 2006. At present energy consumption of the unit GDP and SO2 emissions of Hebei Province improves, but is still higher than the world standards and it is further improved. Tilt of the two curves roughly are the same and almost parallel, which shows that technology of the energy use in Hebei Province are still in low stage, and there is not much improved. High energy consumption is bound to pollute the environment, which requires us to continually improve technology of energy exploitation and utilization and find new energy sources to minimize environmental damage.

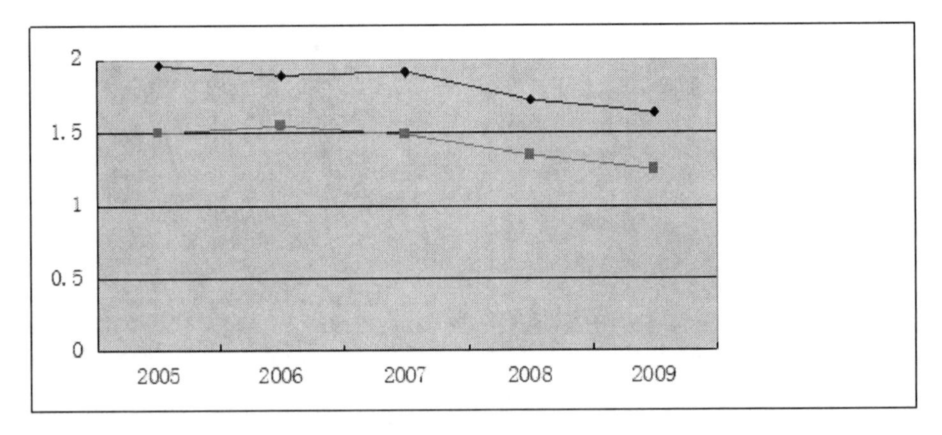

Fig. 4. Energy consumption of per unit GDP and SO2 emissions of Hebei Province from 2005 to 2009(Data from Hebei Statistical Yearbook and Hebei Environment Bulletin in 2009)

In short, in order to achieve coordinate and sustainable development of economic, energy, environment, not only be rapid economic growth, but also more importantly be long-term consideration, and try hard to reduce the unit GDP energy consumption, effectively control SO2 emissions and other harmful substances.

3 Conclusions

This article describes the general condition of economy, energy, environment of Hebei Province and has a detailed analysis to the correlation between the three. We find that there are the trend of coordinate and sustainable development of 3E in Hebei Province, but in the middle some years there will be repeated phenomenon (energy consumption of per unit GDP and SO2 emissions increase analyzed in this article). No doubt that economic growth of Hebei Province is rapid, but energy consumption is excessive, pollutant emissions is also significantly lower than the national discharge standards and is a far cry from the world level. Therefore, in the future development of our province, we must generally take into account the economic, energy, environment, coordinate good the relationship between the three, finally achieve coordinate and sustainable development of our society.

References

1. Jiutian, F., Jianzhong, Y., Honghua, J.: Study on Harmonious Development of Economy and Environment of Shandong Province in the 21st Century.China Population, Resources and Environment (2003)
2. Xie, G.: Improving Environment and Developing Economy are not Contradictory. Reference News, July 25 (2007)
3. Wu, X., Zhejiang, Z.B.: 3E Analysis System. Zhejiang Statistics, (November 2005)
4. Jorgenson, D., Wilcoxen, P.J.: Environmental reg-ulations and U.S. economic growth. Presented at the MIT Workshop on Energy and Environmental Modeling, July 31-August 1 (1989); also published in Rand Journal of Economics, Summer (1990)
5. Xu, Z., Cheng, G.: Economic diversity, dealysis, system engineering theory and practice, 2002velopment capacity and sustainable development of China. Ecological E-economics (March 2002)
6. Kraft, J., Kraft, A.: On the relationship between energy and GNP. The Journal of Energy and Development (1978)
7. Yuan, J.-H., et al.: Population - resources - environment - economy model system for systems analysis. Systems Engineering Theory and Practice (2002)

Study on the Credit Evaluation Model of C2C E-Commerce Websites

Li Shengqi

Zhejiang Agriculture and Forestry University, Hangzhou, Zhejiang 311300, China
Sqli65@163.com

Abstract. Because the current reputation assessment system of the main C2C website is too simple, we can not prevent deceiving on electronic transaction. Focusing on the disadvantages of the evaluation system of existing C2C e-commerce websites, Basing on the four dimensions that product, service, distribution and payment in customer evaluation, this paper conducts analyses respectively from the setting of evaluation grade, the setting of score, the setting of initial credit value, and the credibility of evaluation, proposing an improved credit evaluation model, and conducting a case analysis of the application of the improved evaluation model. It can be seen from the case analysis that the improved evaluation model can better reflect the true credit value of users. Because the reliability of this assessment is composed of the direct and indirect reliability, this model can help prevent sensationalization and slender.

Keywords: e-commerce, C2C, reputation assessment system, model.

1 Introduction

As an important transactional mode in e-commerce, C2C shows a strong development trend in China. Some big domestic C2C shopping websites, such as Taobao, TOM EBay and Paipai have 67.1% market share of online shopping [1]. Even so, e-commerce credit is still the biggest obstacle to the overall development of China e-commerce and C2C e-commerce websites suffer to a greater extent from widespread credit problems. Because participants of C2C are generally individuals and small enterprises with little popularity, both parties of the transaction lack channels of understanding and have no access to the identity and credit status of the trading partners, this means that cyber fraud is more likely to occur. Currently, major C2C e-commerce websites adopt the credit evaluation system to increase user trust of online transactions; however, the existing credit evaluation system is too weak to prevent the lack of credibility and poor reputation of credit. This paper focuses on the disadvantages of the existing credit evaluation system, offers an improved credit evaluation model with higher credibility, and uses examples to validate the model.

2 Analysis of Disadvantages Problems Existing in the Current Evaluation Model

Major domestic C2C e-commerce websites, such as Taobao, TOMEBay and Paipai adopt credit evaluation systems with slight differences, but they are generally similar.

G. Zhiguo et al. (Eds.): WISM 2011, CCIS 238, pp. 150–157, 2011.

In general, the evaluation result values are good, ordinary and bad, respectively corresponding to +1, 0, and -1. The accumulated user score is taken as the credit value of the user [2]. Although analysis of an existing credit evaluation system can evaluate the credit of the buyers and sellers, there are still some disadvantages as follows:

A. The credit evaluation grading system is too simple

The evaluation grades of "good", "ordinary" and "bad" fail to truly reflect the user evaluation. For example, at Taobao, many users complain about sellers' services and products, such as, some products are not packed properly, products are different from those shown in the color pictures. But, due to the fine quality of these products, users still give them high praise.

B. The existing evaluation system neglects the impact of any initial credit value

Some websites only adopt this simple accumulated score principle creating the circumstance where the same result fails to reflect the difference in a user's credit rating. For example, a user gets 100 "good", and another user gets 200 "good" and 100 "bad", reflecting the credit rating of the two users as being not the same, even very different, however, the accumulated credit score through the current system is the same.

C. The existing evaluation system neglects the impact of time weighting in credit evaluation

User credit value is a process of dynamic change, changes in user behavior will lead to changes in credit value, and the credit evaluation of users at different times is also different. For example, the evaluation made by a user 3 days ago is different from that made 45 days ago, But, the currently existing evaluation system cannot reflect this change.

D. The existing evaluation system neglects the impact of transaction times and transaction amounts on credit evaluation

The existing evaluation system neglects the impact of transaction times and transaction amounts, where the effect of a user with 1 transaction is the same as that of a user with 100 transactions. The impact of the evaluation made by users with small transactions is the same with that of users with large transactions and this fails to prevent some users making many small transactions to win a high credit rating and then making fraudulent large transactions when their credit rating is high[3].

E. The existing evaluation system fails to prevent the speculation and defamation of credit

Existing credit evaluation systems assume that the users rating is true and reliable, but all "good" and "bad" evaluations are accounted within the total score which fails to measure the credibility of the evaluation. This accounting method can not prevent some users manipulating their credit value by fraud.

F. The existing evaluation system fails to encourage the third-party services

Literature [4] illustrates the important role that the third party plays in online transactions. Paul A. Pavlou and David Gefen pointed out that the third-party service is an important model of e-commerce, particularly when buyers and sellers come from different societies and under different cultural backgrounds and they have never previously traded with each other. McKnight and Chervany also pointed out the third-party services play an important role in helping buyers to trust sellers and help the completion of online transactions [5]. However, the existing evaluation system only suggests the third-party services without taking any effective measure to encourage third-party services.

3 Improvement of the Credit Evaluation Model

To overcome the disadvantages of the above credit evaluation model, this paper sets out to make improvement in the setting of evaluation grades, the setting of initial credit value, the setting of evaluation score, and the credibility of evaluation.

A. Improvement of evaluation grade
This paper divides user evaluation into four factors: product, service, distribution and payment, with each factor corresponding to one evaluation. One evaluation question is respectively designed for each factor of product, service, distribution and payment. For example, on the aspect of service, the question can be designed as "What about the seller's service attitude?", and the answers to the question are respectively "good", "ordinary" and "bad", respectively corresponding to "+1, 0, -1". In this way, a user can make a relatively objective evaluation of the seller's performance in these aspects during the transaction process.
B. Definition of initial credit value
When both the buyer and the seller applied to log in C2C website for the first time, major C2C shopping websites set up the initial credit value of both the buyer and the seller as zero, which makes the buyer fail to make a judgment of the difference of the seller's credit rating. Accordingly, we hold that the initial credit value of the seller is related to the third-party services the seller adopts.

Definition 1. Assume IS_j is the initial credit value of the seller S_j, there are n third-party services, the seller passes n third-party service authentication. w_i is the weight of No.*i* service authentication in all third-party service authentications. The bigger the weight is, the higher the credibility of the service authentication that the buyer perceives during the transaction process. $f\ (w_i)$ is the score that No.*i* service authentication gets, and the initial credit value of the seller is the sum of the scores that the seller gets in conducting various third-party service authentications.

$$IS_j = \sum_{i=0}^{n} f(w_i) \tag{1}$$

Namely, the initial credit value is related to the authentication it passes, which gets basic identity authentication in the first registration. If means the pass of the third-party authentication, means the pass of identity authentication, and $w_1 > w_2$, if S_1 only passes payment authentication, S_2 only passes identity authentication, and S_3 fails to pass any authentication, then $IS_1 > IS_2 > IS_3$.
C. Setting of user's evaluation score

Definition 2. Assume S_j is No. *j* seller, B_i is No.*i* buyer trading with S_j. T_x is No. *x* transaction between B_i and S_j. $R\ (i,j,x)$ is the evaluation on the credit value of S_j when No. *x* transaction is conducted between B_i and S_j. In 3.a, evaluation has been divided into four factors, $Score$ is the evaluation score obtained by these four evaluation factors, the function is defined as:

$$\begin{aligned} Score &= f(service, product, distribute, payment) \\ &= \alpha * service + \beta * product + \chi * distribute + \delta * payment \end{aligned} \tag{2}$$

Where, service, product, distribute, and payment respectively means the evaluation score of the service factor, product factor, distribution factor and payment factor. Because different categories of products have different requirements on these four factors, so the weight of the factors of different categories of products is also different. If $\alpha, \beta, \chi, \delta$ respectively means the weight value of each corresponding factor in calculating the score, $\alpha, \beta, \chi, \delta, \in [0,1]$ and $\alpha + \beta + \chi + \delta = 1$. The bigger the weight is, the bigger the impact of the factor on the seller's credit during this transaction process. The weight value of $\alpha, \beta, \chi, \delta$ is found according to the statistics of the users in C2C transaction websites. From equation (2), evaluation $R(i,j,x)$ that B_i has on S_j can be reflected as the following function:

$$R(i,j,x) = \begin{cases} 1 & Score > 0 \\ 0 & Score = 0... \\ -1 & Score < 0 \end{cases} \tag{3}$$

From equation (2) and (3), we find that when the user gets "good" evaluation in the four factors, the score is 1; when the user get "bad" evaluation in these four factors, the score is -1; and when the user get 0 in the four factors, the score is 0. If a user's evaluation is (-1, 1, -1, 1), and the corresponding weight of each factor is (0.5, 0.2, 0.2, 0.1), then the score of the user is: $Score$ =0.5*(-1)+0.2*1+0.2*(-1)+1*0.1=-0.4<0, and the credit value of the seller is -1. it can be seen from the above function that if the buyer thinks that the user's quality is satisfactory, but the service and distribution efficiency are not satisfactory, and service plays an important role in this transaction, therefore the seller still gets a score of -1, which avoids the simplicity of previous credit evaluations.

D. Determination of the credibility of user evaluation

The effect of the evaluation made by different users in different situations shall also be different. The evaluation made by a user with multiple transaction experiences shall be more convincing than that made by a user without transaction experience. The evaluation made by the user with large transactions shall be more influential than that made by the user with small transactions. If a buyer has ever traded with multiple sellers, then the evaluation made by the buyer shall be more convincing than that made by the user who has only ever traded with one seller. Accordingly, this can effectively inhibit the speculation and defamation of credit and credit fraud.

Definition 3. The direct credibility Y: after the transaction between B_i and S_j, the degree of credibility of this transaction can be deduced from previous direct transaction times and transaction amounts between B_i and S_j. The function is as follows:

$$Y = k(p * sum + \frac{q}{r} * account).. \tag{4}$$

Where, sum is previous transaction times, $account$ is the total amount of previous transactions, p, q is respectively the factor coefficient of the impact of transaction times and amount on direct credibility, $p, q \in [0,1]$. r is the resolution factor of transaction amount, which ensures that even if the value of $account$ is relatively big, sum and $account$ can impact on Y to the same factor of the order, the value of r is determined on the basis of sum and $account$. k is the regulatory factor, so $Y \in [0,1]$. It can be seen from equation (4) that the greater the transaction times are and the bigger the transaction amount is, the bigger the value of Y is, but the biggest value of can only be 1.

Definition 4. Indirect credibility Z: when B_i is trading with S_j and B_i evaluates on S_j, the credibility of the evaluation made by Bi can be deduced from the transaction condition between B_i and S_j ($i \neq j$). In general, the more the transactions between B_i and other transaction users, the higher the credibility of the evaluation made by B_i. The function is:

$$Z = \rho * num.... \tag{5}$$

Where, $Z \in [0,1]$, ρ is the factor coefficient of the transaction times between B_i and other users on the indirect credibility, $\rho > 0$. num is the number of different transaction users trading with B_i.

Definition 5. The credibility of user's evaluation $C_R(T_x,B_i)$; T_x is No.x transactions conducted by B_i with current seller S_j, $C_R(T_x,B_i)$ is the credibility degree predicted according to the previous transaction conditions among B_i and S_j and other sellers so as to determine the degree of importance of this evaluation on the credit value of the other party. $C_R(T_x,B_i) \in [0,1]$, 0 means that this evaluation is unreliable, and 1 means this evaluation is completely reliable. $C_R(T_x,B_i)$ is composed by direct credibility factors and indirect credibility factors. From equation (4) and (5), we get the following function:

$$C_R(T_x,B_i) = \theta * Y + (1-\theta) * Z \tag{6}$$

Where, $\theta \in [0,1]$, θ is used to control the weight of direct credibility, which has an inverse relationship with the num of the user and other users trading with it. When $num = 0$, namely B_i only trades with S_j, user's credibility $C_R(T_x,B_i)$ is conducted from the previous transaction conditions between B_i and S_j. When the value of keeps increasing, it means that the role of the user's indirect credibility in judging the other user's credibility $C_R(T_x,B_i)$ is bigger and bigger when B_i trades with more and more users. However, the weight of indirect credibility will not exceed a certain value, which shows that the role of indirect credibility is increasing, but it is impossible for it to completely replace the user credibility.

E. Improvement of the evaluation model

Based on the above discussions, we propose an improved credit evaluation model. Considering the impact of time factor, a user's historical evaluation may have expired showing that its role in evaluating the user's current credibility should be reduced. However, the impact of the latest evaluation on credibility should be bigger, and the latest evaluation should better reflect the user's credibility than any previous evaluation. Therefore, we also take time as a weight to evaluate credibility, which definition is as follows:

Definition 6. time weight ω_t is defined as the following function:

$$\omega_t = e^{-\frac{\Delta t}{\lambda}} \tag{7}$$

Where, $\omega_t \in [0,1]$, Δt is time span variable and λ is relevant time adjustment factor to ensure that ω_t is suitable to different situations. For example, if we want to express that the evaluation weight 5 days ago is half of the current evaluation weight (e.g.: the evaluation weight 5 days ago is 0.5, the current evaluation weight is 1, $\Delta t = 5$), then $\lambda = -\frac{5}{\ln(0.5)}$. From equation (1), (3), (6) and (7), we get the definition of the improved credit evaluation model as follows:

Definition 7. *R (i,j,x)* is the total credit evaluation value of the transactions between S_j and B_i, L is the total times of the transaction between B_i and S_j, so we get:

$$R(i,j,x) = \begin{cases} IS_j + \sum_{i=1}^{n} \sum_{x=1}^{L} \left(R(i,j,x) * Cr(T_x, B_i) * \omega_t \right) & i \geq 1 \\ IS_j & i = 0 \end{cases} \tag{8}$$

In the model, seller's credibility is the sum of the initial credit value and user's credit evaluation value. The improved credit evaluation model integrates into one the setting of evaluation grade, the setting of user evaluation score, user evaluation credibility and time factor, etc. It can be seen from equation (8) that when the buyer and seller have never traded with each other, the seller's initial credit value is determined by the third-party services that it conducts. We think that the more all-rounded the third-party services are, the higher his initial credit value is. When the buyer conducts an online transaction with the seller, the seller's credit value is formed from the evaluation made by the buyer after the transaction is over. At this time, the proportion of the initial credit value is smaller and smaller. During the transaction process, the buyer makes his evaluation in terms of the service, quality, distribution and payment of the product which finally constitutes the score of the evaluation. For the credibility evaluated by the buyer, we can get it according to equation (6). From equation (8), the previous evaluation on the seller has certain impact on the seller's credibility. And we think that the more experienced the buyer is, the higher his evaluation credibility is, and the more recent the evaluation is, the better it can reflect the latest credibility of the seller. In addition, it can prevent the speculation and defamation of credit and credit fraud to a certain degree. The product of multiplication of user's credibility, user's evaluation score and time weight constitute the real credit score of the seller made by the user. And the total credit rating of the seller is the sum of the credit scores evaluated by various buyers.

4 Case Analysis of Model Application

To promote the standard uniformity of C2C e-commerce platforms, and for the convenience of users, we think that users can conduct any valid third-party payment authentications and identity authentications which are acceptable on any C2C transaction platform. If S_1 , S_2, S_3 are the sellers selling UV shade cloth, and the current platform encourages two third-party authentications, namely user's identity authentication and the third-party payment authentication, and their respective weight is (0.6, 0.4) and the total initial credit value is 1. S_1 conducts identity authentication and third-party authentication, S_2 conducts third-party authentication, and S_3 conducts no authentication, so their respective initial credit value is 1, 0.4 and 0. Take the transaction record of S_1 as an example to analyze the application of the model, see *Table 1* for transaction record, where the commodity price is 10 yuan/piece and B_i is the user in the transaction. From the credit evaluation system of the current C2C transaction platform, we get the credit score of S_1 is 5, and there is no difference in the effect of user's evaluation. Now, we apply the improved evaluation model (8) to analyze the transaction situations in *Table 1*. According to online statistics, we find that the weight of user evaluation on the four factors of fashion and adornment products is respectively (0.3, 0.4, 0.25, 0.05), and the evaluation score of these transactions on the basis of equation (2) and (3) is (1, 1, 1, -1). In equation (4), assume: $k = 0.5$, $p = 0.2$, $q = 0.8$. And in equation (5), assume $\rho = 0.2$, The credibility of the first evaluation score of B_1: if $r = 10$, and in the first transaction, num=0, so the first transaction of B_1 is Y=0.5*(0.2*0+ *10) =0.4. By equation (5), we get Z=0, and the credibility of the first evaluation score of B_1 is 0.4.

The credibility of the second evaluation score of B_1: if $r = 50$, since num >0 at this time, sum=1, if $\theta = 0.6$,so Y=0.5*(0.2*1+ $\frac{0.8}{50} * 100$) =0.9, Z=0.2*5=1, $Cr(T_2, B_1)$=0.6*0.9+ (1-0.6)*1=0.94. The setting of time weight value: Setting up parameter, we can get the time weight 1 day ago is 1, and the time weight 15 days ago is 0.5. We get that the total evaluation score of B1 on S1 is 0.4*0.5+0.94*1=1.14. Similarly, we can get the evaluation score of B2, B3, and B4 onS1 is respectively 0.4, 0.4, and -0.4.

From equation (8) we get the total credit evaluation score of S_1, namely the sum of the initial credit score and buyers' evaluation on the seller, is 2.54. It can be seen that the credit value of S_1 is less than the credibility obtained before the evaluation system is improved. And when user B_1 trades with S_1 for the first time, the transaction amount and times are less than those of the second time and the evaluation is not the latest evaluation, so the evaluation score is much different from the score of the second evaluation. For the calculation of the evaluation credibility, since the buyer has not previously traded with any other users, the credibility can only be obtained from direct credibility. For the second transaction, since B_1 has gained experience in trading with other sellers, evaluation credibility should be obtained from direct credibility and indirect credibility.

Table 1. Transactions Between BI and S1

Buyer	Current transaction times	Current evaluation	Transaction amount	Transaction time	Number of users trading with other sellers
B_1	1	Good	10 yuan	15 days ago	0
B_2	1	Good	10 yuan	1 day ago	0
B_3	1	Good	10 yuan	1 day ago	0
B_1	2	Good	100 yuan	1 day ago	5
B_4	1	Good (but dissatisfied with service and delivery)	10 yuan	1 day ago	0

5 Conclusion

Focusing on the disadvantages of the evaluation system of existing C2C e-commerce websites, this paper conducts analyses respectively from the setting of evaluation grade, the setting of score, the setting of initial credit value, and the credibility of evaluation, proposing an improved credit evaluation model, and conducting a case analysis of the application of the improved evaluation model. It can be seen from the case analysis that the improved evaluation model can better reflect the true credit value of users, because user evaluation credibility consists of direct and indirect credibility, and so to a certain extent prevents the speculation and defamation of credit and credit fraud.

References

1. CNNIC, The report of internet development in china (2009), `http://www.cnnic.net.cn/,2010-01-15`
2. Abdul-Rahman, A., Hailes, S.: Supporting trust in virtual communities (2000), `http://csdl.computer.org/comp/proceedings/hicss/2000/0493/06/04936007.pdf`
3. Jones, K., Leonard, L.N.K.: Trust in consumer-to-consumer electronic commerce. Information & Management, 45:88-95 (2008)
4. Pavlou, P.A., Gefen, D.: Building Effective Online Marketplaces with Institution-Based Trust. Information Systems Research 15(1), 37–59 (2004)
5. Harrison McKnight, D., Choudhury, V., Kacmar, C.: The Role of Reputation Systems in Reducing Online Auction Fraud. Information Systems Research. 13(3), 334–359 (2002)

The Evolution of Consumer Trust and Satisfaction in Mobile Electronic Commerce

Jiabao Lin and Guanghui Zhang

College of Economics & Management, South China Agricultural University
Guangzhou, China
{linjb,ghzhang}@scau.edu.cn

Abstract. The development of consumer trust in mobile commerce is a dynamic process over time. However, little research has been done on the formation law of this dynamic process. Based on the extended valence theory, self-perception theory and IS expectation-confirmation theory, this study builds a three-stage theoretical model of consumer trust evolution in the context of mobile banking, and mainly analyzes the formation mechanisms of consumer decision-making behavior in the pre-usage phase, the feedback mechanisms of usage behavior in the usage phase and the evaluations mechanisms in the post-usage phase. We posit that trust not only directly affects usage behavior, but also has indirectly influence on it through other variables, usage behavior has significant feedback on the cognitive or psychological factors, users' evaluations have significant impact on satisfaction, satisfaction would prompt the level of user trust in turn which has important effect on the future usage behavior, and ultimately, the evolution of trust forms a round dynamic process, which explain the law of trust evolution.

Keywords: Mobile commerce, Mobile banking, Trust dynamics.

1 Introduction

In the mobile commerce environment, it is important for the success of mobile banking to understand the user's trust dynamic evolvement process, because both non-users and users of mobile banking are concerned about transaction security. Mobile banking as a kind of transactional service and the main obstacles that prevent many consumers from using mobile banking are the uncertainties and risks involved. Trust is the powerful tool to reduce perceived risk and uncertainty [1]. Cultivating consumer trust in mobile commerce is a dynamic process, which extends from initial trust formation to continuous trust development [2]. Although Initial trust is an important indicator of mobile banking success, it does not necessarily lead to the desired managerial outcome unless the trust continues [3]. Previous studies on consumer trust in electronic commerce and mobile commerce mainly consider the antecedent variables of trust based on static cross-sectional data [4, 5]. In nature, the development of consumer trust is an interactive and dynamics evolution process over time. Despite the importance of trust dynamics, this topic has been given little attention in the academic literature. And there has been little academic study on

G. Zhiguo et al. (Eds.): WISM 2011, CCIS 238, pp. 158–165, 2011.

providing insight into trust dynamics in mobile commerce. Furthermore, very little research has addressed the formation mechanism of the dynamic trust process in mobile commerce. Given this general void in the literature, it is necessary to examine consumer trust is how to evolve in mobile commerce over time. The purpose of the study is exploring the dynamic evolution law of the consumer trust in the context of mobile banking. Based on the extended valence theory, self-perception theory and IS expectation-confirmation theory, this study develop a longitudinal model of how users' trust evolve from the perspective of the decision-making process and analyzes the effects of trust on consumers' usage behavior, the feedbacks of usage behavior on the cognitive and psychological factors and the effects of users' evaluations on trust, which explain the consumers' trust evolution law.

2 Theory Background

2.1 Extended Valence Theory

Valence theory originates from economics and psychology and later is applied to marketing field to explain the relationships between consumer behavior and the perceptions of risk and benefit. Peter and Tarpey [6] compare three types of decision-making models: (1) minimization of expected negative utility (perceived risk); (2) maximization of expected positive utility (perceived return); and (3) maximization of expected net utility (net perceived return). The results of their study indicate that the net perceived return model could explain more variance in brand preference than the other two models, which is the principle of valence theory. The valence theory suggests that there are positive and negative consumer perceptions toward products, namely as perceived risk and perceived benefit, and they are the two main aspects of consumer decision-making behaviors which are to maximize the net value (the subtraction between benefit and risk). Intuitively, the valence theory is a good model as it considers both positive and negative decision-making attributes.

In the electronic commerce or mobile commerce environment, trust is a crucial factor affecting consumer behaviors [7]. Therefore, this study integrates trust and the valence theory, and formats the extended valence theory. This model is shown in Figure 1. We suggest trust not only directly affect the usage behavior, but also has indirectly impacts on it through perceived risk and perceived benefit.

2.2 Self-perception Theory

The self-perception theory posits that the usage behavior of users is their evaluation base for products or services, which means The more usage there is, the more favorable the user evaluations [8]. The self-perception theory also suggests that people do not form specific evaluations on routine behavior until they are asked to do so. Later, researchers take the self-perception theory as the theoretical basis of the feedback mechanism. In IS filed, scholars have empirical examined this theory and many studies find that past usage behavior will have effects on the motivation and evaluation of the future. Based on a meta-analysis, Ouellette and Wood [9] find that past behavior is significantly related to attitudes, subjective norms, perceived behavioral control, and intention. In the context of personalized portal, past usage behavior has significantly positive effects on perceived usefulness, perceived ease of use, and usage intention [10].

2.3 IS Expectation-Confirmation Theory

The logical framework of expectation confirmation theory is as following: Consumers form specific the expectations of products or services before purchase. When consumers use the products or services for a period of time, they will form the performance perceptions. Then, they compare pre-purchase expectations and post-purchase performance perceptions to measure the extent that they expect the confirmation. If the expectations of consumers are greater than the actual performance, that is unconfirmed. On the contrary, if the expectations of consumers are less than or equal to the actual performance, that is confirmed. The extent to confirmation and the pre-purchase expectations determine their level of satisfaction. Finally, the level of satisfaction consumer satisfaction with products or services determines their intention to buy again [11]. Bhattacherjee [12] apply the expectation confirmation theory to the IS filed, posit IS expectation confirmation theory and suggest that the IS continuance usage and the repurchase of consumers are similar to each other. The IS expectation confirmation theory posits that IS continuance intention is determined by three constructs: expectation of the IS, confirmation of expectation following the actual usage and perceived usefulness.

3 Research Model

3.1 Pre-usage Phase (T1)

Trust is a complex construct, because previous researchers have defined it from different perspectives and disciplines, but do not form a unified definition. Trust is one party's psychology expectations that another party do not engage opportunistic behavior [13]. It is also the willingness of a party to be vulnerable to the actions of another party [14]. Some studies adopt the conceptualization of trust as a set of specific trust beliefs which is the beneficial features that a party think another party has, including a party’s perception of another party’s ability, benevolence, and integrity[15]. Others view that trust is the whole belief that another party is trusting, namely as trusting intention, meaning that one party intends to rely on another party, although has no the ability to monitor or control that party [16]. In this study, taking the trusting belief perspective, our perspective is consistent with the definition of trust in online trading by Lee [17]. Applying their view to mobile commerce, the high level of trust will increase consumer intention and behavior. In the context of mobile banking, as the transactions of mobile banking take place in virtual network, consumers are worried about mobile banking may not be able to fulfill its responsibilities. Thus, consumer trust is an important factor of affecting usage behavior of mobile banking. This study proposes the following hypothesis:

H1: Consumer trust in mobile banking (T1) has a positive effect on usage behavior. Perceived risk is used to explain consumer behavior in early days, and later many studies explore the effect of perceived risk on traditional consumer behavior. In our study, perceived risk is defined as using mobile banking will bring the potential and uncertain negative value of subjective beliefs. The risk of mobile commerce mainly contains security/privacy risk and financial risk.

Security/privacy risk is defined as a potential loss due to fraud or a hacker compromising the security of a mobile banking user. Mobile Web viruses, in particular by placing a malicious hacker or phishing Trojan program, access to customer account information and password, leading to the risk. Both fraud and hacker intrusion not only lead to users' monetary loss, but also violate users' privacy, a major concern of many Mobile Internet users. Many consumers believe that they are vulnerable to identity theft while using mobile banking services.

Financial risk is defined as the potential for monetary loss due to transaction error or bank account misuse. Mobile banking is a kind of the services based on transaction and involves payment, money transfer and other financial business. Many customers are afraid of losing money while performing transactions or transferring money over the Mobile Internet. At present mobile banking transactions lack the assurance provided in traditional setting through formal proceedings and receipts. Thus, consumers usually have difficulties in asking for compensation when transaction errors occur. In short, many consumers think that using mobile bank will exist risk which makes them vulnerable, and dare not use it.

In the electronic commerce environment, prior research shows that trust is a determinant factor reducing the perceived risk and uncertainty. If consumers have the high level of trust in mobile banking, they will think that mobile banking does not violate its commitments. This indicates that a high level of trust will reduce the perceived risk of consumers. On the contrary, if the level of trust is low, consumers are usually concerned that mobile banking would discharge of its responsibilities, therefore, produce high levels of perceived risk. Based on the above discussions, this study posits the following hypotheses:

H2a: Consumer trust in mobile banking (T1) has a negative effect on perceived risk.

H2b: Perceived risk has a negative effect on the usage of mobile banking.

Mobile banking is the innovation and extension of online banking. As small and portable mobile terminal and extensive coverage of wireless networks, consumers can use mobile banking anytime and anywhere. Mobile banking allows consumers do not have to wait in line or go to the bank's counter service, and it can be extended the banking business to user side. Perceived benefit of mobile banking divides into two types: one is the direct benefit and the other is the indirect benefit. The direct benefit is defined as the tangible benefits that users can feel, such as low transaction costs. At present, to attract more customers, mobile banking does not charge commission charge and provide preferential price for money transferring and remittance. The lower commission charge is, the more people transfer from offline banking and online banking to mobile banking. Therefore, mobile banking can improve the corporate image. The direct benefit is defined as those the benefit that is intangible and difficult to measure, for instance, mobile banking users can enjoy 24-hour service and really obtain the features with the need to use, which saves time cost. In addition, Mobile banking not only has the basic financial business that offline banking provides, but also has the personal financial services such as mobile stock, bank-securities transfer, and the transactions of gold, allowing users to understand the market and accumulate wealth. Perceived benefit will motivate consumers to use mobile banking, as it can bring them many gains. However, these gains are achieved on the premise that consumers believe mobile banking fulfills its obligations. The trusted customers have confidence in the obligations of mobile banking, and further have confidence in obtaining benefits, which indicates a higher level of trust will promote greater benefits. Conversely, if consumers only have a low level of trust, they will doubt that whether mobile banking can fulfill its responsibilities. Therefore, this may produce a lower level of perceived benefit. In electronic commerce, previous studies find that consumer trust significantly affects perceived benefit and perceived benefit positively influences the intention of online shopping. Given the above analysis, expanding the background to mobile commerce, we propose the following hypotheses:

H3a: Consumer trust in mobile banking (T1) has a positive effect on perceived benefit.

H3b: Perceived benefit has a positive effect on the usage of mobile banking.

3.2 Usage Phase (T2)

Based on trust, perceived risk and benefit, consumers will form comprehensive evaluation which determines whether they use mobile banking. As long as consumers begin to use mobile banking, it indicates they enter the usage phase. The self-perception theory suggests previous usage behavior has a feedback effect on consumer future beliefs. According to the self-perception theory, we posit the following hypotheses:

H4: The usage behavior of mobile banking has a positive effect on confirmation.

H5: The usage behavior of mobile banking has a positive effect on satisfaction.

H6: The usage behavior of mobile banking has a positive effect on perceived usefulness.

3.3 Post-Usage Phase (T3)

The post-purchase process is very different from the pre-purchase process primarily because in the post-purchase phase consumers have substantial and direct prior experience to draw on. In other words, in the post-purchase evaluation process, mobile banking will be evaluated in the context of the consumer's prior expectations and the actual performance of the service as perceived after its usage. In marketing, expectation-confirmation theory suggests that consumer satisfaction is a state of psychology and comes from the comparison between the consumer's prior expectations and the actual performance of the service. If the performance is greater than or equal to the expectation, consumers are satisfied. if the performance is less than expected, that is unconfirmed and consumers are not satisfied [11]. In the field of information systems, IS expectation-confirmation theory suggests that customer satisfaction is determined by two factors: one is confirmation, the other is perceived usefulness [12]. Confirmation is formed when the performance after the usage is greater than or equal to the prior expectation in mobile banking. Conversely, that is not confirmed. Confirmation positively affects user satisfaction because confirmation reflects the realization of expected benefit. Perceived usefulness is a core variable of the technology acceptance model. The technology acceptance model finds that perceived usefulness plays a important role throughout the usage stage. In the IS expectation-confirmation theory, perceived is the representative for the post-usage performance. Previous research on consumer behavior suggests that higher post-usage performance can produce higher satisfaction. Thus, perceived usefulness has a positive effect on the satisfaction of mobile banking. In addition, user confirmation can adjust perceived usefulness of mobile banking, especially when the initial level of usefulness is relatively low, because they're not sure using mobile banking what can obtain. When users experience confirmation, the level of perceived usefulness will rise. Based on the above analysis, this study proposes the following hypotheses:

H7: User confirmation has a positive effect on the satisfaction of mobile banking.

H8: Perceived usefulness has a positive effect on the satisfaction of mobile banking.

Based on the extended valence theory, self-perception theory and IS expectation-confirmation theory, this study builds a three-stage (pre-usage, usage and post-usage) theoretical model of consumer trust evolution in the context of mobile banking. The research model is shown in Figure 1. In this model, we posit that trust in pre-usage phase not only has a direct effect on usage, but also has indirect effect on it through perceived benefit and risk. We also assume that the actual usage of mobile banking positively affects confirmation, perceived usefulness, and satisfaction, and satisfaction has a positive impact on trust in post-usage phase.

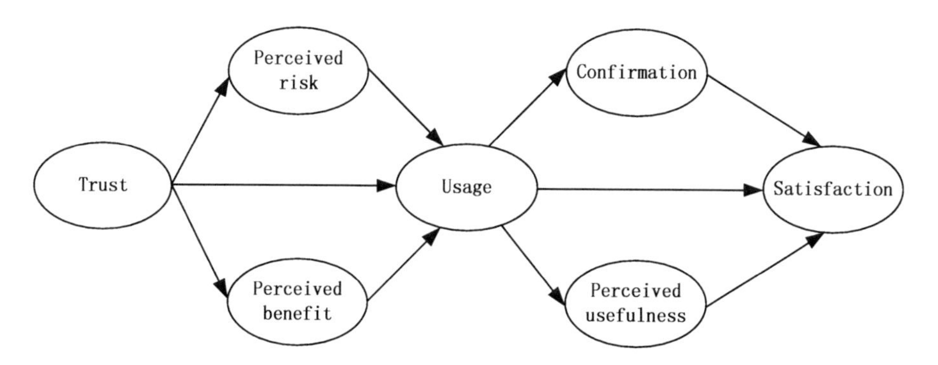

Fig. 1. Research model

4 Conclusions

This study builds a three-stage theoretical model of trust evolution in mobile commerce and explains the law of the dynamics of trust evolution. We assume that in pre-usage stage, trust not only directly affects usage behavior, but also has indirectly influence on it through other variables; in usage stage, usage behavior has significant feedback effects on the cognitive or psychological factors in the future; in post-usage stage, consumer evaluations have significant impact on trust, which forms new trust and further will enter a new round cycle, reflecting the law of trust evolution. The empirical examination will be done in future research.

Acknowledgment. This work was partially supported by a grant from the Soft Science Foundation of Guangdong (2011B070400011), a grant from the Presidential Foundation of South China Agricultural University (4700-K10057), grants from the National Natural Science Foundation of China (70971049, 70731001), a grant from NSFC/RGC (71061160505), and a grant from TD-SCDMA Joint Innovation Lab, Hubei Mobile Co., China Mobile Group.

References

1. Kim, D.J., Ferrin, D.L., Rao, H.R.: A trust-based consumer decision-making model in electronic commerce: the role of trust, perceived risk, and their antecedents. Decision Support Systems 44(2), 544–564 (2008)
2. Siau, K., Shen, Z.: Building customer trust in mobile commerce. Communications of the ACM 46(4), 91–94 (2003)
3. Vatanasombut, B., Igbaria, M., Stylianou, A.C., Rodgers, W.: Information systems continuance intention of web-based applications customers: The case of online banking. Information & Management 45(7), 419–428 (2008)
4. Hu, X., Wu, G., Wu, Y., Zhang, H.: The effects of web assurance seals on consumers' initial trust in an online vendor: A functional perspective. Decision Support Systems 48(2), 407–418 (2010)
5. Vance, A., Elie-dit-cosaque, C., Straub, D.W.: Examining trust in information technology artifacts: The effects of system quality and culture. Journal of Management Information Systems 24(4), 73–100 (2008)
6. Peter, J.P., Tarpey, L.X.: A comparative analysis of three consumer decisions strategies. Journal of Consumer Research 2(1), 29–37 (1975)
7. Jin, C., Cheng, Z., Yunjie, X.: The role of mutual trust in building members' loyalty to a C2C platform provider. International Journal of Electronic Commerce 14(1), 147–171 (2009)
8. Bem, D.: Advances in experimental social psychology. Academic Press, New York (1972)
9. Ouellette, J., Wood, W.: Habit and intention in everyday life: The multiple processes by which past behavior predicts future behavior. Psychological Bulletin 124, 54–74 (1998)
10. Kim, S.S., Malhotra, N.K.: A longitudinal model of continued IS use: An integrative view of four mechanisms underlying postadoption phenomena. Management Science 51(5), 741–755 (2005)
11. Oliver, R., Burke, R.: Expectation processes in satisfaction formation: A field study. Journal of Service Research 1(3), 196 (1999)
12. Bhattacherjee, A.: Understanding information systems continuance: An expectation-confirmation model. MIS Quarterly, 351–370 (2001a)
13. Rousseau, D., Sitkin, S., Burt, R., Camerer, C.: Not so different after all: a cross-discipline view of trust. Academy of Management Review 23(3), 393–404 (1998)

14. Mayer, R.C., Davis, J.H., Schoorman, F.D.: An integrative model of organizational trust. Academy of Management Review 20(3), 709–734 (1995)
15. Gefen, D., Karahanna, E., Straub, D.W.: Trust and TAM in online shopping: an integrated model. MIS Quarterly 27(1), 51–90 (2003a)
16. Moorman, C., Zaltman, G., Deshpande, R.: Relationships between providers and users of market research: the dynamics of trust within and between organizations. Journal of Marketing Research 29(3), 314–328 (1992)
17. Lee, M.C.: Predicting and explaining the adoption of online trading: an empirical study in Taiwan. Decision Support Systems 47(2), 133–142 (2009)

Impediments to E-Commerce Adoption as Perceived by Businesses

Xibao Zhang

Department of International Economics and Trade, Qingdao University
xibao.zhang@qdu.edu.cn

Abstract. In this study the underlying dimensions of e-commerce impediments as perceived by businesses are investigated. Statistical analysis of empirical data from a questionnaire survey of managers of Chinese businesses resulted in three meaningful e-commerce impediments dimensions being uncovered. They are the Unfamiliarity, Cost/Benefit, and Product Suitability dimensions, which represent what companies perceive as aspects inhibit them from adopting e-commerce. In addition, the limitations of this study and directions for future research are also discussed.

Keywords: electronic commerce, impediments, inhibitors, Chinese businesses.

1 Introduction

The impediments to e-commerce uptake by businesses have been dealt with in past research. Reference [1], in their study of New Zealand firms that were already online at the time of their study, found that the major barriers are mainly security-related. The samples of [2] and [3] included both online and offline firms. Reference [2] found that, in addition to security, the top five impediments also include 'Time', 'Making it a priority', 'Cost', and 'Not convinced that benefits are real'. Reference [3], on the other hand, found that security is not a significant barrier; the significant barriers include 'lack of time' and 'Does not lead to more efficiency or low costs'. Reference [4] also reported similar findings. In addition, other scholars, notably [5-8], also reported diverging and yet somewhat related results. On the consumer side, there also has been research on impediments to e-commerce adoption [e.g., 9]. However, the focus of this study is on e-commerce adoption impediments as perceived by businesses. In classifying these findings, we use [10]'s framework which groups impediments to e-commerce into three categories, financial bottlenecks, customer confidence issues, and organizational issues (Table 1).

These impediments, however, have been proposed or identified by the researchers, and may not necessarily represent what businesses perceive as the impediments to e-commerce adoption. In other words, there is lack of research on business perceptions of e-commerce impediments, especially dimensions underlying the specific impediments proposed or identified in previous research. Therefore, the purpose of this paper is to find out if there are any theoretically meaningful underlying constructs or dimensions of e-commerce impediments from a business perspective.

G. Zhiguo et al. (Eds.): WISM 2011, CCIS 238, pp. 166–170, 2011.

Table 1. Major Impediments to E-commerce as reported in previous research

Category	Impediment	Source
Financial bottlenecks	Cost	[2]
	Initial set-up costs	[4]
	Does not lead to more efficiency or lower costs	[3]
Customer confidence issues	Enforceability of contracts negotiated over the network	[1]
	Small e-commerce market	[3]
	Does not lead to more sales	[4]
	Enforcements of contracts concluded over the Internet	[3]
Organizational issues	Decreased productivity through frivolous use	[1]
	Fear of unknown	[2]
	Making it a priority	[2]
	Not convinced benefits are real	[2]
	Lack of dedicated resources	[4]
	Lack of market awareness	[4]
	Lack of IT infrastructure	[4]
	Lack of time	[3]
Security issues	Guarantee of message delivery	[1]
	Tampering with network messages	[1]
	Unauthorized access to internal networks	[1, 3]
	Interception of network messages	[1]
	Verification of authorship of messages	[1]
	Security	[2]
	Security of financial information transmitted over the Internet	[3]
	Security issues	[4]
	Interception of messages by third parties	[3]
Other	Lack of support from top government	[4]

2 Methodology

The empirical data of this study come from a questionnaire survey of Chinese managers. The respondents were first asked to indicate firm size, which is broken into five categories on an ordinal scale of 1 through 5: 1 (employing 1-5 people), 2 (6-20 employees), 3 (21-50 employees), 4 (51-200 employees), and 5 (over 200 employees). They were then asked to rank, on a 7-point scale, a list of e-commerce impediments adapted from [2], which includes Cost, Time, Making it a priority, Fear of the unknown, Don't/Didn't understand the terminology, Security concerns (privacy, confidentiality and authentication), No knowledge of courses available (and which ones were reputable), Am not/ was not aware of the benefits, Am not/ was not convinced that benefits are real, Convinced that staff will abuse it, Is not/was not possible to connect to the Internet, and Not willing to take the risk with such a new technology. The questionnaire was administered to a sample of students in executive training classes in Qingdao, China. The students were asked to participate in the survey, and those who agreed were further instructed on how to complete the questionnaire. A total of 83 usable responses were collected. The collected data were then entered into a Microsoft Excel file and analyzed with SPSS.

The specific SPSS procedure used is CATPCA (categorical principal component analysis) so as to find out if there are any theoretically meaningful underlying constructs or dimensions. CATPCA is a type of principal components analysis that includes nonlinear optimal scaling, and is suited not only for categorical data, but also for ordinal and nominal multivariate data [11].

3 Results and Discussion

The sample is skewed toward large firms, with 10 firms (11.66% of the sample) employing 21-50 people, 16 firms (19.28%) with 51-200 employees, and the remaining 57 firms (68.67%) employing over 200 people.

As Tables 2 and 3 show, the CATPCA procedure resulted in three meaningful dimensions. The first dimension can be called the Unfamiliarity dimension, with Fear of the unknown, Unfamiliarity with terminology, No Internet connection, and Perceived risk as its principal components. Apparently this dimension is associated with individuals and companies that are still not familiar with the Internet and e-commerce resulting from no connections to the Internet, and therefore still perceive the Internet and e-commerce to be highly risky. This is interesting considering that there is a rather high percentage of Chinese firms using the Internet [12]. It also shows that education is should be directed toward these people so as to familiarize them with the Internet and e-commerce, thereby overcoming this obstacle.

Dimension 2, on the other hand, can be termed the Cost/Benefit dimension, with Cost, Priority, Unawareness of benefits, and Doubts about whether its benefits are real, as its principal components.

The last dimension can be considered as the Suitability dimension because it includes Products unsuitable for e-commerce and Doubts about whether its benefits are real, as its major components. This dimension represents an impediment that is very difficult to overcome since products are perceived not to be suitable to sell over the Internet.

4 Limitations and Directions for Future Research

This study represents an effort in identifying underlying dimensions of impediments to e-commerce adoption as perceived by Chinese businesses. It is notable that of the myriad impediments or inhibitors proposed by academics and/or identified by practitioners, three meaningful underlying dimensions have been uncovered in this study. These dimensions are valuable in that they can be used in future research so as to simplify theoretical deliberations.

Apparently the 14 specific impediments that have been used to collect first-hand data in this study have a strong influence on the dimensions subsequently uncovered. And as such they do not constitute an exhaustive list. One direction for future research is to include more impediment measures, which could well lead to more meaningful underlying dimensions to be discovered.

Table 2. Model Summary

Dimension	Cronbach's Alpha	Variance Accounted For	
		Total (Eigenvalue)	% of Variance
1	.612	2.318	16.554
2	.463	1.753	12.525
3	.322	1.426	10.185
Total	.881[a]	5.497	39.263

a. Total Cronbach's Alpha is based on the total Eigenvalue.

Table 3. Component Loadings

	Dimension		
	1	2	3
Cost	-.273	.709	.105
Time	-.282	-.024	-.416
Priority	-.245	.653	.207
Fear of unknown	.654	.364	-.178
Terminology	.652	.039	-.242
Security	-.605	-.078	-.207
No courses	-.180	.291	-.448
Benefits unaware	.190	.601	-.056
Benefits real?	-.030	.225	.746
Staff abuse	-.615	.028	-.218
Noconnection	.411	.069	-.155
Risk	.514	-.013	-.052
Biz unrelated	.042	-.333	.071
Products unsuitable	.037	-.265	.475

References

1. Abell, W., Lim, L.: Business use of the Internet in New Zealand: An exploratory study (1996)
2. Moussi, C., Davey, B.: Internet-based electronic commerce: Perceived benefits and inhibitors. In: ACIS 2000. Queensland University of Technology, Brisbane (2000)
3. Walczuch, R., Van Braven, G., Lundgren, H.: Internet Adoption Barriers for Small Firms in The Netherlands. European Management Journal 18(5), 561–572 (2000)
4. National Office for the Information Economy (NOIE), The Current State of Play, in The Current State of Play - November 2000, NOIE, Canberra (2000), http://www.noie.gov.au/projects/information_economy/ecommerce_analysis/iestats/StateofPlayNov2000/index.htm
5. Ah-Wong, J., et al.: E-commerce progress: enablers, inhibitors and the short-term future. European Business Journal 13(2), 98–107 (2001)
6. Dubelaar, C., Sohal, A., Savic, V.: Benefits, impediments and critical success factors in B2C E-business adoption Technovation 25(11), 1251–1262 (2005)
7. Kshetri, N.: Barriers to e-commerce and competitive business models in developing countries: A case study. Electronic Commerce Research and Applications 6(4), 443–452 (2007)
8. Molla, A., Licker, P.S.: eCommerce adoption in developing countries: a model and instrument. Information & Management, 42(6), 877–899 (2005)
9. Ancker, B.: Drivers and inhibitors to e-commerce adoption: Exploring the rationality of consumer in the electronic marketplace (2003)
10. The European Union, The development of e-commerce in the European Union: A general assessment (2002)
11. Meulman, J.J., Van der Kooij, A.J., Heiser, W.J.: Principal Components Analysis with Nonlinear Optimal Scaling Transformations for Ordinal and Nominal Data(). In: Kaplan, D. (ed.) The Sage Handbook of Quantitative Methodology for the Social Sciences. Sage Publications, Thousand Oaks (2004)
12. Zhang, X., Moussi, C.: Level of Internet use by Chinese businesses: A preliminary study. Electronic Commerce Research and Applications 6(4), 453–461 (2007)

The Design and Implementation of Earthquake Service System in Sichuan Province

Xin Fei and Shujun Song

Institute of Geo-Spatial Information Science and Technology
University of Electronic Science and Technology, Chengdu, China
fei_xin19860312@163.com

Abstract. This paper deeply describes the methodology of Attenuation Model of Seismic Intensity in terms of the characteristics of Sichuan regions and illustrates the method for establishing and modifying Seismic Intensity Effect Fields in the GIS-based system of earthquake service. The proposed method gives a good generation of intensity fields. So the accuracy of generating the spatial distribution scope of damage is improved, especially, via three approaches of modification. In order to achieve the share of earthquake information, the framework of system is designed for three-tier architecture consisting of data service, middleware, and application service tier. Basic spatial data related to the earthquake and attribute data are stored in seismic database in the data service tier to meet the requirements on the client side, such as information querying, analysis and evaluation, etc. The methods of data processing and shortage are proposed in this paper, which are the key steps to make an accurate evaluation for disaster loss and casualties.

Keywords: seismic intensity effect fields, framework, database, evaluation.

1 Introduction

In natural disasters, seismic hazard is one of the most serious disasters causing casualties and economic losses. Application of GIS established in seismic service system is the important and effective way to earthquake disaster reduction. It permeates all aspects such as earthquake prediction, earthquake risk assessment, earthquake aided decision support, etc. However, traditional GIS technology is based on the single-machine environment, so there are some limitations to their application including the spatial data sharing, low efficiency, the small scope of application and so on.

With the development of computer network, application of GIS has been a trend that GIS provides users with the functions of spatial data to browse, query and analyze via the network. Web GIS technology, which makes use of Web technology to expand and improve the geographic information system (GIS), is a new technology. It is widely applied to the seismic hazard assessment, inquiring information service published, aided decision support and other aspects. The appearance of the technology that makes the application of seismic systems more extensive and flexible, accelerating the seismic information prediction, assessment and the acquirement, plays an important role in making rapid decisions. So it reduces the casualties and economic loss

G. Zhiguo et al. (Eds.): WISM 2011, CCIS 238, pp. 171–177, 2011.

effectively. Yao Baohua applied WebGIS technology to emergency decision system, who established three-tier architecture consisting of data service, middleware, and expression service tiers based on B/S mode[1], and realized the integration of GIS technology with Web technology .

2 System Structure

The B/S architecture is applied to the GIS system for earthquake disaster reduction in this paper, which consists of data service, middleware, and application service tiers.

Based on basic geographic data and attribute data related to the earthquake, map management, spatial analysis and querying, achieve earthquake damage prediction and evaluation，updating and publishing information rapidly. Figure 1 illustrates the system architecture. The functions and design standard of each tier are as follows:

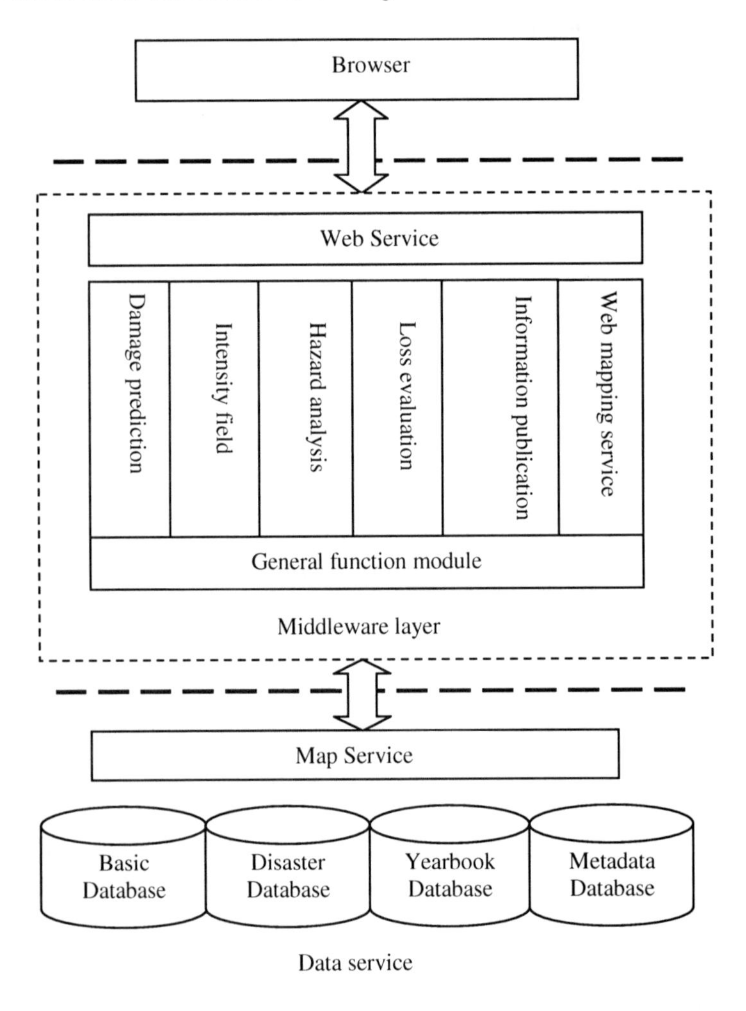

Fig. 1. System general structure

(1) The data service is mainly responsible for storing a variety of data related to earthquake disaster in seismic database. It is the basis of other two tiers.

(2) The middleware tier is responsible for processing business logic parts, which connects data service tier with application service tier for mutual communication. Application requests from the application service tier are performed in the data service tier and computing and analyzing results are returned to the application service tier through the middleware tier. In order to meet the requirements of earthquake system for prediction and assessment, several function components including hazard analysis, loss evaluation are designed in the tier to realize applications synthetically.

(3) The application service tier provides the interface functions for users to use the interactive mode. Several modules are designed in this tier, such as seismic hazard prediction module, loss assessment module, web mapping service module, information publication module and so on, especially, loss assessment module including building loss, economic loss and casualties loss.

3 Basic Analysis of Models for Earthquake Disaster Prediction

3.1 Attenuation Models of Seismic Intensity

This paper has chose Sichuan zone as a deep research area. Based on collecting historical earthquake data in this area, the chart of earthquake intensity lines is displayed for significant ellipse in Sichuan and adjacent regions.

The attenuation models of seismic intensity applied for the area are the mathematical foundations for Earthquake Disaster Prediction, which are applicable to the characteristics of Sichuan regions. The models are used to predict the degree of earthquake damage, outlining earthquake effect fields, to evaluate loss results in different intensity areas. So it improves efficiency for earthquake emergency response and aided decision support in the earthquake disaster prevention and reduction. Elliptical model is illustrated by the following equation (1).

$$\begin{cases} I_a = d_{a1} + d_{a2}M + d_{a3}\lg(R_a + R_{a0}) \\ I_b = d_{b1} + d_{b2}M + d_{b3}\lg(R_b + R_{b0}) \end{cases} \quad (1)$$

Here, M stands for the magnitude; R stands for the epicenter distance (unit: km); R_0 stands for preset item for distance; d_1, d_2 and d_3 stand for regression coefficients; a, b stands for long axis and short axis respectively.

Based on collecting historical earthquake data in Sichuan regions, the attenuation relationship of seismic intensity is showed in Table 1 [2]:

Table 1. Attenuation Relationship of Seismic Intensity

Region	Axis Direction	d_1	d_2	d_3	R_0	σ
Sichuan Province	Long Axis	4.0293	1.3003	-3.6404	10.0	0.45
	Short Axis	2.3816	1.3003	-2.8573	5.0	0.45
	Average Axis	3.3727	1.2755	-3.2858	7.0	0.42

3.2 The Method of Outlining Earthquake Effect Fields

Generally speaking, the outlining methods include the direct input method, the historical earthquake method, as well as the parameters method [3]. The paper is mainly focused on the direct input method, in which the method describes outlining earthquake effect fields according to the longitude, the latitude, the magnitude, as well as the attenuation direction. The location for earthquake happening is fixed on the map, combining relevant information with the formula of earthquake intensity attenuation, and then the earthquake effect fields are outlined. The historical earthquake method is based on earthquake cases in history and the parameters method adopts the probability methodology to outline the earthquake effect fields.

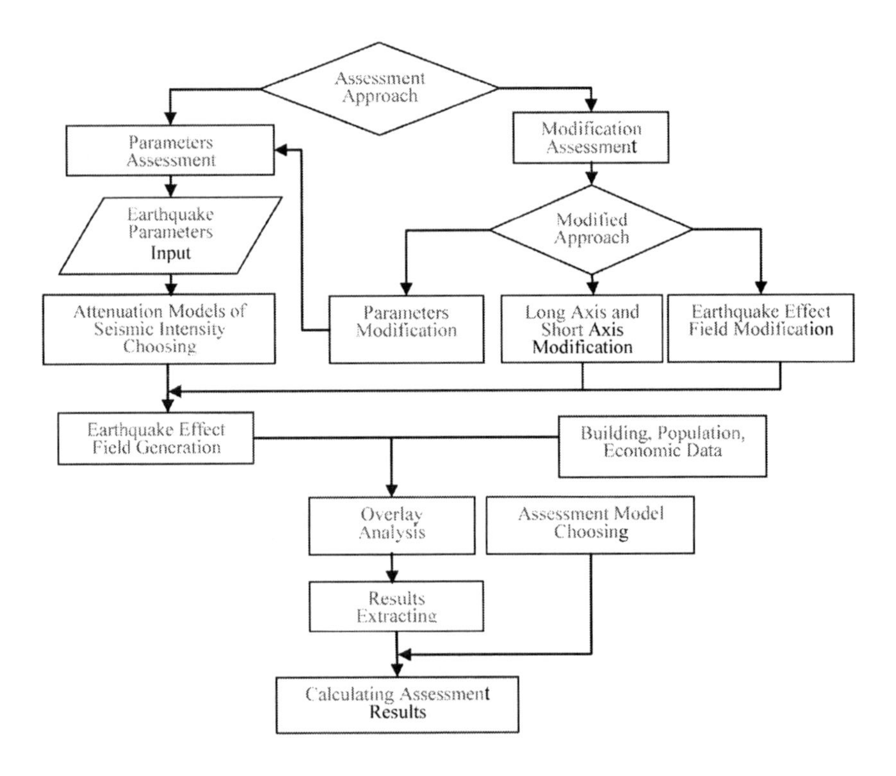

Fig. 2. The flowchart of earthquake effect fields

3.3 The Modification Method of Earthquake Effect Fields

Due to the error of seismic parameters and the inaccuracy of attenuation models, this paper provides three correction methods: parameters amendment, long axis and axis amendment and the amendment of earthquake intensity fields. The parameters amendment is described as correcting earthquake parameters. The long axis and short axis amendment is described as outlining the earthquake effect fields based on the value known of long axis, short axis and major axis direction. The amendment of earthquake intensity fields is described as outlining the earthquake effect fields according to the geological conditions in practice, and then stores the earthquake effect fields in the earthquake database.

The flowchart of modifying earthquake effect fields is showed in Figure 2.

4 Database Establishment

4.1 Data Collection and Processing

In the database of earthquake disaster prediction and reduction, the sources of data are divided into three parts.

(1) Governments provide data related to the earthquake, including fundamental geographic information data, potential disaster data, emergency plans of governments, lifeline data and so on.

(2) The yearbook releases part of data, including population, economic data and buildings data in various regions.

(3) The earthquake system has existing data, including seismic zoning maps, administrative maps, geological maps, as well as other data.

After collecting various data, processing data in unified regulations and technical standards, all kinds of electronic maps are conversed and analyzed. Data from different sources is transformed into unified coordinate system. Spatial data and attribute data are matched, especially establishing accurate topological relations [4].

So data processing includes digital input, editing, establishing the topological relations, coordinate transformation, and unified spatial data formats.

4.2 Data Storage

After finishing data processing, Geodatabase model is adopted for data storage. Data is stored in database using ArcCatalog through the channel of ArcSDE. Feature Dataset and Feature Class are established for spatial data storage, and then various data are imported into database in terms of their types. Attribute data is stored in tables, in relation to spatial data via the fields.

4.3 Database Type

The database for the earthquake service system is composed of four sub-databases.

Basic Database. It includes spatial data, which consists of basic geographical maps, geological maps, potential seismic hazards data, hazards distribution data, etc, and attribute data such as economic data, population, weather and building areas stored in attribute tables.

Earthquake Disaster Database. It mainly stores the calculated result data of assessment and analysis, the middle results data and disaster data.

Yearbook Database. It stores part of data, which reflects economy, population distribution situation in various regions, making up for the shortage of basic database.

Metadata Database. The database stores data, which describes various types of data.

4.4 Functionality of the System

One of the main functions of the earthquake service system is to predict seismic hazard and evaluate earthquake loss quickly. The earthquake intensity effect fields are quickly calculated and drew by inputting the parameters of earthquake, such as the epicenter location consisting of longitude and latitude, earthquake magnitude, attenuation direction, based on the local attenuation model of seismic intensity after the occurrence of an earthquake. Basic seismic maps are overlaid with the earthquake intensity effect fields to extract the lists of buildings, lifeline data, potential seismic hazards data and resident population in every intensity zones. With these results and their corresponding seismic vulnerability matrixes, the buildings destruction, economic loss and casualties caused by the earthquake are calculated in terms of analysis models based on disaster loss evaluation models [5]. And then the evaluating results will be queried and spatially visualized on the client side such as thematic maps. In addition, the functions of the system include information release, web mapping service, database management and so on. Information release supports disaster data publication and downloading the assessment results in terms of different needs of users. Web mapping service includes maps querying and maps analysis modules which provide shortest path analysis of distributed supplies and rescue. The results are displayed on web clients such as statistical graph, tables. The functions of the system have been displayed in Figure3.

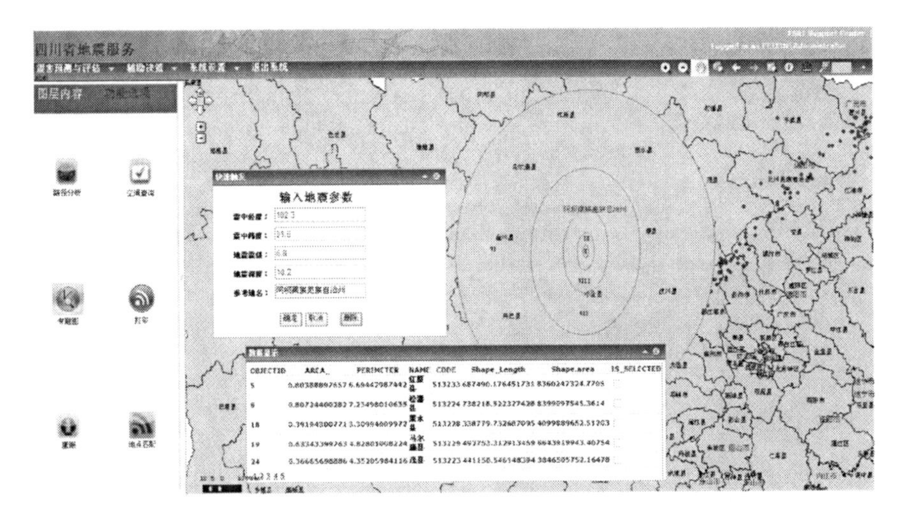

Fig. 3. The functions of the system

5 Conclusions

This paper has illustrated the attenuation model of seismic intensity, analyzing its realization process. Especially, several dynamic modification methods have been put forward, making outlining earthquake intensity effect fields more accurate and efficient, and the earthquake intensity effect fields have topology structure and information of longitude and latitude. It is convenient to evaluate economic losses and casualties, so the results of evaluation and analysis can be published on the web clients rapidly, achieving the share of earthquake information.

Acknowledgments. Thank teachers and students for their help in the laboratory, they help me solve many problems in study.

References

1. Yao, B., Lu, H., Shi, W.: A WebGIS Based General Framework of Information Management and Aided Decision Making System for Earthquake Disaster Reduction. In: Geoscience and Remote Sensing Symposium, pp. 1568–1571 (2006)
2. Lei, J., Gao, M., Yü, Y.: Seismic Motion Attenuation Relations In Sichuan And Adjacent Areas. Acta Seismologica Sinica 29, 500–511 (2007)
3. Nie, S.: Design and Application of the Seismic Influence Field based on GIS. Journal of Basic Science and Engineering 16, 546–555 (2008)
4. Peng, J., Long, X., Han, L.: Application of GIS to the Construction of Basic Database for Earthquake Emergency Response. Plateau Earthquake Research 22 (2010)
5. Tang, A., Xie, L., Tao, X.: Based Urban Earthquake Emergency Respose System. Seismlogical Research of Northeast China 17 (2001)

The Preliminary Study on 3D Terrain Visualization

Xiuming Jia and Qiang Guo

College of Mining Technology, Taiyuan University of Technology,
030024 Taiyuan, China
xiumingjia@163.com, guoqshenshi1@yahoo.com.cn

Abstract. With more and more concerned of the 3D geographic information system, the 3D application system receives quick support from the application units because it is more realistic than the 2D in the visual. The author has taken the geological topographic maps and high resolution images in Dongyao Village, Wutai County as the basic data, generated the 3D topographic map on Dongyao Village area, by editing the vector data, doing some remote sensing image geometric correction, generating DEM, and doing a series of spatial analysis on this 3D landscape. This paper describes the main methods and processes that making the 3D terrain visualization and realizing roam by using ERDAS IMAGINE and ARCGIS software.

Keywords: ERDAS IMAGINE, the 3D landscape, DEM.

1 Introduction

Since the release of Google Earth, the 3D Geographic Information System (3DGIS) is widely concerned, it becomes a hotspot on research and application, and gets a rapid development. Liked 2DGIS, 3DGIS needs spatial data processing with the most basic functions, such as data acquisition, data organization, data manipulation, data analysis, data performance etc. Compared with 2DGIS, 3DGIS has the following advantages: (1) More intuitive display of spatial information. Compared with 2DGIS, 3DGIS provides a more abundant and more realistic platform in displaying of spatial information, which to visualize the difficult spatial information. In this way it makes GIS more popular in public, in order to make accurate and fast judgments ordinary people can also combine with their own relevant experiences to understand GIS. Doubtlessly, 3DGIS has a unique advantage in visualization. (2) Multi-dimensional spatial analysis is more powerful. The spatial analysis of 3DGIS is more powerful, many functions that can not be achieved in 2DGIS can be done in 3DGIS, for example, submergence analysis, geological analysis, sunshine analysis, spatial diffusion analysis, Visibility Analysis and other advanced spatial analysis. 3D data itself can be reduced to 2D, so 3DGIS naturally contains 2DGIS's spatial analysis function. 3DGIS has powerful multi-dimensional spatial analysis function, which fully embodies the characteristics and advantages of GIS. Base on remote sensing comprehensive prospecting information in Dongyao Village, Wutai County, the author used the raw data of the project, edited the vector data, corrected the remote sensing image, generated DEM, thus generated the 3D topographic map on Dongyao village area, and did a series of spatial analysis on this 3D landscape.

G. Zhiguo et al. (Eds.): WISM 2011, CCIS 238, pp. 178–184, 2011.

2 The Raw Data Preprocessing and DEM Generation

2.1 Software Selection

This topic chose the US ERDAS IMAGINE software. ERDAS is a world-class remote sensing image processing and remote sensing and GIS integration software provider, its product ERDAS IMAGINE system is the leading ones of the remote sensing industry, which can handle the remote sensing images including satellite images (Landsat and SPOT, etc.), aerial photographs and radar images. The system not only provides image enhancement, filtering, geometric correction, integration and other simple basic applications, but also provides powerful tools. Furthermore, ERDAS IMAGINE also provides a wealth of expansion modules, for example, Vector, Radar, Virtual GIS, OrthoMAX, OrthoBASE etc. In this paper, Writer used the core module of ERDAS IMAGINE to get simple 3D landscape observation. Through the input of DEM, image, vector, 3D models, and other information, you can create, display and analyze the 3D landscape of the study area, and do the associated 3D analysis of the landscape.

2.2 DEM Generation

Digital elevation model (DEM for short) is expressed as the digital models of topographic features in space distribution, can be considered as another expressional mode of physiognomy, it has been widely used in the scene modeling. The DEM data sources and collection methods are: (1) measuring from the ground directly, such as using GPS, total station, field measurement etc. (2) Obtaining by photogrammetric means according to aviation or space images, such as the 3D coordinates of observation instrument, analytical mapping, digital photogrammetry etc. (3) Collecting from the existed topographic map, such as reading grid point method, tracking by the digitizer, semi-automatic acquisition by the scanner, and DEM was generated by interpolation. Currently, the Interpolation Algorithm of DEM is used by contour lines and elevation points to establish triangular irregular network (TIN for short). Then interpolating linear and bilinear based on the TIN to building DEM.

The writer got the contour line and the contour identification information by editing the geological topographic maps (scale 1:5000) in Dongyao Village, Wutai County. In the MapGis 6.7 software environment, through editing the micro-contour lines and the broken lines of the contour line by using the Line editing tools, made sure the micro-contour lines smooth and did not intersect, then inputted the elevation information based contour line information by elevation automatic assignment tool, as shown in Figure 1.

Raw data of this topic was the geological topographic map, it lacked the data projection information, whose coordinates recorded was the screen coordinates, so the data was required to be further edited in ARCGIS. By using the entire graph transformation tools of MapGis 6.7, the contour data was converted to a scale of 1:1000, using the file conversion tool to convert to SHAPE files which ARCGIS supported. In the ArcMap environment, using the Define Projection tool in ArcTool box redefined the projection information of the contours. Opened the QuickBird Remote Sensing of this area, used the Georeferencing tools for geometric correction of images, and made sure it to match with the vector data, as shown in Figure 2. In the ArcMap, using the Add Feature to TIN tool in3D Analyst to form TIN. Because

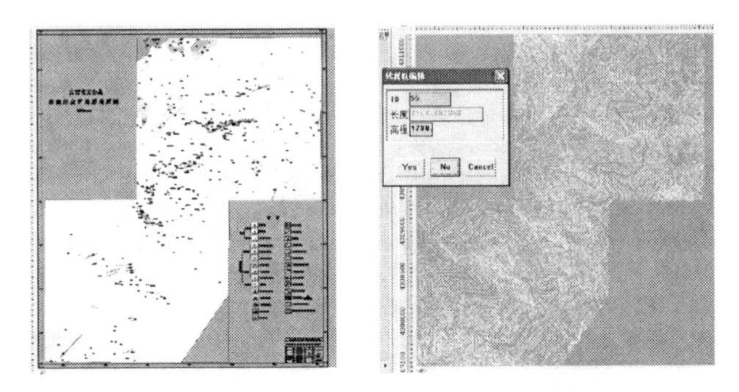

Fig. 1. The contour and elevation information into the treatment

the region was irregular polygon, there were some errors in the edge of the region. It had to cut the wrong TIN by using the clip tools. Then DEM was generated by using the TIN to Raster tools, as shown in Figure 3.

Fig. 2. The vector data and image matching

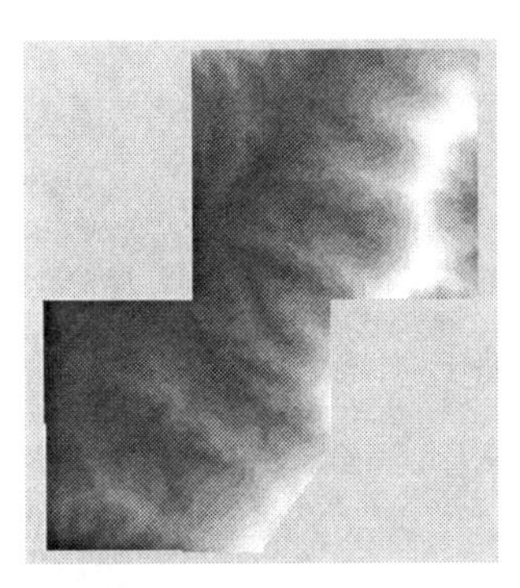

Fig. 3. DEM generation

3 Visualization of 3D Terrain

The authenticity of the terrain is the important component of terrain visualization. In order to improve the authenticity of the terrain, it always can add attached texture image attached to achieve. As the real texture remote sensing digital image processing of the surface topography, it has rich content in the terrain surface stacking texture image (such as aviation image, space imaging, digital image) and high authenticity, thus it has become an ideal terrain of the effective methods to improve the authenticity, has been widely applied in highly realistic 3D topography production [1].

3.1 Display of the 3D Terrain in the ArcScene

To test the accuracy of DEM data, in the ArcScene environment, the writer imported the files of the line, DEM data, and Quickbird image data into the region. After setting the Base Heights of each file, simple 3D terrain view was generated. The projection wall was generated by setting the files of the line, as shown in Figure 4.

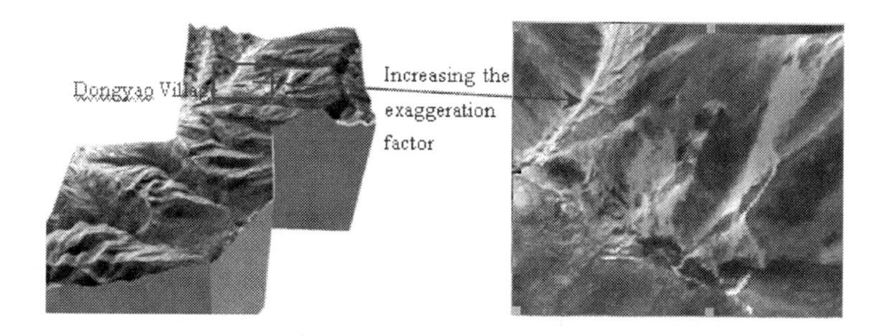

Fig. 4. Display of 3D terrain in the ArcScene

3.2 Display of 3D Terrain in the ERDAS IMAGINE

ERDAS IMAGINE VirtualGIS is a 3D visualization tools. VirtualGIS Scene is composed of digital elevation model and remote sensing images with the same map projection and coordinate system [2]. In VirtualGIS window, opening the DEM and image files in turn, increased note on the text layer, and saved the VirtualGIS Engineering. You can edit VirtualGIS Scene, for example, adjusting the solar source bits by setting the solar azimuth, solar altitude and light intensity and other parameters used the menu item of Sun Position; adjusting the visual characteristics of Virtual, including DEM display characteristics, background display properties, 3D roaming feature, 3D display characteristics, note symbol properties and so on. The adjusted Virtual Scene was shown in Figure 5.

Fig. 5. Display of the 3D terrain in the Virtual

4 The Spatial Analysis in ERDAS IMAGINE

In the window of the VirtualGIS which has already been created, you can perform a variety of property data analysis such as the vector, the marker, the analysis of heavy fog, the analysis of flood inundation, the threat analysis, visibility analysis etc., through stacking various attribute data layer such as vector layer, annontation layer, water layer, mist layer, model layer, model layer, intervisibility layer. The writer carried out analysis of flood inundation, slope and aspect analysis, 3D roaming display, according to the specific circumstances.

4.1 Analysis of Flood Inundation

ERDAS IMAGINE provides a large number of analytical tools to complete the monitoring of flood disasters and loss estimates. In VirtualGIS window stacks water Layer (Overlay Water Layer) for the inundation situation analysis, and the system provides two analysis modes (Fill Entire Scene and Create Fill Area) to operate. After the layer of flood overlap the VirtualGIS scene, the menu of Water has already existed in the VirtualGIS windows menu, we can observe the situation that overrun and submerge of flood and so on by setting flood height and flood display characteristic in the Water menu, was shown in Figure 6.

4.2 Slope Analysis and Aspect Analysis

Each triangle net of TIN triangulation nets constitutes a plane, a point in the surface must be contained in a certain network of triangle, it means that the point is contained in a particular plane as well. So the slope of the point is defined by the angle between its place and the horizontal plane [4]. Through analyze the slope in the area research we could conclude the relationship between the rainfall intensity and the soil and water conservation, which play an important role in monitoring the soil erosion.

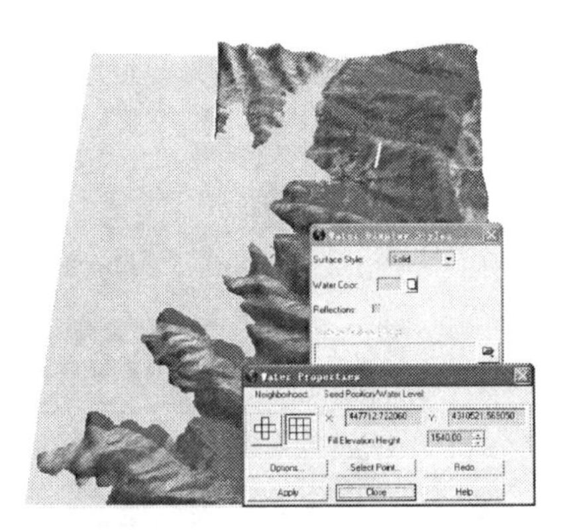

Fig. 6. Analysis of flood inundation

A certain point in the surface of TIN is defined as the slope in the triangle plane which contains that point. That is the triangle slope in the direction of planar found the direction of projection mean [4]. Making the slope of the study area as reference, we could make more scientific and reasonable arrangement in distributing those plants who love sun or hate.

4.3 Threat and Intervisibility Analysis

The threat and intervisibility analytic function is the analysis of intervisibility relation between different observers and observation scope. In 3D virtual window, setting observer and creating the observer's visual airspace range according to some conditions, can do the visibility and threat analysis. In Intervisibility module of ERDAS IMAGINE, you can set the observation point by using the Observer Attribute tool, and get the results of analysis, as shown in Figure 7.

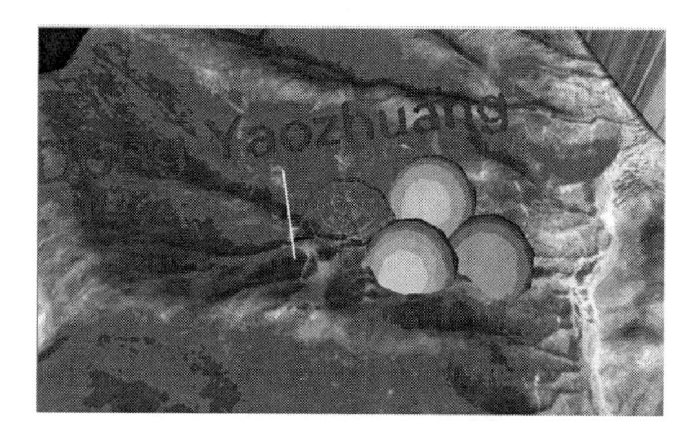

Fig. 7. Threat and intervisibility analysis

4.4 The 3D Roaming Display

There were many kinds of navigation mode in VirtualGIS, the virtual navigation was carried out by selecting the navigation mode that system provided. In the VirtualGIS, according to the need to define the flight route, and flying along the defined route in virtual 3D environment. The writer created a curve as flight route in the IMAGINE 2D window. In order to express more figuratively, we need to edit flight route, including the height set, route characteristics set, flight patterns set etc of the flight route. And then performing the flight operations, controlled the flying beginning and ending, as shown in Figure 8.

Fig. 8. 3D roaming displaying

VirtualGIS provides the three dimensional animation production tools, the VirtualGIS engineering which contained the flight line converted to a 3D roaming animation by using those tools. Using a variety of media players, 3D animation files, which have been generated, can be broadcasted without the ERDAS IMAGINE as shown in Figure 9.

Fig. 9. 3D roaming displaying

5 Conclusion

3D terrain visualization has important application value in earth science research [3]. It has important significance on the dynamic, image, multi-angle, round, multi-level description. This study carried out the task which was based on geological topographic map and high resolution images in Dongyao Village, Wutai County. Due to the research conditions and study time restriction, there is no more in-depth research on 3D visualization, but the experimental results still has strong theoretical and practical value. The author achieved 3D terrain visualization, which based on the IMAGINE ERDAS software system. It could simulate the scene for some simple operations, and didn't need complicated programming. It easily solved many real life specific problems, the method has maneuverability, and the application scope will greatly widen.

The high precision of 3D image animation provided the real and objective ground landscape image for the macro observer, and helped them to make the right decisions. This experiment was the extended part of remote sensing comprehensive prospecting information in Dongyao Village, Wutai County, can make the results more realistic.

References

1. Zhao, J., Du, Q., Meng, F., Wang, H.: Overview of Large-scale Realistic 3D Terrain Technology. Geology of Chemical Minerals 28(4), 245–250 (2006)
2. Dang, A., Jia, H., Chen, X., Zhang, J.: ERDAS IMAGINE Remote Sensing Image Processing Tutorial. Tsinghua University Press, Beijing (2010)
3. Liu, J., Yang, G.: 3D Terrain Visualization of ERDAS And its Applications. Journal of China West Normal University(Natural Sciences) 29(3), 307–312 (2008) (references)
4. Tang, G., Yang, X.: GIS Spatial Analysis of Experimental Tutorial. Science Press, Beijing (2010)

The Methods and Applications for Geospatial Database Updating Based on Feature Extraction and Change Detection from Remote Sensing Imagery

Guohong Yao[1,2,*] and Jin Zhang[3,**]

[1] Institute of Geodesy and Geophysics, CAS, 340 Xudong Street, Wuhan 430077, China
[2] Graduate School of CAS, A19 Yuquan Road, Beijing 100049, China
yaogh118@126.com
[3] Taiyuan University of Technology, 79 West Yingze Street, Taiyuan 030024, Shanxi, China
zjgps@163.com

Abstract. Change detection of the spatial object and geospatial data updating is an important research for geospatial information science, which also guarantee the applications of GIS data. Applying change detection method to remote sensing data to analyze geospatial objects, combine with the method of object-oriented image classification to extract features and analyze change information, which fulfilled the needs of updating the geospatial databases with remote sensing imagery, has a great significance for the ongoing digital city construction and national conditions monitoring. This paper describes the key technology of updating the geospatial database with the remote sensing imagery. Using the object-oriented feature extraction and classification, the updated thematic information was obtained satisfactorily.

Keywords: Feature Extraction, Change Detection, Geospatial Database.

1 Introduction

The up-to-date state of geospatial data is the soul of Geographic information systems (GIS). Having established the basic version of geospatial database, surveying and mapping departments focus on the sustainable updating of geospatial dataset [1].

The techniques of remote sensing data acquisition tend to be multi-sensor, multi-platform, multi-angle with the imagery of high-spatial resolution, high-spectral resolution, and high-temporal resolution [2]. It is one of essential methods for geospatial database updating to extract information from remote sensing imagery.

Recently, spatial information cognition, object-oriented change information extraction, auto-generation of multi-resolution mapping data and version management of geospatial dataset have become the hot research spot for domestic and overseas [3].

---

* Guohong Yao (1968-), Associate Professor, Ph.D candidate of CAS. Synthetic Geographic Information Center of Shanxi.
** Corresponding author: Jin Zhang (1963-), Ph.D & Professor, Taiyuan University of Technology.

G. Zhiguo et al. (Eds.): WISM 2011, CCIS 238, pp. 185–191, 2011.

2 Test Area and Data Sets

The test site locates around PingShun country in Shanxi province, covering an area of 25km^2. Mountains are the dominant terrain, with plains and hills distributing between them. Features that are extracted in this thesis including settlements, transport, water, vegetation and other objects.

The multi-temporal and multi-source remote sensing imagery used are QuickBird image data collected on 18 August 2009 and ADS40 image collected on 16 June 2010, as Fig. 1 shows.

Fig. 1. The QuickBird image (Left) and The ADS40 image (Right)

3 The Updating of Geospatial Database Using Remote Sensing Imagery

3.1 Tasks and Key Point

The updating of geospatial database aims to determine the change of location and attribute of objects, basing on all kinds of current information available. The new version of data and change information is obtained by replacement, emendation and coordination of data-relation to update database.

Lately, new requirements for updating service have been put forward, including achieving latest products of specific elements from specific region within a few days or dozens of days, dealing with the emergencies and satisfying urgent need for major projects.

Having built the fundamental geo-information database, surveying and mapping departments are on their way to updating major elements within major regions as required. Administrative boundaries, new roads (e.g. expressways, national roads, railways), settlements above the country level, water system should be renewed every one or two years.

3.2 The Methods of Updating Geospatial Database with Remote Sensing Imagery

Remote sensing technique is the only method that could provide dynamic observation for large area, with spatial continuity and time sequentiality, providing multi-temporal imagery for feature detection and monitoring [4]. Analyzing the frequency and amplitude of features to determine change objects and its range, with the latest remote sensing imagery and other up-to-date information is the road geospatial database updating must get through.

The conventional approach of geospatial database updating relying on man-machine interactive interpretation is time-consuming with high labor intensity. The result lacks inheritability, transitivity and could not ensure high accuracy.

This thesis utilizes multi-scale segmentation and object-oriented analysis to test the effectiveness and reliability of extracting features and change detection. The image preprocessing of multi-source and multi-temporal high resolution remote sensing imagery includes compositing, enhancement, rectification, correction and fusion. We are searching for rapid intelligent geospatial database updating techniques suitable for large-scale production. Fig. 2 reveals the working flow of geospatial database updating with imagery.

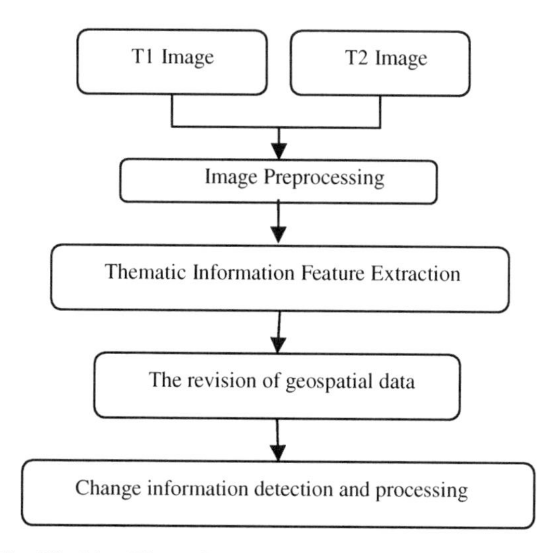

Fig. 2. The Working Flow of Updating Geospatial Database with Imagery

3.2.1 Image Preprocessing

Since image acquisition conditions differ, preprocessing is needed before change detection to make the spectrum and position of the same features from different image consistent. Radiometric correction and geometry correction are the central steps [5].

3.2.2 Feature Extraction of Thematic Information

Traditional classifications like Maximum Likelihood method are pixel-based techniques, which lack space structure and texture information. Those ways obtain discrete results, which could not represent the boundary and area features [6]. Multi-scale image segmentation proposed by Martin BAATZ and Arno SCHÄPE makes object-oriented image analysis technology got rapid development.

Object-oriented method takes advantage of spatial and spectrum features to segment images into objects consist of homogeneous pixels. It can overcome the shortcomings of pixel-based methods, and improve the accuracy of classification [7]. Fig. 3 exhibits the working flow of object-oriented feature extraction and classification.

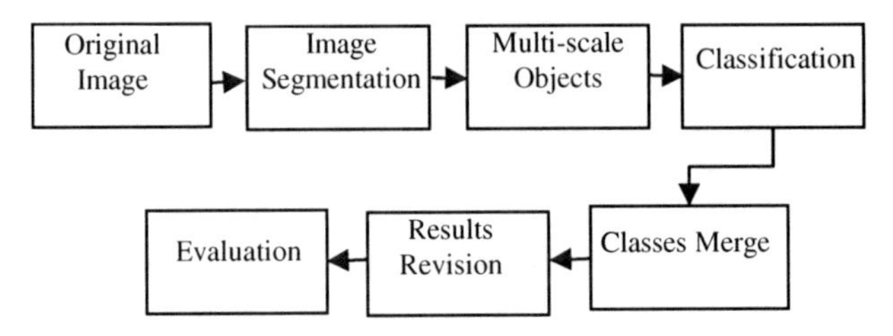

Fig. 3. The Working Flow of Object-oriented Feature Extraction and Classification

According to the object types of the test area, we combine visual interpretation with on-the-spot investigation, and choose the classes to be updated including road, settlement, water, grassland, farm land and forest land.

Image segmentation is the key to object-oriented classification and information extraction, which influence the accuracy of results [8]. The parameters including waveband, scale, color, shape should be properly set to form mutually disjoint, homogeneous objects with rich semantic information. Fig. 4 reveals the image after multi-scale segmentation.

The imagery is classified using supervised method after being well segmented. The classification process contains defining of classification system and determining of classification rules. According to the features to be updated, hierarchical classification levels are built. Firstly, use the mean gray value of near-infrared band to classify the image into water and non-water. Secondly, classify the non-water class into vegetation and non-vegetation with the normalized vegetation index (NDVI). Last but not the least, subdivide the vegetation class into grass and tree using maximum variance and subdivide the non-vegetation class into settlements and roads with length-width ratio of shape parameter. The object-oriented classified image using eCognition software is represented in Fig. 5.

Fig. 4. The Image after Multi-Scale Segmentation

Fig. 5. The Classification Result of 2009 (Left) and 2010(Right)

3.2.3 Change Information Detection and Processing

The results of image taken in different time should be optimized after segmentation and classification. Spectral difference segmentation is used to calculate mean gray value of adjacent objects in the same level. The objects of later image are merged on basis of the threshold set. For the former image, adjacent objects of the same class are combined into a larger object using merge region algorithm. The change information of features are detected basing on context Information and class-related features as well as existence of super objects [9].

Changing information acquired by comparison of classification results is revealed in different color. The results can be export in report or figure forms as needed. Fig. 6 shows the change information of ADS40 image in 2010 basing on QuickBird image of 2009. Fig. 7 shows the change information vector export result.

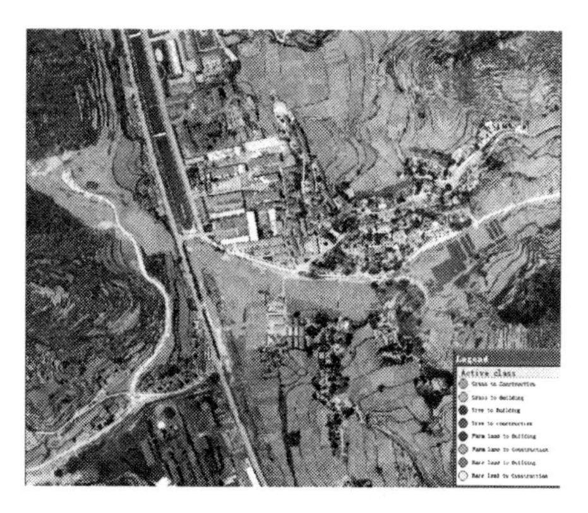

Fig. 6. The Change Information of ADS40 Image Basing on QuickBird Image

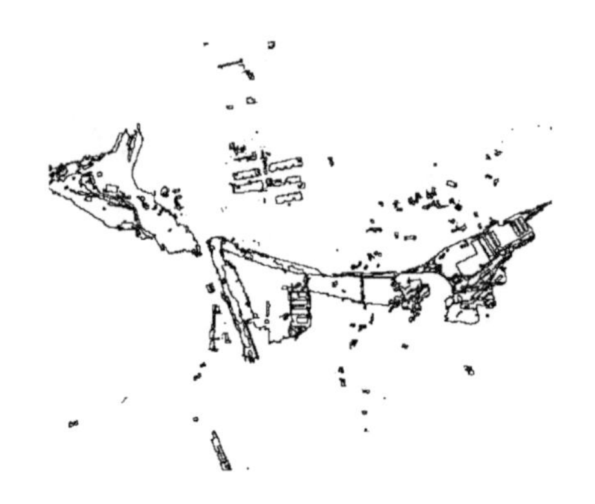

Fig. 7. The Change Information Vector Export

3.2.4 Geospatial Database Updating

On the basis of changing information, the geographic elements could be updated accordingly [10]. The added and reduced features should be treated separately, and the mode of processing and procedure differs from each other.

The reasons of features decreasing contain disappear of elements like the vanishing of pool, and partial disappearance of elements, like the change of shape and diminishment of area. For the first case, elements can be deleted from original vector data. While for the second case, the part should be deleted from geographic entity.

The reasons of features increasing contain augment of partial elements like the change caused by expansion of residents, and new elements like new channels. For the first case, it should be merged to the original elements to keep integrity, and for the second case, the new element are added directly.

Generally, for the objects changed partially, there will be no change of its property. For the overall changed objects, further interpretation or field investigation is essential.

Finally, relationship of adjacent spatial elements should be rectified. The processing of rebuilding Topology relationship, meeting edge and updating geospatial database by layers is among the post-processing of the process.

4 Conclusion

From what has been discussed above we may safely draw conclusions as follows:

1) Remote sensing imagery could offer abundant surface information without delay, and have become the major source of geospatial database updating.

2) It is a highly effective and reliable way using multi-resolution segmentation and object-oriented classification methods to intelligently updating geospatial database on the basis of multi-temporal and high-resolution remote sensing imagery. In this thesis we extract ground features like settlements, traffic and water system.

3) Various of spatial elements are involved in the updating of geospatial database. On-site verification after the updating work is indispensable since the relationship of adjacent spatial elements should be rectified embodying topological relation, coordinate spatial relation and matching of attributes.

Acknowledgment. This work is partially supported by the National Natural Science Foundation of China. The Grant No. is 40771175.

References

1. Chen, J.: Key Issues of Continuous Updating of Geo-spatial Databases. Geomatics World 2(5), 1–5 (2004)
2. Li, D.: Change Detection from Remote Sensing Images. Geomatics and Information Science of Wuhan University 28(5), 7–12 (2003)
3. Zhang, J.: Theory and Key Technology of Updating Geospatial Database Based on Multi-resource Spatial Data. Science & Technology Review 23(8), 71–75 (2005)
4. Fang, Y.: Technique for Auto Change Detection and Updating of Topographic Map Using Hyperspectral Image. Bulletin of Surveying and Mapping 7, 51–53 (2007)
5. Zhong, J.: Change Detection Based on Multitemporal Remote Sensing Image. [A Dissertation]. National University of Defense Technology, Changsha (2005)
6. Walter, V.: Object-based Classification of Remote Sensing Data for Change Detection. ISPRS Journal of Photogrammetry and Remote Sensing 58(3-4), 165–225 (2004)
7. Baatz, M., Schäpe, A.: Multiresolution Segmentation:An Optimization Approach for High Quality Multi-Scale Image Segmentation. Journal of Photogrammetry and Remote Sensing 58(3-4), 12–23 (2000)
8. Jian, C.: Research on the Scale of 1:10 000 Digital Maps Revision Based on High Resolution Remote Sensing Data. Science of Surveying and Mapping (supplement), 52–54 (2006)
9. Champion, N.: 2D Change Detection from High Resolution Aerial Images and Correlation Digital Surface Models. In: International Archives of the Photogrammetry, Remote Sensing and Spatial Information Sciences XXXVI, Part 3/W49A, pp.197–202 (2007)
10. Woodsford, P.A.: Spatial Database Update and Validation—Current Advances in Theory and Technology. In: ISPRS Archives XXXVI, Part 4/W54, pp. 1–7 (2007)

Research on Software Buffer Overflow Flaw Model and Test Technology

Junfeng Wang

Distance Education Research Center of Shaanxi Province
32, HanGuang North Road(Middle), Xi'an, Shaanxi Province, P.C 710068
Tel.: 86-029-82069192; Fax: 86-029-82069222
wangjynfeng@sina.com

Abstract. This paper summarized the development of software remote buffer overflow flaw model and test technique concerned. Based on the analysis of the cause and principle of software remote buffer overflow, a software remote buffer overflow theory model and a logical analysis technique of network protocol session sequence were brought forth. Through introduction to black-box test method, an I/O test technique based on software remote buffer overflow model was presented to solve the following key problems, such as, position technique of software remote buffer overflow doubtful sites, analysis technique of error injection contents and detection technique of software remote buffer overflow. At last, an application system was completed, and experiment results show that the I/O test technique based on software remote buffer overflow model can effectively find and position the potential remote buffer-overflow flaws in system software.

Keywords: software remote buffer-overflow flaw, software remote buffer overflow model, I/O test technique.

1 Introduction

The remote buffer overflow is a great danger of software security flaws, commonly found in Windows, Unix / Linux and other popular operating systems and the software running on, such as browsers, media players, messaging tools, office software, Web and FTP servers, etc. Criminals use software remote buffer overflow to start attacking, such like implant Trojans, spread Malwares and Spams, control target computer and build up Botnet, causing extremely threat for internet security. In 2007, Microsoft released 69 security bulletins, 56 of them are about software remote buffer overflow, reached 81% of total. High-risk Vulnerabilities got 42, 41 of them are about software remote buffer overflow, reached 97.6% of it. Recently there is no effective way to test and found the potential software remote buffer overflow of software, application requirements are urgent.

There is two ways to prevent software remote buffer overflow attack: The one is, by patch or upgrade to fix vulnerabilities that have been found, this is a passive prevention mechanism, another one is using test technique to find and fix the potential flaw that might be exist, this is dynamic prevention mechanism. Combine two of them, will be

G. Zhiguo et al. (Eds.): WISM 2011, CCIS 238, pp. 192–202, 2011.

more available Coping with the threat of software remote buffer overflow. So the research of software remote buffer overflow test technique has important theoretical Significance and application Value.

The contributions of this paper are build up software remote buffer overflow model, analysis the basic from and principle of software remote buffer overflow, brought out the relative(related) test technique based on software remote buffer overflow model, provides a wealth of software security evaluation techniques for Network Information Security Evaluation Department.

2 Related Works

2.1 Software Security Flaws

In International information security, software security flaw usually referred to as software vulnerability, which means those software security problems could used by attacker, mainly include the flowing two parts: bug and flaw. Commonly bug exists in the code, generated in coding phase. e can use Code Scanner to detect errors in the code. Software security flaws may arise in all phases of software production, can cause software insecurity, usually it is difficult to delete by code scan. Common software flaws are: shared object, mutual trust, Unprotected data channel(Internal and external channel), error or failure of the access control mechanisms, lack of security audit, log or incorrect log, Timing(especially in multithreaded system), Buffer Overflow and other problems. In order to standardize the name of the software security flaws, to exchange data more easily between independent software security flaws and different security tools. MITRE company of USA brought out and build up a CVE(Common Vulnerabilities and Exposures) dictionary, the purpose is provide a unique identification and a piece of description, unified description on security flaws, convenient for communion exchange and data sharing between different developers and users. the similar security flaw dictionary are CWE(Common Weakness Enumeration)etc..

2.2 Software Security Flaws Model

Software security flaws model refer to a mathematical model, in the different phases of software development, build up for conveniently analyze the security flaw might exist in the software. In the analysis requirements phase, according to the different needs of software security, proposed software security flaw analysis model[1]. In the testing phase, proposed software security flaw testing model[2]. In software maintenance phase, proposed software security flaw detect model[2], due to the causes of software security flaw, some one proposed other software security flaw models[3].

2.3 Software Remote Buffer Overflow Test Technique

The software remote buffer overflow is one of the commonest software security flaws. According to American NIST statistics in recent years, software remote buffer overflow hold about 10%-20% of all software security flaw. So, the research of software remote buffer overflow test technique has important theoretical Significance and application Value. Currently, foreign existing technology primarily is white-box testing technique, this technique can be divided into static and dynamic testing, these two testing techniques research status quo are as fellow.

The main idea of static testing technique is: By scanning and analyzing source code, to identify remote buffer overflow mode, complete model-based security flaw testing. ITS4[4] construct a security threat model structure database, use lexical analysis to pattern match the source cod. This kinds of detection technology with easy to achieve, high efficiency algorithm and other characteristic, but does not consider the information of syntax and semantics level, easy appeared mistakes and failure of reporting. David Evans and others[5] developed LCLint, a tool use specification to check the security of cod, its main idea is: Use the specification file that coded by C source code files and a series of LCL language as input, then automatically check the inconsistency between source files, specification files and programming rules, output the corresponding warning report, Compare with general Program analysis tools, LCLint can check the abstract border and Illegal use global variable problems, but not all of the software production has LCL coded specification files. David Larochelle et al[6] proposed a plan using constraint analysis technique to resolve the problem of security flaw testing: Depend on semantic annotation of program to construct constraint set, use constraint analysis to judgment software remote buffer overflow, this plan is a improvement of LCLint based annotation assistant detection tools, the performance almost same as the general compiler. Yichen Xie etc[7]developed a detection tool called ARCHER: First os all through source parsing generate AST, then use tool construct corresponding program flow control char, have path-sensitive analysis on the chart. In Send mail, Linux and some open source system test, ARCHER has good results. Nurit Dor etc [8] developed a buffer detector prototype system called CSSV, it is committed to find all buffer overflows of the source code, the main idea is to introduce the concept of contract at the first, then construct to annotation version program with the contract and source code, to proceed the Pointer analysis of annotation version program, reflect the result as integral analysis to resolve. The problem of this tool is: only can find local buffer overflow, and has higher false alarm rate. David Wagner[9], in his PhD thesis transformed the software remote buffer overflow detection to integer range analysis mode to process, construct a effective algorithm based on graph theory, analyze the data range of related variable of program, the main idea is Statically analyze for the constraints and pointer dependencies of source code and transfer it to integer range problems to resolve. And according to this technique developed a prototype tool called BOON, but in recent years there is no new research published. Vinod Ganapathy etc [10] based on linear programming and static analysis theory proposed a lightweight detection algorithm, this algorithm use software tool to generate some pointer info and AST, then generate some linear constraints, use linear programming idea to accomplish range analysis, find out possibly occurred software remote buffer overflow, but this algorithm mainly round the busses software tool difficult to implement, has some Limitation.

In the dynamic testing technology, generally by using extend the GNU C compiler method to achieve, teat the returned address of function call's exception or not to judgment the software remote buffer overflow occurred or not in software processing. Eric Haugh etc[11] introduced a series statements such as assert into source code, trace the buffer of the memory, experiment results show that method proposed from the programming Normative angel by the programmer, its only can find some software remote buffer overflow that already existed. Kyung-suk etc[12] proposed a dynamic detection for C language library function calls caused buffer overflow problems, through extend the GNU C compiler, makes compiled program maintaining local variables, function parameters, global variables and other type of buffer mode info in the Heap, and record the buffer size that allocated in the Heap, according to this information, to judge the overflow event occurred or not during the program running. StackGuard [13] insert a "Canary" between return address and buffer in memory, by check "Canary" string to judge the return address was modified or not before call return. StackShield [14] brought forth to keep the return address in a safety place, and recover the return address before call returned. These two kinds of tools are more focus on precautions of remote buffer overflow attack, but not analyze the remote buffer overflow itself [15][16][17].

At present, Domestic research in this area is still at an early stage, mainly focus on define the concept of remote buffer overflow, generate cause analysis and precaution method and others [18][19], the research include by analyze C and other source programs to precaution the generation of buffer overflow vulnerability. use existed buffer overflow vulnerability to proceed attack analysis, such as Windows RPC attack, Microsoft IFRAME attack and other attacks.

In short, the main testing technology in domestic and overseas are whit box testing, the advantage is easy to achieve and high efficiency, disadvantage is high false alarm rate, and need the source code of the tested software, couldn't have test for the numerous nor open source busyness software and system programs. For non source code software, use simulation attack method to test and verify the known buffer overflow vulnerability, but couldn't test and found unknown buffer overflow vulnerability, couldn't achieve the initiative correct and management purpose of software security. From above we can see, in the software remote buffer overflow security flow testing technique, black box based testing technique has more significance of research.

2.4 Software Remote Buffer Overflow Model

Software remote buffer overflow model mainly use for describing the basic from of software remote buffer overflow and generation mechanism, it can be divided into two forms, remote stack overflow and heap overflow. The remote stack overflow is common and easy attacked by attacker, the model as shown in fig 1(a). If remote attacker inject super-long data into native local variable corresponding buffer, then the spare data will along the address space increase direction respectively cover stack frame pointer, EBP, functions return addresses, local variable of calling function even more. If covered content are EIP that will be overwrite the return address, make program execution sequences has changed, that is easier attacked by attacker.

The remote heap overflow is more complicated than remote stack overflow, the model as shown in fig 1(b). If the data transmitted through the network exceeds the size of a remote heap buffer, will cause a remote heap overflow. The spill point could be appeared in static data area, unformatted data area, heap area and so on. Windows heap controller usually use two-way link list to manage spear memory blocks, by cover the heap management data, when heap controller re-allocation or released these blocks with the damaged heap management data, that also cause remote heap buffer overflow occurred.

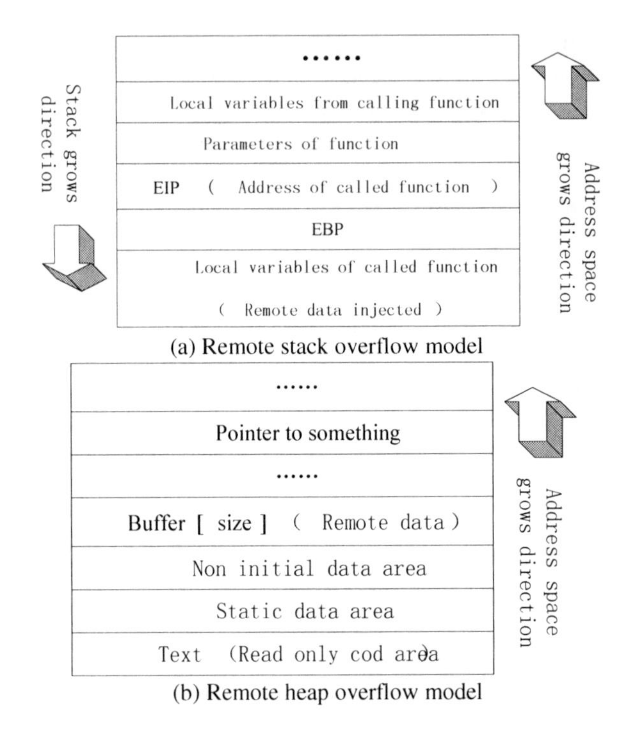

Fig. 1. Software remote buffer overflow model

2.5 I/O Testing Model

I/O testing model was proposed for the test of software remote buffer overflow model, As shown in fig 2. It mainly consist of three parts: I(Input) represents local software input for remote software are detected. P(Program)represents the process procedure of remote software are tested. O(Output) represents the respond of remote software for input (I).

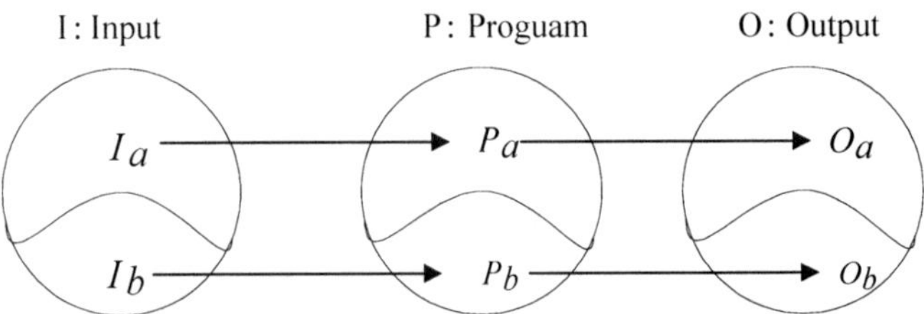

I_a :normally remote data input

P_a :remote software process procedure

O_a : I_a input respond of remote software

I_b :based on fault remote data input, called fault injection.

P_b :remote software process procedure based on I_b input

O_b : I_b input respond of remote software

Fig. 2. I/O testing model

During establish the I/O testing model, considered form the different angle of developer and attacker, to have a detailed classification of input (I), program (P) and remote software respond (O). If there are some fault input can cause memory management exception of remote software, that means the software has doubtful point of remote buffer overflow.

3 Software Remote Buffer Overflow Test Technique

Since use a large number of business, sharing and system software does not provide source code, these source code unknown software, mainly use error injection based black box testing techniques to test and found the potential remote buffer overflow which exists in the software. So research black box testing techniques has more reality significance. The basic idea of black box testing based software remote buffer overflow testing technique in this article is: use network data capture technology to acquire system session data of software, analyze the sequence Logic of session. Based on that, analyze and find out doubtful error injection point. according to the characteristic of error injection point and session sequence Logic, analyze the contents of the error injection, use script builder automatically generate error injection script, finally use I/O test model to test and verify for doubtful error injection point, and report the result of the test. The software remote buffer overflow testing technique system model as shown in fig 3.

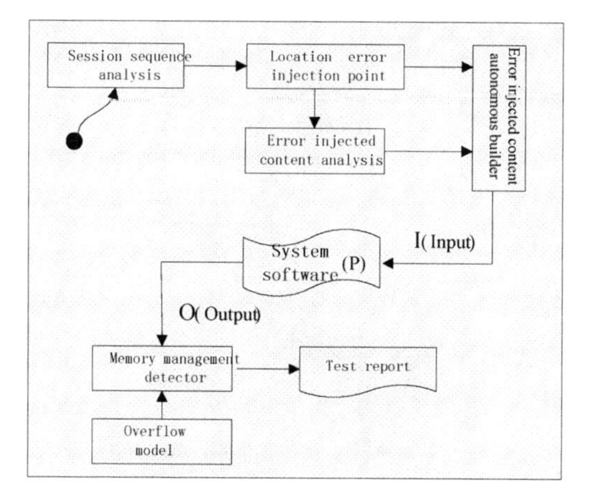

Fig. 3. The software remote buffer overflow testing system model

3.1 Sequence Analysis of Network Protocol Session

Network protocol temporal logic analysis technology is mainly used to analyze the network session flow for the software to judge the software has be used error injection point or not. Mainly include : (1) Real-time capture and storage technology of network flow: by monitor the process of software network session, real-time Capture network session packet, and storage in the database. (2) Sequence analysis technology of network protocol: according to the relationship of sequence to arrange the session packets which captured in different time, construct different time frame session subset, have statistical analysis for different session subset, get sequence relationship of session, provide basis for analyzing and positioning of fault injection point.

The real-time capture and storage technology: In order to resolve the real-time of data packets capture and storage, avoid packet loss, in this paper, use two different thread pool to receive and process data packets: receiver thread pool and storage thread pool. Receiver thread pool keep sending read request to network driver layer, in order to avoid buffer overflow. Storage thread pool in charge of monitor the accomplishment of these read request thread: when check out a read request accomplished, from storage thread pool wake up a slept thread to accomplish the storage task. By this kind of processing, will effectively resolve packets loss. in addition, by setup different priority for different thread, enhance efficiency of packets capture.

3.2 Temporal Analysis Technology of Network Protocol

By scan the source address and destination address of packets to judge it is session data or not. and according to the sequence of received time to arrange. if it is TCP session, because the relationship of session packets serial number as fallow:

Send : a packet with SEQ=the received ACK
serial number at last time
Receive: a packet with ACK= the received SEQ at last time+ received TCP data length+1.

Then by relationship of packets serial number to confirm time sequence relationship of session. The point of Tcp session assembly is to resolve three problems: (1) confirm valid data: before confirm valid data, put it in to buffer, until capture confirm packets. (2) capture and identify SYN and FIN packets, by analyze the IP header of captured packets to resolve. (3) Processing discord, retransmission, error and packets loss. If it is UDP session, according to IP address and port number of each other, filter the session UDP packets from captured network flow. Then according to the serial number of UDP packets arrange, to confirm the time sequence relationship of session.

3.3 I/O Based Testing Technology

First of all I/O(Input/Output) model based Testing Technology by using Error injection point analyzing and positioning technology to trigger the potential candidate of remote buffer flow, then analyze the characteristic of the candidate, chose relevant Error injection content for further testing, in test, continuously testing buffer exception events, by testing result to confirm that is overflow point or not. Mainly include: error injection point analyzing and positioning technology, Error injection content analyze technology and detection technique of software remote buffer overflow.

Error injection point analyzing and positioning technology: After get sequential logic of tested network session, need confirm the position of single-point error injection in session. According to the characteristic of black-box test, most of the buffer overflow triggered by input data to the tested software by users in normal network session. So analyze the network session of tested software, obtain the specific position of normal data or control information inputted by users in network session protocol, to confirm the error injection point of buffer overflow, because most of data input by users are ASCII or other printable character (like Chinese characters), so chose the data segment of network session protocol and the content are ASSII or other printable character (like Chinese characters) as the position of error injection. then use error injector simulate the normal network session process between users and the tested software, and put designed error injection content to the error injection point of the session packets. at last, send the session packets to the tested software by network, to achieve black-box test based on error injection. Therefore, this paper brought forth a heuristic algorithm of error injection point, the specific describe of algorithm as follow. Suppose P stands for the program need to analyze, x stands for the input data of program, Σ stands for all possible inputted data set of program P ,Q stands for normal inputted data set of Σ, $Q^{'}$ stands for abnormal input data set of Σ, $Q^{''}$ stands for specific inputted data set, L stands for possible existed software remote buffer overflow doubtful site, $Pred$ stands for the criterion of memory management exception and program execution procedure has been changed, processing every doubtful site L of P as following algorithm.

Step1: count=0.

Step2: Randomly select from Q, $Q^{'}$ and $Q^{''}$ to input x or a serial number, if program P occurred exception during process X, according to data state of program P accomplished processing to establish a set, called Z.

Step3: Modify a single variable "a", generate a new set $Z^{'}$, and makes program accomplish processing on $Z^{'}$:

Step4: If the output of P satisfied $Pred$,then count++ ;

Step5: Repeat Step2 4, until all input test set accomplish, the repeat number record as n .

Step6: For every position Z, under the P ,Q was given, use $1-count/n$ as Safety Index of evaluation.

The analysis technology of error injection content: According to the characteristic of error injection, chose the suitable content, in order to effective trigger software remote buffer overflow occurred. Error injection content divide into two types random and designed injection. The choice of to use random or designed injection depends on specific software system. In buffer overflow attack based on pre-designed content, usually need precise design the content of error injection(shell code). The target of this paper is to trigger the buffer overflow flaw of the tested software to achieve the test of flaw, injection content needn't design precise shell code, only need to generate software remote buffer overflow. According to above, this paper use super-long string as error injection content. Because there is no way to confirm how long of the string as a trigger for buffer overflow in a certain injection point, so this paper use super-long string generate technique based on active learning. If there is no buffer overflow occurred after inject super-long string, then followed a certain algorithm generate longer string to inject. When the string length or injections times over a certain threshold, then deem this injection point hasn't buffer overflow flaw. So the paper proposed a algorithm of error injection content analyze, the specific describe as the followings.

Step1: From character library of error injection(C) random chose a character c ($c \in C$),as error injection character. initial length setup to. L=10,and NUM1 = | C | -1, $C=C\backslash c$.

Step2: L=L+10.

Step3: select a p_e randomly from the set of doubtful injection point($p_e \in P$, P are the set of doubtful injection point) NUM2 = | P | -1, $P=P\backslash p_e$.

Step4: according to automatic script builder of error inject put L times C inject from p_e followed I/O tested model,

If memory management exception or process procedure changed, then record L, c and p_e,

Else GOTO Step2.

Step5: If NUM1≠0 or L > 250, then GOTO Step1.

Step6: If NUM2≠0, then GOTO Step3.

Buffer overflow events detection technique: In black-box testing system, use distributed architecture to test the tested object of remote detection site from control panel, monitor the testing process by a detection agent running on the same detection site, real time detect there is buffer overflow event triggered or not after error inject, to test and verify software remote buffer overflow exist in software. According to software remote buffer overflow model, here need to research the real time detection technique of stack overflow event and heap overflow event. To detect software remote buffer overflow in software, first capability must be have is find out buffer overflow event occurred in process. In this paper by load memory management detector to identify and capture overflow event occurred. ◦ Memory management detector uses Win32 to debug API capture these exception event, mainly use two core API: Wait For Debug Event and Continue Debug Event.

In the initialization phase, memory management detector by call the Create Process function or Debug Active Process function with DEBUG_PROCESS label to gain the debugging right for detected process, then by call Wait For Debug Event function wait for debug event. When the process try to access inaccessible memory, run into illegal instruction or occur other exception will generate EXCEPTION_DEBUG_EVENT. After memory management detector detected EXCEPTION_DEBUG_EVENT by analysis on the type of exception, thread context, call stack, Memory status and other relevant information to judge buffer overflow occurred or not. If buffer over flow is detected, process end, else call Continue Debug Event inform the OS exception is disposed or need next exception mechanism to process

The practice of buffer overflow event detection: According to software remote buffer overflow theoretical model and I/O test model implement a prototype system. This system use distributed system structure, consist of control panel and detection node. the process of tested software and memory management detection process resident in detection nod. on one hand use control panel to simulate client (or sever) of tested software, send error injection content to the tested process on detection node by error injection script builder, on the other hand send control information to the detection process on detection node, to control the execution of memory management detection process. By testing a famous domestic OA software, this prototype system has found two new software remote buffer overflow point and confirmed by software test group of the OA software part, the practice shows the testing method proposed by this paper is valuable.

4 Conclusion

Buffer overflow is an important software security flaw, which includes local buffer overflow and remote buffer overflow. The software remote buffer overflow security could be found and used by attacker, to implement remote attack for the host system in network. Currently the testing of software security flaw mainly uses white-box testing technique, that need to be provided by the source code of tested Software. For the large number of system and application software that does not provide source code, the only way is to use black-box testing technique. According to the research based on software remote buffer overflow theoretical model and testing model, the paper proposed an I/O testing technique based on error injection, and carry out a prototype system, experiment results show that the prototype system can effectively find and position the potential remote buffer-overflow flaws in system software.

References

[1] Wang, T.t., Han, W., Wang, H.: Analyze model of softerware based on security flaw required. Computer sinc. 34(9) (2007)

[2] Kallen, M.J., van Noortwijk, J.M.: Optimal periodic onspection of a deteriation process with sequential condition states. Int. J. Pressel Vessel Piping (in prcss) (this issue)

[3] Byers, D., Ardi, S., Shahmehri, N., Duma, C.: Modeling Software Vulnerabilities With Vulnerability Cause Graphs. In: The 22nd IEEE International Conference on Software Maintenance (2006)

[4] Viega, J., Bloch, J.T., Kohno, T., McCraw, G.: ITS4: A Static Vulnerability Scanner for C and C++ Code. Annual Computer Security Applications Conference (December 2000)

[5] Evans, D., Guttag, J., Horning, J., Meng, Y.: LCLint: A Tool for Using Specification to Check Code. In: SIGSOFT Symposium on the Foundations of Software Engineering (December 1994)

[6] larochelle, D., Evans, D.: Statically detecting linkly buffer overflow vulnerabilities. In: USENIX Security Symposium, Washington, D.C. (August 2001)

[7] Xie, Y., Chou, A., Engler, D.: ARCHER: Using Symbolic, Pathsensitive Analysis to Detect Memory Access Errors. In: ESES/FSE 2003, Helsinki, Finland, September 1-5 (2003)

[8] Dor, N., Rodeh, M., Sagiv, S.: CSSV: towards a realistic tool for statically detecting all buffer overflows. In: C.PLDI 2003, pp. 155–167 (2003)

[9] Wagner, D.: Static Analysis and Comuter Security: New Technique for Software Assurance. PHD Dissertation, Fall (2000)

[10] Ganapathy, V., Jha, S., Chandler, D., Melski, D., Vitek, D.: Buffer Overrun Detection using Linear Programming and Static Analysis. In: CCS 2003, Washington, DC, USA, October 27-30 (2003)

[11] Hsugh, E., Bishop, M.: Testing C Programs for Buffer Overflow Vulnerabilities. In: The 10th Annual Network and Distributed System Security Symposium Catamaran Resort Hotel San Diego, California (February 2003)

[12] Lhee, K.-s., Chapin, S.J.: Type-Assisted Dynamic Buffer Over-flow Detection. In: USENIX Security Symposium, pp. 81–88 (2002)
[13] Cowan, C., Pu, C., Maier, D., hinton, H., Walpole, J., Bakke, P., Beattie, S., Grier, A., Wagle, P., Zhang, Q.: Stackguard: Automatic adaptive detection and prevention of buffer-overflow attacks. In: Proceedings of the 7th USENIX Security Symposium, San Antonio, TX, pp. 63–77. USENIX (January 1998)
[14] StackShield, http://www.angelfire.com/sk/stackshield
[15] Baratllo, A., Singh, N., Tsai, T.: Transparent runtime defense aganist stack smashing attacks. In: Proceedings of the 2000 USENIX Annual Technical Conference, pp. 251–262. USENIX, San Jose (2000)
[16] PaX, http://pageexec.virtualave.net
[17] SolarDesigner, Non-executable stack patch, http://www.openwall.com/linux
[18] Niu, L.B., Liu, M.R.: Research on software defects classification. Chinese Journal of Application Research of Computers 21(6) (2004)
[19] Ye, Y.Q., Li, H., Zheng, Y.F., Hong, X., Zheng, D.: Analysis of buffer overflow in binary files. Chinese Journal of Computer Engineering 32(16) (2006)

Research and Realization of Security Model Based on Web Services

Shujun Pei and Deyun Chen

Computer Science and Technology Harbin University of Science and Technology
150080 Harbin, China
{peisj,chendy}@hrbust.edu.cn

Abstract. With wide use of Web service technology in distributed system, the security issue becomes increasingly prominent. Based on an in-depth study of WS-Security and WS-Policy, the paper designed a Web service security model and implemented it by using WSE service component under the C# .NET environment. This model realized the end to end security of SOAP message, improves the interoperability of WEB service and greatly meets the need of security requirement of WEB service application environment.

Keywords: WS- security, WS-policy, WSE component, .Net.

1 Introduction

With the increasing maturity of Web service technology, many Web-based distributed application systems choose Web service system which is simple, standard and with interoperability property as its basic key technology. And secure communication is the basis for the wide use of service in the Internet.

Web service communication uses SOAP message as its transmission carrier, a service call may span multiple intermediate nodes (SOAP server, Web services, etc.); therefore, Web service secure communication must enable the end to end security of SOAP message, and meet the special requirement of Web service application, such as message local encryption and decryption, etc. However, the transport layer based SSL and network layer based IPsec and other common point to point secure communication mechanism is the key issue for the further development. On the other hand, different application needs different service secure requirement. In order to improve the interoperability, the secure policy document of Web service needs XML to describe to combine with the existing Web service technology. In addition, each user and each role owes different permissions of Web service. To establish a flexible, expandable visiting control model is another key issue [1].

The background of this research is the second period Project of traffic management information system. The project adopts Web service technology to solve business integration and data integration problems among Web-based heterogeneous distributed systems. According to the safety requirement of the project, the paper puts forward WS-Security [2] and WS-Policy [3] based secure model, enabling confidentiality, integrity and non-reputation of layer SOAP message, realizing authentication and authorization of Web service, improving interoperability of Web service and finally achieving the model.

G. Zhiguo et al. (Eds.): WISM 2011, CCIS 238, pp. 203–210, 2011.

2 The Related Security Technology of Web Service

In order to solve the secure communication problem of Web service, Microsoft and IBM define the Web service secure model together. The model uses WS-Security regulation as its core, guaranteeing the security of SOAP information. It guarantees the security of upper application system through defining WS-Policy, WS-Trust regulations and so on.

2.1 WS-Security Regulation

WS-Security regulation puts forward a set of standard SOAP security extension methods. Through the message encryption, digital signatures and certification, it ensures secure end to end transmission of messages [4].

1) Authentication: to authenticate users identity by passing the Security Token through SOAP header.

2) Message encryption: to use the XML-Encryption technology to guarantee the confidentiality of message transmission.

3) Message Digital Signature: to use the XML Signature technology to realize the integrity and consistency of message.

2.2 WS-Policy and WS-Security Policy Regulation

1) WS-Policy. it defines Web service policy model and policy reference containment mechanism. Policy is expressed by policy expression. Other Web service regulation can use the framework that is provided by WS-Policy to define relevant policy mechanism to describe corresponding service request, preference and capability [5].

2) WS-Security Policy. It uses WS-Policy regulation as basis for the definition of XML element which can be used to specify the security policies. And the element explicitly specified policy. Put it another way, WS-Security appoints how to use WS-Policy language to express WS-Security policy. These policies specify the specific security needs for the special circumstances, such as the adoption of authentication, which digital signature algorithm the client receives, what encryption algorithm the server supports, etc [6]. The secure policy XML element that is defined by WS-Security Policy includes:

< Security Token>: specify the type of Security Token.
<Integrity>: specify the means of digital signature.
<Confidentiality>: specify the encryption algorithm.
<Visibility>: specify the message fragments which need decryption.

3 Design of Web Service Security Model

Traffic management information system is divided into two sub-networks (internal network and external network). The internal network mainly faces the information system managers and the user types of the service are single; and the internal network adopts non-public secure technology SSL, so the security characteristic it has reduces the security requirement to Web service, for example, it can encrypt and sign without X.509 certificate. On the other hand, the security requirement of external network requires Web secure system must reach the security level of message-level. In addition, in specific Web methods, the methods to check, submit and modify the information (such as to check record or the submission of traffic information) have different requirement towards information security. The former may not need to encrypt the information or signature on it to avoid the depletion of system resource and affect the performance; while the latter may need to guarantee the confidentiality and integrity during the process of message transmission, enabling it is without monitoring, tampering and other problems of this kind. Therefore, in order to improve the interoperability, service applicant must acquire some relevant security information of the service which is provided by service provider by certain means, such as encryption algorithm, signature algorithm, authentication, etc. Here, we adopt security policy document of Web service which can well meet the requirement.

3.1 Secure Communication Mechanism

Before the introduction of a whole communication process, we first give the symbols involved and their expressions. They are shown in Table 1.

Table 1. Secure communication mechanism symbol table

Symbol	Description
S	Web service provider
U	Web service consumer
{P}	Web service security policy document
Lx	X.509 certificate of entity X
$K[x],K^{-1}[x]$	Public keys and private keys of entity X
{m}(E=K[x])	message m that is encrypted by public key K[x]
$\{m\}(E=k^{-1}[x])$	message m that is encrypted by private key K[x]
$\{m\}(E=k_1,S=K_2)$	message m that is encrypted by key K1, signed by key K2.
A-- >B:{m}	entity A sends messages to B
A-- >B:{P}	entity B acquires security policy document of A
$V[\{m\}(S=K^{-1}[x])]$	authenticate the digital signature of message m which is signed by key K-1.
$DE[\{m\}(E=K),K^{-1}]$	to decrypt the message which is encrypted by public key K by using key K-1.

Instance of secure communication process (ignore using timestamp to prevent replay attack process):

$(1).\ S \rightarrow u : \{p\}$

$(2).\ U \rightarrow s : \{m1(s = [Lu], E = [Ls], Lu)\}$

$(3).\ S : DE[\{m1\}(E = K[Ls])], V[\{m1\}(S = K[Lu])]$

(4). $S \rightarrow U : \{m2\}(S = K[Ls], E = K[Lu], Ls) \mid$

(5). $U : DE[\{m2\}(E = K[Lu])]V[\{m2\}(S = K[Ls])]$

Introduction to the communication process

1) Before the secure communication process, service applicant acquire secure policy document that corresponding to Web service provided by service server through SOAP message or other means.

2) Service applicant signs and encrypt the very part that need secure protect according to secure requirement, signing the whole message by the use the private keys of secure applicant, encrypting the whole message by the use of public keys of service provider. Finally, send the result and certificate to service provider.

3) After receiving the request message, service provider first get the certificate of service applicant in the message, to verify its effectiveness and pass it. To authenticate the effectiveness of message signature by taking advantage of the public key that provided by certificate and decrypt the whole message.

4) If the security authentication is effective, then establish the whole communication process, realizing the call to service. And then send the corresponding message which is after security procession to service applicant according to the request of secure policy document.

5) The service applicant receives the security processed message. If it passed the effective security authentication, then to sign, verify and decrypt the message according to the request of security policy document, and acquire the original response message at last.

The above mentioned communication processes are all set to have timeout mechanism, if timeout occurs, then re-run the entire communication process. If there is fatal error during communication process, for instance, the failure of authentication, decryption, or failure of verification, then stop communication and send out wrong response message.

3.2 Secure Model Design

1) Secure Model System Structure. The secure model system structure is shown in the Fig.1. as follows:

According to the security requirement of Web service of this project, we construct a security model system of Web service by combining WS-Security and WS-Policy. The basis of security of this system is the extension which is SOAP header based. They realize SOAP encryption, SOAP signature, SOAP authentication and authorization, SOAP security attribute extension (timestamp). For visiting control, we designed the visiting controller of Web service, realization of this controller is mainly based on X-RBAC model, dynamically control the visiting limitation of users to Web service. The legacy system includes a private CA center which facilitates the implementation of security model, and is responsible for the key, certificate generation and management. In addition, security log also ensures the model system with audit capabilities. Security policy manager is responsible for the management of security policy documents and security policy library holds particular security policy document corresponding to the file.

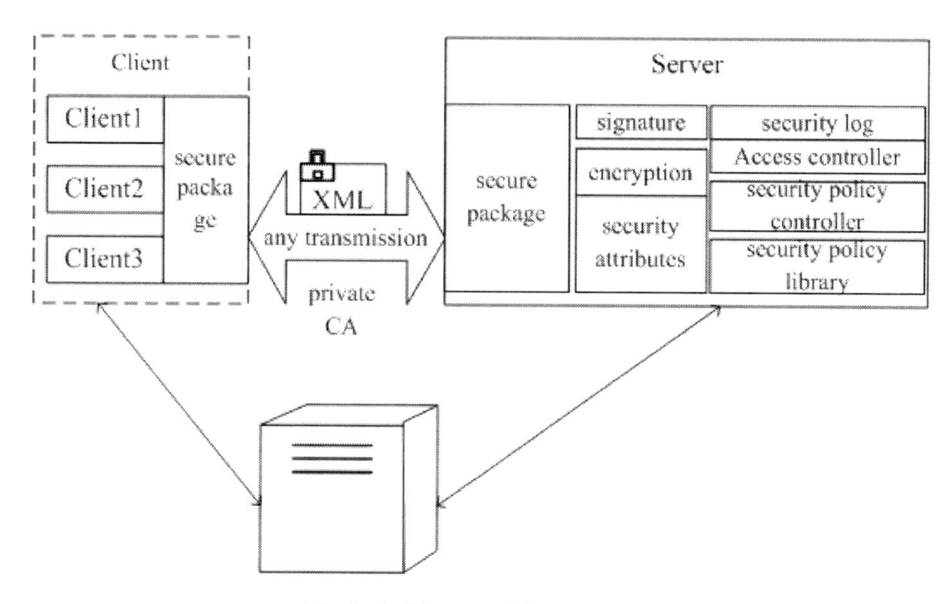

Fig. 1. Architecture of Secure Model

2) Composition and Function of the Security Model. Security model includes Security package (Security Utility), access controller and security policy manager. Security Package is composed of a number of SOAP message filter. The package function principle is shown in Fig.2.

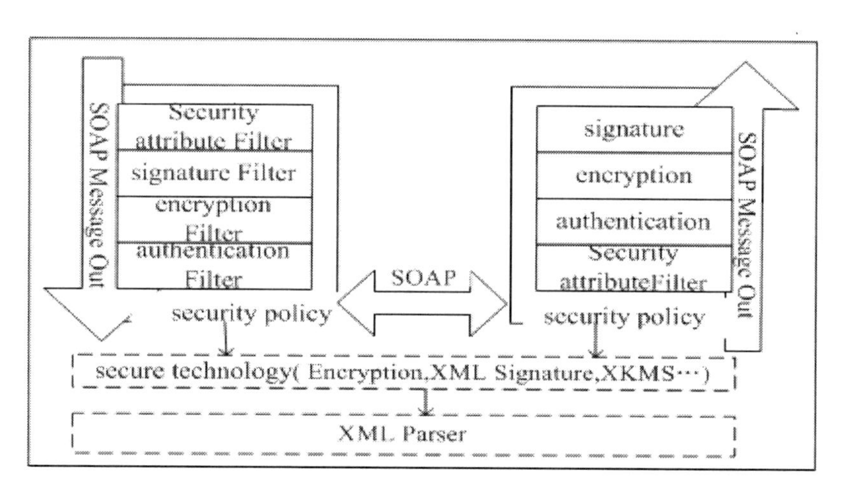

Fig. 2. Function and principal of security model

Main functions of Security package are as follows:

a) Message Encryption: to realize the encryption and decryption function of SOAP message based on XML Encryption. Management of key is based on XKMS technology.

b) Message Signature: to realize signature and signature verification function of SOAP message base on SML Signature. Management of key is also based on XKMS technology.

c) Certification Information Processing: to achieve authentication on users identity through Username/Password Token or X509 certificate security token which is supported by WS-Security.

d) Additional Security Attributes: the additional security attributes are mainly to prevent the replay attack to Web service which is achieved by timestamp mechanism. Of course, additional security attributes can be customized for additional expansion of the security needs of users, achieving a more secure and reliable SOAP message.

4 Implementation of Web Service Security Model

The whole web service security model is developed and implemented by Microsoft .NET platform, C# language. .NET platform strongly support Web service, among which the Web Service Enhancements (WSE) 2.0 of Web service support the advanced Web service protocol such as WS-Security and WS-Policy, it is the basis of achieving security model.

4.1 Secure Policy Document

Web service providers customize their security needs based on security mechanism of WS-Security protocol, such as the support for security token types, encryption algorithms, signature means and so on. And then generate corresponding security policy documents by the using of WS-Security Policy according to security requirement. Finally, bind it together with Web service by WS-Policy. The examples are as follows:

```
<Policy Document>
<Mappings>
<! Web service which is associated with security policy >
<map to =”http://www.yahu.com/demservice.asmx”>
<default policy=””policy-GUID>
</map>
</mappings>
<Policies>
<wse: policy wsu: id=”policy-GUID”>
<wsse: integrity wsp: usage=” wsp: Required”>
<Wsse: tokeninfo>
< ! Description of security token >
<Security Token>
<!  Signature by using X.509v3 certificate >
<wsse: tokenTye>wsse: X.509v3</wsse: tokenTye>
</security Token>
</security Token>
</wsse: tokeninfo>
</wsse: integrity>
</wsp: policy>
</policies>
</policy Document>
```

4.2 Security Package

1) Authenticate Filter: the sample is authenticated by Username/Password Token.

```
///server side // inherit IPassword Provider class
Public class wsepasswordprovider: ippasswordprovide {
GetPassword (uToken) // acquire user's verification code
GetSoapMessage () //acquire SOAP message
Foreach (uTokens) {Validate () ;} // traverse user token, verify identity
/// client add user name and password into SOAP message
AddSoapMessage (uName, uPass);
```

2) Encrypt Filter: code sample uses X.509 certificate encryption and decryption SOAP message, ignoring the logic judgment.

```
GetSoapMessage (); // acquire SOAP message
GetCert1 (); //acquire certificate Cert1
X509SecurityToken (sCert); // using certificate to establish a security token
EncryptedData (sToken); //encryption message
```

3) Signature Filter: code sample uses X.509 certificate signature SOAP message, ignoring the logic judgment.

```
GetCert2 (); // acquire certificate Cert2
X509SecurityToken ("cert2"); // using certificate to establish a security token
GetSoapMessage (sToken); // acquire SOAP message
Message Signature (Stoken); //signature message
```

4) Security Attributes Filter:

```
GetSoapMessage (); // acquire output SOAP message
///SOAP message output
TimestampOutputFilter (); define timestamp output filter
ProcessMessage (env); // add timestamp
/// SOAP message output
TimestampOutputFilter (); // define timestamp output filter
ProcessMessage (env); //read timestamp, throw the exception if it extends
```

5 Conclusion

This paper put forward Web service secure communication solution which is WS-Security and WS-Policy based, safe, interoperable and practical. It is flexible, extensible and easy to be maintained. The security model is not only suitable for the Intranet environment of traffic information management system, but also applies to the Internet environment Web service application system, solving communication security of typical Web service application mode. We will focus on more complex Web service application scenarios in future, such as after calling a series of operations, an application program will exchange a large number of SOAP messages with some other application program, put it another way--the research of how to establish the shared security context of a certain type between two application programs.

References

1 IBM Corporation and Microsoft Corporation. Security in a web Services world: A Proposed Architecture and Roadmap-A joint security whitepaper IBM Corporation (July 2008), http://www-106.ibm.com/developerworks/library/ws-secmap/
2 WIKIPDIA.Public-keycryptography (January 2008), http://en.wikipeda.org/wiki/Public-key_cryptography
3 Freier, A.O., Karlton, P., Kocher, P.C.: The SSL Protocol version3.0, Netscape Communications, http://wp.netscape.com/eng/ssl3/ssl-toc.html
4 Fensel, D., Bussler, C., Maedche, A.: Semantic web enabled web services. In: Proceedings of the First International Semantic Web Conference on the Semantic Web, Sardinia, Italy, June 9-12 (2009)
5 Budak Arpinar, I., Zhang, R., Leman-Mezaetal, B.A.: Ontology driven web services composition platform. Information Systems and E-Business Management 3(2), 175–199 (2009)
6 Benatallah, B., Casati, F.: Special issue on Web services. Distributed and Parallel Databases 12(2/3), 115–116 (2008)

Design and Implementation of the Uniform Identity Authentication System Based on LDAP

Jian Lin[1], Hua Yan[1], and Bing Wu[2]

[1] Center of information management and development,
Taiyuan University of Technology, Taiyuan, China
[2] Research Institute of Mechanic-electronic Engineering,
Taiyuan University of Technology, Taiyuan, China
{linjian,yanhua,wubing}@tyut.edu.cn

Abstract. This paper describes the current characteristics of the campus network authentication technology at first. And then combined with the actual situation of our university, a campus network unified authentication system based on LDAP protocol is designed. Based on comparison of the Information Model, Named Model, Function Model and Security model and other models, a model of unified authentication framework based on Agent Model is constructed. It is implemented that all applications share one user's authentication information in the campus network. The unified authentication system solution solves the different application systems cannot be recognized the authentication problems with each other.

Keywords: LDAP, Uniform Identity Authentication System, Directory service.

1 Introduction

The digital campus is the ultimate goal of university informatization system. The network is the infrastructure of digital campus. With the expanding of network applications, all kinds of management systems based on network have widely been applied in teaching activities at present. Due to the drawbacks of different systems independent authentication of application, it makes the centralized management of using campus network difficult to realize. Such as user information lacks effective sharing, it cannot unify update and serious redundancy of information. Information systems are more and more difficult to manage because of the serious inconsistency of user's identity. When the same user enters different application system, they must memory and adopt a large number of different login and identity information, and systematic safety cannot be guaranteed. Therefore it is an important step to establish a uniform certification system to manage unified network user information, authentication and authorization according to the different circumstances.

Establishing a unified identity authentication system, the user must first certify his identity from the authentication system before he wants to access network resources. In this way, it can easily achieve the unified management of users.

Uniform Identity Authentication and Single Sign-On are the basic framework of digital campus. Single Sign-On should include the following three points: 1) All

G. Zhiguo et al. (Eds.): WISM 2011, CCIS 238, pp. 211–217, 2011.

application system share a Uniform Identity Authentication System. 2) All application system can identify and withdraw authenticated information. 3) Application system can recognize users who have logged in the systems and can automatic analyze their information, then it can be completed the single sign-on functionality.

There are some difficulties to achieve a unified single sign-on authentication because of the various units on campus applications, network services and other factors. Therefore, the first step is how to use the same user account login different systems. This paper presents an identity authentication platform model of digital campus based on LDAP directory services. This model can be used to solve that users log in different application system with the same account in campus network.

2 Overview of LDAP Protocol Model

Lightweight Directory Access Protocol (LDAP) [1] service has very clear purpose to simplify the complexity of the directory X.500 in order to reduce development costs, while adapting to the needs of Internet. On large-scale network systems using Transmission control protocol (TCP)/Internet protocol (IP), it is a simplified version based on x.500 standard. LDAP service has many great features, such as providing quick and advanced search, quick response and hierarchy view of data. It also can be utilized to many different applications. We can apply LDAP to design a Uniform Identity Authentication System.

LDAP Information Model [2] is defined a kind of data type and information unit storage in the directory, the basic information unit of directory is entry - Object's collection of properties according to rules. Each property has a type and value. Directory service is a method to realize information management and service interface. A directory service consists of the directory database and a set of Access control protocol. It is a special database, Directory information is all kinds of resources in the network environment, they are storage in hierarchical in the directory tree structure. We can store, access, manage and use the information.

The directory database has the following characteristics: distribution, quick and flexible searching, platform-independent and security. As services for integrating management information of catalogue in the network, directory services are one of the important underlying technologies of network system.

LDAP Named Model describes the way of organizing and referencing LDAP data, namely entries localization way in the LDAP directory. It describes that all entries store and arrange in an inverted tree structure, as shown in Fig. 1. Users can access thus entries in any position of directory tree through Standard Named Model.

LDAP Function Model describes the way of operating directory through using LDAP. It contains many important function models offered by LDAP service [3].

- Search and read operation: This function contains reading, searching and listing operation. Reading operation is to be able to read the data that you specify. Searching operation can help end-entity obtain the desirable data by searching the whole database according to the criteria defined by end-entity. Restricting search operation can scope the specific search area to help LDAP server not only reducing the server load but also preventing any ambiguous search.

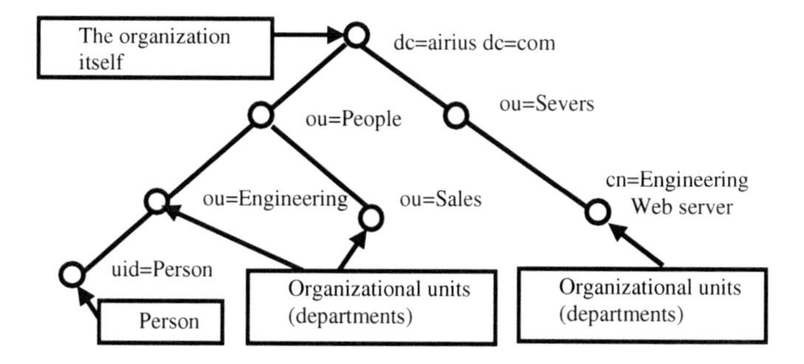

Fig. 1. The organization and reference mode of LDAP model

- Modification operation: This function contains modification, addition, deletion and modifying Relative Distinguished Name (RDN) operations. Modification operation allows end-entity to modify its secure data. Addition operation can add a new end-entity to the database. Deletion operation can delete a cause after end-entity's privilege being expired or revoke the cancellation resulted from cheating or something else. Modifying RDN operation helps end-entity to modify the associated distinguished name (DN) components.
- Authentication operation: This helps initiate a session and prove its identity to the directory. Authentication contains bind, unbind and abandon operations, and these operations support several authentication methods, such as simple clear-text password and public-key based authentication. And LDAPv3 can support stronger authentication service.

The built-in extensibility of the LDAP architecture makes it easier to modify the protocol for new situations and to meet end-entities' needs. Therefore, it is convenient to add what we want. We can create attributes like access control lists (ACL) to manage the end-entities and improve the security of the system.

3 Ldap Security Model Architecture

In the network system, it is important that we can confirm the identity of the communication. When we confirm correct identity, System allows for communications.

We can identity authentication secure channel and access control through using LDAP Security Model [3]. With the purpose of LDAP Security Model is providing a framework in order to preventing an unauthorized access to directory date. There are three Authentication mechanisms Architecture:

- Identity authentication: there are three Authentication mechanisms in the LDAP, including anonymity authentication, simple authentication and SSL authentication. Users don't need certification in the anonymity authentication, it be used the method of full public only.
- Communication security: LDAP provide a communication security based SSL/TLS [4]. SSL/TLS is an information security technology based PKI. It is widely used in Internet now. Although SSL protocol can establish a secure

channel between Web server and browsers, it cannot protect the end system from threats if end system does not have enough secure protection.

- Access Control: LDAP access control are flexible and plenty. It can realize access control based Access control strategy procedure in LDAP. A role is a management and access control policies for semantic construction. System manager creates the roles. Every end-entity has a role and is bound by this role to follow the rules of the role. Whenever it wants to access the resources in this system, it must authenticate and identify himself. Then the system checks if the role of this end-entity has the permission to access this resource.

This approach can simplify the management of the system and increase the security of this system.

4 Uniform Identity Authentication System Based on LDAP

We have designed a Uniform Identity Authentication System based LDAP architecture. A method based on Agent Model is used in this system. The user management of the model is a Centralized mode so that the system has a larger expansibility and security [5]. Agent-based model is shown in Fig. 2.

Unified authentication based on agent model, its main job is to develop client and server side applications to provide a uniform interface. Certificate server is the most important device. It is responsible for issuing user's electronic identity and his authentication and identity management. It does not rely on IP of host or operating system. It does not ensure that all hosts on the network's physical security, and assuming the transmission of data packets can be randomly stolen tampering.

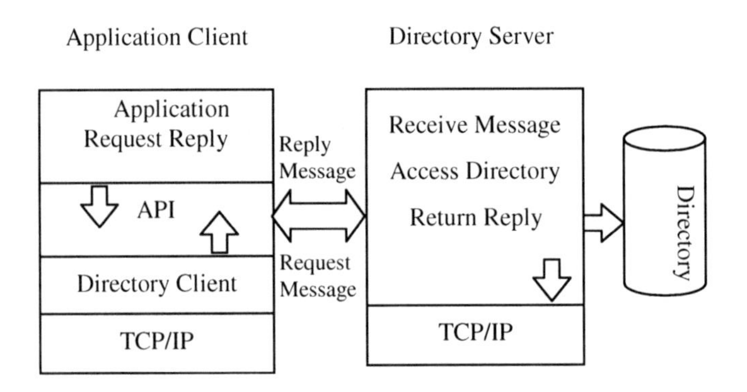

Fig. 2. The diagram of interaction between client and server

In Fig. 2, the uniform identity authentication system is consists of three parts: client, certificate server and application server. The authentication server plays the role of agent, because all authentications are done through it. Before accessing the application resources, all the clients send their identity authentication information to the authentication server. In order to improve the security of the system, it can be used mutual authentication mode. After the user authentication, the certificate server returned to the user an electronic identity label. The user can visit application servers according to this electronic identity label. If the electronic identity is illegal, the application server will refuse to provide services.

- A client must send a request to the server to establish a session, at first. The request is sent to the server through TCP, Port number is 389. So the client sends a series of operation request to the server, and the server response it.
- The server returns a response in accordance with the client's request. In some special cases (user settings), client can send to the server a number of consecutive requests without waiting for response from the server.
- Meanwhile, the server can handle according to the actual situation of sending a response to the client.
- Note: Applications do not need direct access to the directory but by a series of API functions to generate requests and send it to the directory server.
- When requested directory information is not local directory, directory server will send request to other serve in order to collect directory information.
- After the directory services have been established, it can provide services through the port, the client can also access data through this port.

The uniform identity authentication system model based on agent-based authentication is designed, as shown in Fig. 3.

- All kinds of application Server and LDAP server are connected through the campus network.
- Two LDAP servers are set to provide authentication services with Dual-machine backup, and synchronized with a heartbeat cable. The prime LDAP Server works usually, and the backup software is used RSYNC based on Linux. RSYNC uses Remote-Update protocol ,it allow RSYNC transmitting incremental value of two groups file rather than file itself, so RSYNC can improve the transmission efficiency.
- Storage device is used Fiber SAN (Storage Area Network). Two LDAP servers and storage connect through the fiber. There are directory information and others network application data in this Storage. Application system accesses the LDAP server and storage through NAS.

The client should configure LDAP parameters according to requirements, such as LDAP server IP, port number, authentication model, encrypted transmission, SSL Model, LDAP type and other parameters. Application systems access LDAP server according to these parameters. There are many application systems in this architecture at present.

Fig. 4 shows the directory hierarchy structure of our school in LdapBrowsers software.

The administrator of the LDAP server can modify or delete end-entity's attribute. And the punishment for end-entity who deceives or violates the protocol can be done by deleting or modifying its attribute.

LDAP start up SSL/TLS service through StartTLS 5. First, SSL protocol offered Data confidentiality by Information encryption. Secondly, Encryption also can ensure the consistency of the data. Finally, Secrets has important role in the security verification. This system has higher security. When a user needs to have LDAP server authenticate its identity, he must encrypt its identity and password first and send the encrypted data and its name certificate to LDAP server to do verification.

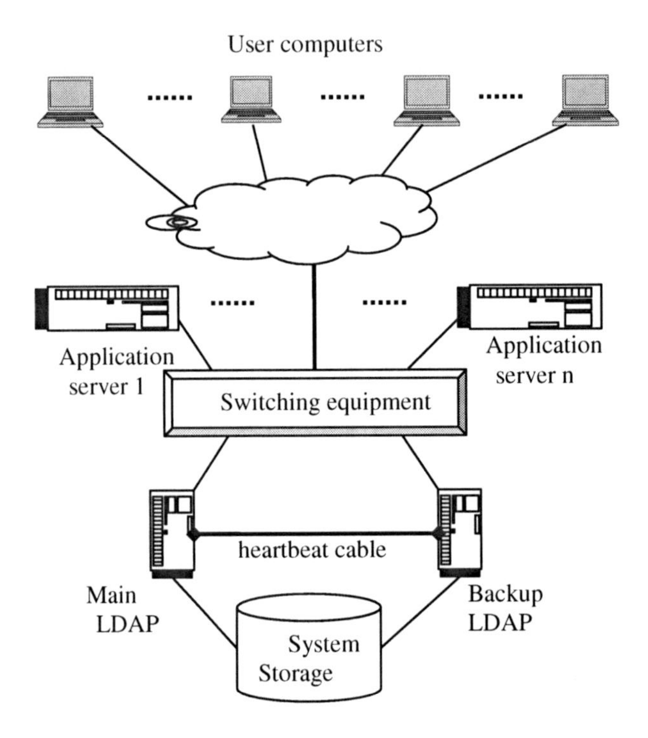

Fig. 3. The topology of uniform identity authentication system of TYUT

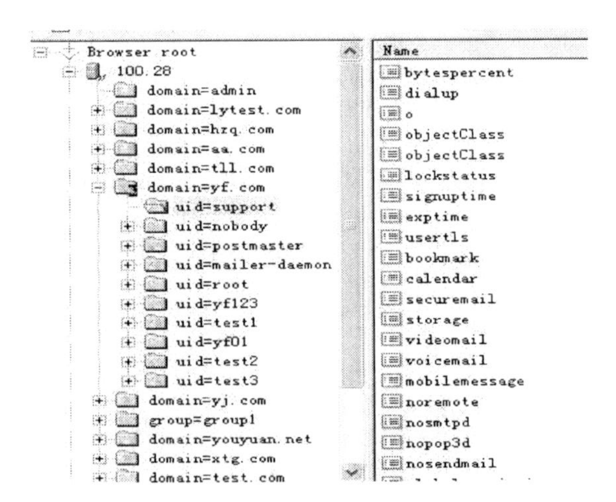

Fig. 4. Software directory hierarchy structure of the LDAPBrowsers

5 Conclusion

This paper solves two questions of Single Sign-On, namely all application systems share an identity authentication system and all application system can identification and extraction Authentication information. In the future, we can use Kerberos protocol to realize that application system will recognize users who have logged in the systems and can automatic analyze users so that the uniform identity authentication system can be the single sign-on.

Kerberos [6] is a distributed authentication service system, based on KDC and Needham.schroeder method. The overall program of Kerberos is third-party authentication service trusted. Kerberos can coordinate client and server to do identity authentication. This architecture assumes that users can guarantee the safety of Kerberos system itself. After authentication success, the authentication system generates unified authentication ticket, returned to the user. In order to determine the validity of ticket, system Parity it . These clients and servers exchange authentication information, thus completing the high-level Single Sign-On.

References

1. Carrington, C., Speed, T., Ellis., J., et al.: More on LDAP. In: Enterprise Directory and Security Implementation Guide, pp. 35–40 (2002)
2. Dong, L.-w., Wang, W.-y.: Study on Unified Resource Management Architecture Supporting SSO. J. Journal of Computer Applications 26, 1146–1147 (2005)
3. Yeh, Y.-S., Lai, W.-S., Cheng, C.-J.: Applying lightweight directory access protocol service on session certification authority. Computer Networks 38, 675–692 (2002)
4. Zeilenga, K.: Named Subordinate References in Lightweight Directory Access Protocol (LDAP) Directories. RFC 3296 (July 2002)
5. Liu, C., Xu, H., Yang, Z.: The Research and Realization of an Uniform Identity Authentication System on Campus Based on LDAP and SOAP. J. Computer Applications and Software 23(12), 59–60, 80 (2006)
6. Ashraf, E., Tarek, S.S., Zaki, M.: Design of an enhancement for SSL/TLS protocols. J. Computers & Security 25(4), 297–306 (2006)

3DES Implementation Based on FPGA

Fang Ren[1], Leihua Chen[1], and Tao Zhang[2]

[1] Meteorological Science and Technology Service Center Shaanxi Meteorological Service
Shaanxi, China, 710014
[2] Lighting-proof Technology Center Shangqiu Meteorological Service
Henan, China, 476000
renfang829200@163.com

Abstract. In order to meet the demand of plenty continuous encrypting-deciphering, and meet the demand of enhancing the security of encrypting-deciphering algorithm, the fundamental technologies such as pipeline technology and finite state machine (FSM) are applied, 3DES encryption algorithm's encryption chip's circuit based on FPGA are designed and realized. On the platform of FPGA of Xilinx Virtex4 series, the ISE 10.1 development kits are used to realize the simulation confirmation and the logic synthesis. The result indicates that the 3DES cryptographic system's speed is able to achieve 860.660Mbps and the encrypting-deciphering speed is greatly enhanced. The design could be used in network security products and other security equipment extensively.

Keywords: 3DE, FPGA, finite state machine, pipeline, VHDL.

1 Introduction

Modern Block Encryption Algorithm [1] (such as DES [2], 3DES [3] [4], AES) can be realized by either software or hardware, software encryption accomplishes encrypting function by circulating encryption software on a host. Besides occupying resource of host, software encryption computes slower and has lower security than hardware encryption, on some occasions demanding high speed data transmission; it fails to fulfill the demand. Hardware encryption accomplishes encrypting function by hardware encryption device which is independent of host system, and all the memory and computation of key data is accomplished inside by hardware, it do not occupy resource of host, computes faster, has higher security, stability, compatibility.

3DES (168bit key) ensures its security by enhancing its complexity of algorithm while keep original system without great modification for it is based on DES algorithm on the base layer. But the basic defect is slowness of software encryption, it often implies bit operations such as transposition, position-change, exclusive OR etc., so hardware encryption has its special advantage. Hereafter, accomplishing by hardware encryption and optimizing 3DES on hardware platform becomes a new hot search point. FPGA [5] is flexible and effective, therefore, 3 grade pipeline technology and finite state machine (FSM) are applied to realize finally the encryption system in modularized design method, and the system was simulated, synthesized, and optimized to enhance its entire performance.

G. Zhiguo et al. (Eds.): WISM 2011, CCIS 238, pp. 218–224, 2011.

2 Description of DES and 3DES Algorithm

2.1 DES Algorithm

Since its disclosure, DES suffered kinds of attacks, has gone through tests over long period of time, some new algorithms derives from it, has extensive applications. The entire system is public; security of DES system is dependent absolutely on confidentiality of keys. Plaintext and cipher text processed by DES is 64 bit, a 64 bit key is applied, and the effective length is 54 bit.

DES encryption's procedures can be summarized as: initial transposition, 16 rounds circulating iteration, inverse initial transposition. Through an initial transposition, plaintext is divided to half left and half right, then processed in 16 rounds exactly same operating, the plaintext finally becomes 64bit ciphertext through the last transposition. Every round operating includes extension transposition, S-box alternation P-box transposition and twice exclusive OR, besides, there is a round key (subkey) in every round operating.

The course of DES decryption is similar to the course of encryption, in decryption, the same algorithm is implied. The difference is that the sequence of application of the round key is reversed. Fig.1 is the flowchart of DES encryption algorithm.

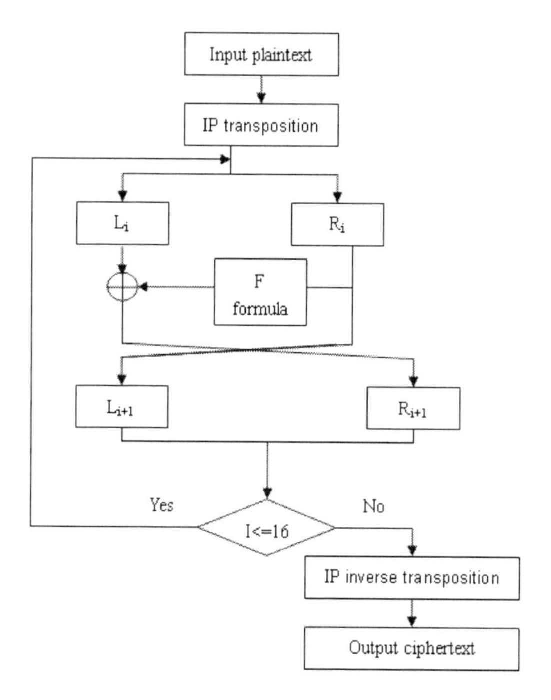

Fig. 1. Flowchart of DES encryption algorithm

2.2 3DES Algorithm

Triple-DES is a DES algorithm's reform with more security. 3DES's keys include three 64bit keys, K_1, K_2, K_3, and once DES encrypting algorithm was implied as its basic module. 3DES divides 64bit plaintext into blocks and then encrypts, take $E_K(x)$ and $D_K(X)$ representing encryption and decryption to a 64bit data which is divided into blocks by DES, the key is K, 64bit plaintext is X, 64bit cipher text is C, then the program of encryption and decryption can be expressed by equations below. Fig.2 is the flowchart of 3DES encryption and decryption algorithm.

Equation (1) is encryption course of 3DES.

$$C = E_{K_3}(D_{K_2}(E_{K_1}(X))) \tag{1}$$

Equation (2) is decryption course of 3DES:

$$X = D_{K_1}(E_{K_2}D_{K_3}(C))) \tag{2}$$

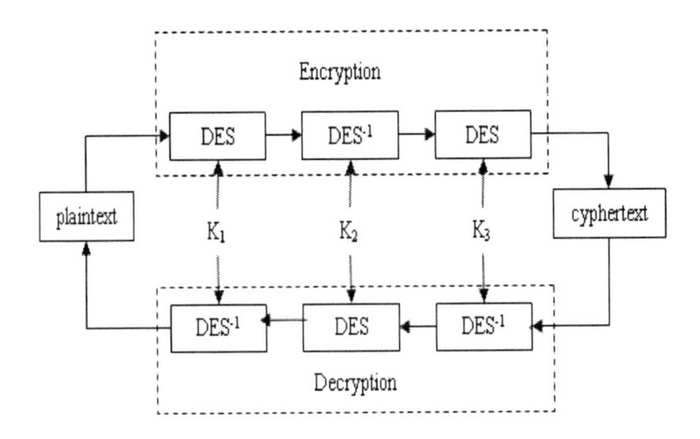

Fig. 2. Flowchart of 3DES encryption and decryption algorithm

To attain higher security, the three keys should be different, it's equal essentially to encrypt with a 168bit key. In case of data which requires not high security, K_3 can be same to K_1, and then the key's effective length is 112bit.

3 FPGA's Design and Realization

3.1 Scheme of Design

Principle of the design is to enhance the designed frequency while consuming as less resource as possible. Therefore, 3 grade pipeline technology and finite state machine (FSM) are applied in the system, under the premise that the system occupies resource rationally, they enhance the system's performance as far as possible. After being synthesized and optimized, the system's speed of process of data is enhanced further.

Based on design's goal above, study and analysis of 3DES algorithm, modularization design method is implied to construct integral structure of 3DES system's operating modular. Fig.3 is integral structure of the system.

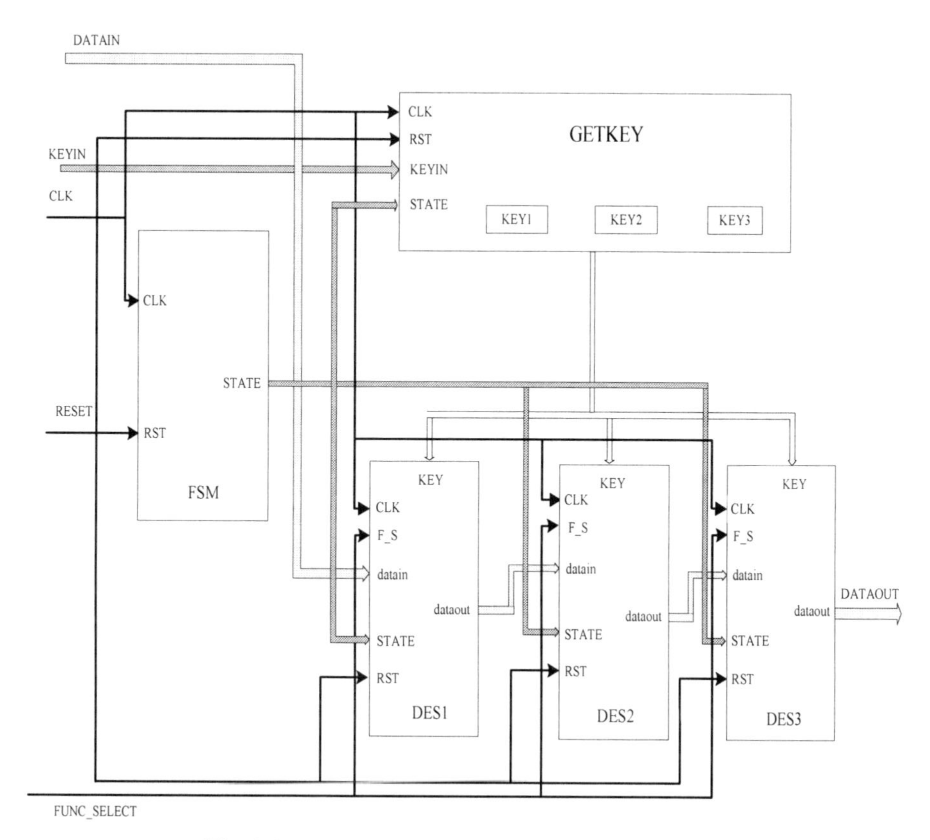

Fig. 3. Integral structure of the 3DES algorithm system

In the system, FSM takes charge of all the state's transformation, is core of the whole system. Under control of the controlling signal, GETKEY accomplishes reception of initial key, results in subkeys which operating modular demands in every round iteration, and stores it up. DES1, DES2, DES3 control the processing procedure according to states, and complete pipelining approach of encryption or decryption.

3.2 Application of Key Technology

Design of finite state machine (FSM): FSM has a simple structure of program, can be controlled flexibly, and is able to eliminate burs. The progresses execute collaterally, the computing speed is enhanced effectively. Fault tolerant technique was applied, so it has great reliability. The hardware resource is spared, for the developed hardware is simple. The control of the whole encryption circuit's state is accomplished by FSM.

In the initial state, FSM produces signal to control input of key, and input of plaintext or cyphertext, in this way, the call of key modular and DES operating modular is achieved. The FSM have two states, WaitKeyState (state waiting for key's input) and WaitDataState(state waiting for plaintext's input). The states of the system's top layer modular shift as Fig.4.

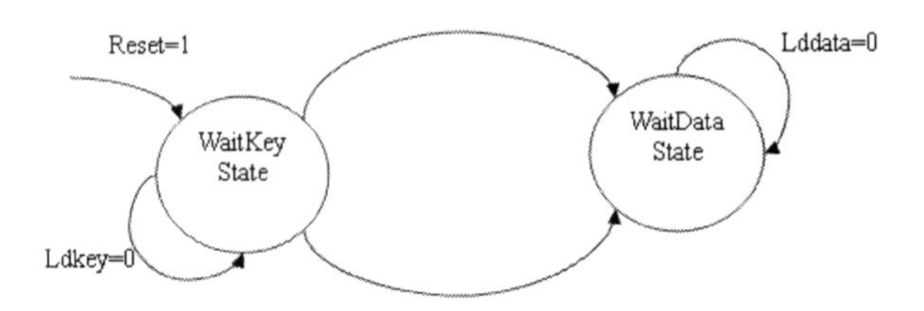

Fig. 4. The states of the system's top layer modular shift

Supposing sixteen 48bit keys are applied in 16 rounds circulating computation, the procedure is complicated; accordingly, it is difficult to control the system. Therefore, FSM design is adopted to complete the control of every subkey matching its relevant round number and the control of data-processing procedure. The FSM designed 5 states, they are waitkey (wait for key's input), waitdata (wait for data's input), initialround (initial round), reapeatround (reapeat round), finalround (final round). The FSM uses a 4bit counter roundcounter to record round number. Fig.5 is illustration on 16 rounds computation and state shift of subkey's choice.

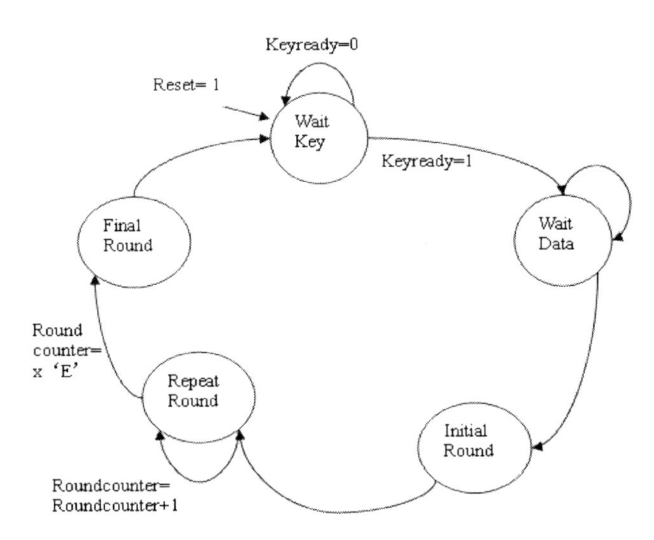

Fig. 5. 16 rounds computation and state shift of subkey's choice

Design of pipeline: Design of pipeline is to separate assembled logic systemically, and to insert memories between parts, to store up the mid-data, in order to enhance data processing speed.

In comprehensive view of capability and resource occupied, 3 grade pipeline's form is applied, every DES acts as a grade of pipeline. 16 rounds iteration operating is completed. Key-birthing modular generates subkeys required in 16 rounds iteration operating to advance. FSM is applied to control every subkey matching its relevant round number. Since operation of encryption or decryption on the first block of data, from the 49th circle on, i.e. after 48 circles' delay, every 16 clock circles results in an output of 64bit encrypted or decrypted data.

Design on key modular: In the design, design of key-birthing machine is realized independently of DES round operation function. Under control of the controlling signal, the modular accomplishes reception of initial key, and generates subkeys which operating modular demands in every round iteration.

In former design, time-sharing multiplex of a key- birthing modular applied by three keys, it makes system's speed of encryption or decryption slower than copying multiple operating modular collaterally. Considering this, copying 3 key modular collaterally is applied to reduce the time consumed by grouped keys' time-sharing wait and the time consumed in waiting for subkey in every round operating.

Subkey of every round of encryption or decryption in DES is generated by circulating left shift or right shift, and every round of shift has a certain shift bit. Every round number and value of initial key relative to every key can be deduced by integrate analysis, irrespective to subkeys. Hence, valuing directly and generating to advance are applied in the design's key-birthing. Round number is controlled by a counter, to select a relevant subkey. All the subkeys are generated in the same time collaterally, are ready to advance, selected by key-selecting signal, conveyed to corresponding operation modular.

The system's capability is enhanced greatly by way of copying 3 key modular collaterally and generating all the subkeys to advance.

4 Result of Simulation

After accomplished design of the system, to find out its defect early, simulation confirmation is issued. Simulation confirmation here adopts the method of carrying bottle-up, and the confirmation is divided into modular-level and system-level two parts. Limited to space, only image of the result of system-level confirmation is provided here.

In order to descry result of the confirmation directly, we input 3 initial keys is the same as: ox00 00 00 00 00 00 00 00, data_in[0:63] input: ox80 00 00 00 00 00 00 00.As Fig.6, kinds of signal and situation of data' input are demonstrated. Encryption operation is applied to the inputted plaintext, cyphertext is outputted at the 49th circle. In the Fig.7, if the output signal is ture, the cyphertext outputted is: ox95 f8 a5 dd 31 d9 00. The result of simulation is correct.

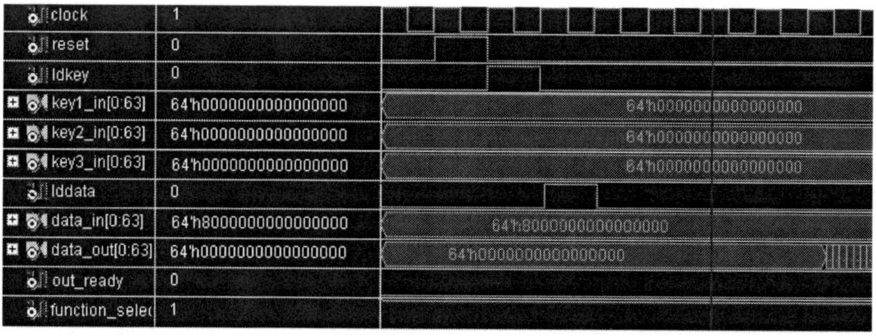

Fig. 6. 3Des_top's oscillograph of simulation confirmation on encryption

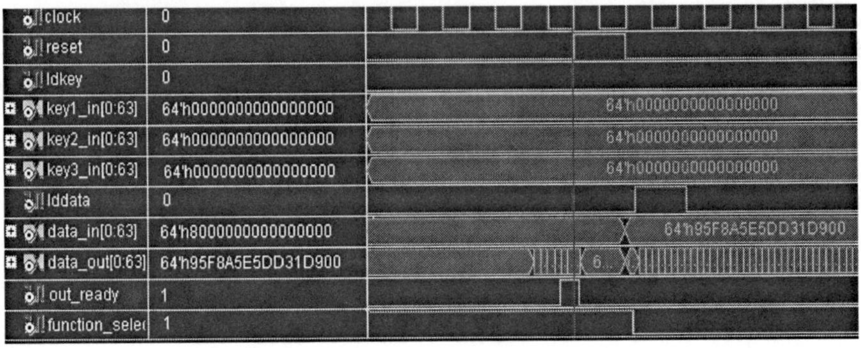

Fig. 7. 3Des_top's oscillograph of simulation confirmation on encryption (continue to former image)

In the simulation confirmation of decryption operation, take the outputted cyphertext from the system formerly as plaintext in decryption operation, keep the keys the same, the overcome outputted by the system is same to the plaintext inputted in encryption operation, the result is also correct. (Image omitted)

5 Conclusion

The algorithm of encryption is realized mainly applying Xilinx corporation's xc4vlx25-12-ff676 components of Virtex4 series as platform. VHDL codes are translated and edited, integrated, adapted, and simulated in ISE10.1 environments. According to the results of analysis and synthesis, the highest operating frequency of the system's clock is 215.165MHz, the shortest establishing time is 3.454ns, and the longest maintaining time is 3.793ns. The design can be implied extensively in network security products and other security devices.

References

1. Yang Bo. "Modern cryptology," .Peiking: press house of Tsinghua university, 200:3146:148-149.
2. Pan, W.: development and realization of DES encryption chip. Jinan university Master's paper (June 2008)
3. Dang, Z., Zhang, G.: Design and realization of IP core in high speed3_DES algorithm. CPLA University of Information Science and Technology, Master's paper (April 2007)
4. Shao, J., He, Z.: High speed realization of 3DES' encryption algorithm, based on FPGA. Modern Electronic Technology 27(21), 55–57 (2004)
5. Li, Y., Song, R., Lei, J., Du, J.: Xilink FPGA Foundation of design (Version of VHDL). Press house of Xidian university, Xi'an (February 2008)

A New Log Audit Model to Improve the Effectiveness of Network Security Research

Guofeng Wu, Liliang Zheng, and Dingzhi Lu

School of Computer and Information, HeFei University of Technology, HeFei 230009, China
{wgf2867,hnzl1}@126.com, dingzhilu5258@163.com

Abstract. The network logging important events during the operation of information on log analysis, found that the required security audit events and the law is the fundamental purpose. Based on the existing network security measures analysis, a new model of network security audit log is proposed. Dynamic increase of the model rules of the crisis on the network, according to the specific needs of some of the rules of the new additions and changes. At the same time adding in the log of the log classification system, greatly improving the efficiency of the audit log.

Keywords: log acquisition, audit, log classification, rule base.

1 Introduction

The rapid developments of the network, not only bring convenience to people, but also bring a lot of security risks. To enhance the security of the system, we usually inserted firewall or network isolation and information exchange system into the inside and outside the network, additional IDS (Intrusion Detection System) [1] within the network. This security solutions just use to before invasion or invading thing, they are very difficult to deal with the invaded thing.

Audit system as a firewall system and an intrusion detection is added. You can review the log to find the existing danger or the risk will occur.

Log audit system[2] is a distributed, cross-platform network information security system; it supports real-time collection of information, this information comes from various network devices ,servers ,databases and various specific business system which produce the log , message ,status ,etc. On the basis of real-time analysis, we can find intrusion [3] and issue various real-time alarms from that monitoring the operational status of various hardware and software systems. Provide a historical log of data mining [4] and correlation analysis. Managers can get accurate, detailed statistical analysis of data and exception reports from the visual interface and reports. The log audit system can help managers to find security leak, take valid measure, boost secure grade.

2 Improved Log Audit Module

Event log can generally be divided into: the operating system logs, security device logs, application logs and network device logs. Security audit system must be unifying the log format before using these data.

G. Zhiguo et al. (Eds.): WISM 2011, CCIS 238, pp. 225–232, 2011.

2.1 Unified Log Format

Since uniform format log is the premise of the log audit system, otherwise the system will appear the problem of matching difference when auditing various logs. So it's very important that the log format should be unified. There are two general approaches to unified log:

1) We can take same characters of various logs; we can take same characters of various logs and take them as the message of unifying log. This method ignores the part property of the original log, which make the particle size of the audit bigger and rougher.

2) By using a language, translate into the different kinds of logs to a uniform format. This method is by defining a data structure of structure type, which makes the data noted in different logs map to the corresponding data structure.

In this paper, while considering the above method, by a careful division of the log, we can make uniform log more efficiency. First, determine the type of collectible log whether it is a Syslog log. If they are then stored directly (Because the main object of this system is the Syslog log); Otherwise, by using the other rules to deal with the log format. The specific log unified process as shown in Fig. 1.

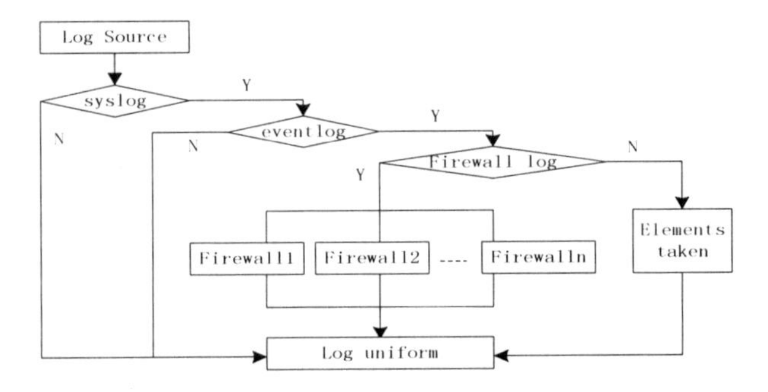

Fig. 1. Flowchart of log uniform

2.2 Log Classification

Considering the large number of logs generated, it's particularly necessary that how to find valuable information from a large number logs. So in this paper a kind of log classification system is proposed, then make a scan classify of the log which are unified. The specific classified criteria are as follows:

(1) Original IP/MAC address: By searching the original address, we can clearly see the operation status of the original host.

(2) Destination IP/MAC address: By searching the Destination address, we can see the details of network dealing with the host network.

(3) Time segment: By analyzing a certain time segment of the log, we can find the potential dangerous.

(4) Protocol character [5] segment: By classifying the protocol character, we can see the operational status of this protocol in current network.

Above four points are important for log audit system, some specific cases can be extended. The log classification map as shown in Fig. 2.

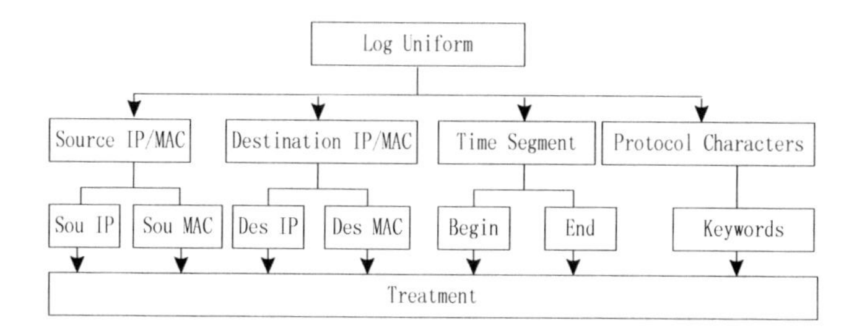

Fig. 2. Log classification map

2.3 Establishment of the Rule Base

We used the method based on feature, namely analysis of various invasive methods, feature extraction. Characteristics of various invasive methods form the rule base, then let log information compare with the characteristics of various intrusions during invasion detection analysis in the log information. The aim is that using characteristics to distinguish between normal network and intrusion behavior. Therefore, Invasion of feature extraction, the quality of the intrusion rule base to establish, will directly affects the ability to detect intrusion detection and the level of detection efficiency.

First, we must depth analysis of various protocols, and then know the various protocols rules and the operation rules. Particularly carefully study of some important agreement, such as IP, TCP, ICMP, FTP, TELNET, HTTP, SMTP, and DNS. In addition, it is important to collect all kinds of attacker [6], and then study a variety of attack patterns and the invasion strategy. By using formal and random process method, we can abstract and model the intrusion process on the basis of a large number of network intrusion case studying. The aim is extracting the feature of traces and process. This method has the advantage of high accuracy detection and disadvantage of known attacks just being detecting, so the system do nothing for unknown types of attacks. Various rules in the rule base structure table as shown in Table. 1. The table of part rules in rule base as shown in Table.2.

Table 1. Structure table of rule base

List name	Data structure	Length
Rule number	Char	4
Rule name	Varchar	20
Protocol	Varchar	4
Severity grade	Varchar	2
Time period	Varchar	20
Limit value	Int	4

Table 2. Part rule table of rule base

Rule umber	Rule name	protocol	Severity grade	Time value	Limit value
0001	land	IP	middling		
0002	Ping of death	ICMP	high	1s	100
0003	Syn flood	UDP	high	1s	400
0004	Smurf	ICMP	high	1s	100
0005	Arp cheat	ARP	high	1s	400
...	...	...	...	...	...

Intrusion detection process: first, read the log data, and then package them into one-dimensional chain structure. At the same time, every rule of rule base should packaged the into one-dimensional chain table. Then, each log of list table matches with rule protocol. If they are IP protocol, then use IP rule to detect. Else if they are TCP protocol, then use TCP rule to detect. This action will skip some functions which are not necessary to test, greatly improving the detection efficiency. During the relevant agreements detection, each of the relevant test should be detected, and implement the corresponding detection function. The following detection algorithm is about the ICMP protocol:

```
Transfer log data
begin
if log_protocol = ICMP then check ICMP_attack
While check_rules <> NULL
if rule = smurf then smurf_check
...
if rule = ping of death then ping of death check
goto next rule
end
```

2.4 The Process of Safety Audit

Adopt the log audit method based on rule base, first, let log information match with rule base to gain the security thing. Then, use security incidents to remove the repeated one. And through the correlation analysis of security event excavate intrinsically linked to achieve the log audit function. Finally, the system quantitatively assesses the security threats of node theory.

Audit model presented in this paper is divided into five modules: The first is the unity module of log format, through this step, the log classification system and analysis system can be more accurate treatment of the corresponding log. The second is the log classification system, using this system can be summarized on the log processing, which is more conducive to view and analyze the log. The third is the rule base management system, which can dynamically update the rules of the various crises. The fourth is a log analysis system, which is the most important part in the log audit. It can be analyzed with the log of the current situation and the current crisis, which give warning and suggestions for management in time. The last module is warning which can clearly understand the current network operations. The Security Audit Flowchart as shown in Fig. 3.

State of the log is divided into three levels, which are representing the different treatment conditions and store in different tables. They are the following three: Red Alert (this log contains dangerous), Yellow alert (this log contains dubious information), Green alert (security log which is no risk). Red alert log, the system will automatically give treatment suggestions. Yellow alert log, the system will give alarm. Green alert log, the system will do nothing. The system must find the log information of suspicious or dangerous from vast amounts of data. And then use ring, E-mail or other means to alarm, notify the administrator to take response measures to bug fixes.

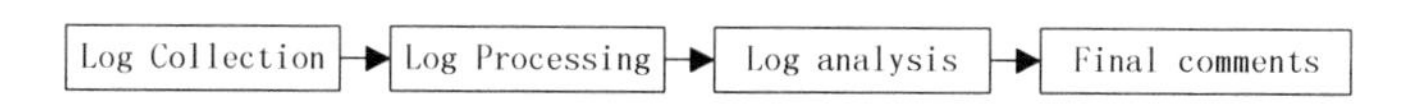

Fig. 3. Security Audit Flowchart

2.5 The Method Log Audit

Security strategy can use the method which is real-time audit combination of post audit. Real-time audit, when the system is running, which is that let collected log data match with the set rules [7]. If the rule Match success, the corresponding alarm of process information will given by a certain way, and then the corresponding audit data is store into the real-time audit database. Otherwise, this log event is the safe operation. Post audit is a centralized data processing behavior, since the audit process in real time a single log data is focused, for many associated operation is not able to make accurate judgments. History audit make up this lack, for some events associated with the continuous which can make a more accurate judgments. At the same time, the correlative result is store into the history audit database.

The heart of the security audit system is audit rule base [8], self-learning ability was added in the rules of rule base, which can capture the characteristics of new crisis, and then write into rule base. The so-called self-learning refers to the rule base can approach the records administrator and the actual data mode automatically generate new rules, which can better adapted to practical needs. This greatly improved the ability to adapt to the self-audit log, namely the adaptive rule base mainly refers to the security administrator based on the evaluation of the actual audit report, dynamic adjustment of the credibility of each rule and some other parameters. Rule matching speed is fast, rule itself has self-learning and adaptive capacity which is a good security audit system should be considered. The following is a brief description of log audit method, we can achieve self-learning me, time, target);

Where name ,time and target are security events'parameters;

```
If relate(Event)=threat   Then
    insert the relate(Event) into rules;
Else goto next;
Let VoT represent the value of theoretic threat;
VoT = evaluate(Event);
Return VoT.
END
```

By the visual interface and reports to provide accurate and detailed data of statistical analysis and exception analysis, which can assist managers to detect security leaks, and take effective measures to improve the security level which makes the security audit system more human and more integrity.

3 Design of System

The most important module of audit system are as follows: log collection, log combination, check the log and the management of log. The test data source of the system is from the Linktrust syslogserver.

3.1 Log Collection

The audit object of the log audit system is all kings of logs, so the integrality and the timely of the log collection directly affect the veracity of log audit. The system throw off some functions to reduce the resource consume, which can protect the system timely get the integrate logs. The most key technique is the using of SOCKET. After the analysis of the syslog protocol trait collect log. By analysis of the syslog protocol trait collect log. The part code as follows:

```
unsafe private void Receive(byte [] buf int len)
  {......
PacketArrivedEventArgs e=new PacketArrivedEventArgs()
fixed(byte *fixed_buf = buf)
     {
        SyslogHeader * head = (SyslogHeader *) fixed_buf;
        e.HeaderLength=(uint)(head->syslog_verlen & 0x0F) << 2;
         temp_protocol = head->syslog_protocol;
         switch(temp_protocol)
         {
   case 1: e.Protocol="ICMP";     break;
   case 2: e.Protocol="IGMP";     break;
   case 3: e.Protocol="TCP";      break;
   case 4: e.Protocol="UDP";      break;
   case 5: e.Protocol="SYSLOG";     break;
   default: e.Protocol= "UNKNOWN"; break;
          }
        ......
  e.PacketLength =(uint)len;
  e.MessageLength =(uint)len - e.HeaderLength;
  e.ReceiveBuffer=buf;
 Array.Copy(buf,0,e.IPHeaderBuffer,0,(int)e.HeaderLength);
 Array.Copy(buf,(int)e.HeaderLength,e.MessageBuffer,0,(int)e.MessageLength);
  }
 OnPacketArrival(e);
}
```

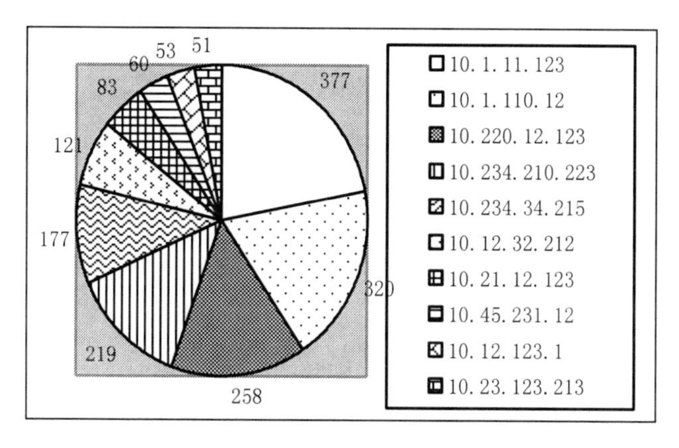

Fig. 4. Visit the server the IP of top 10 map

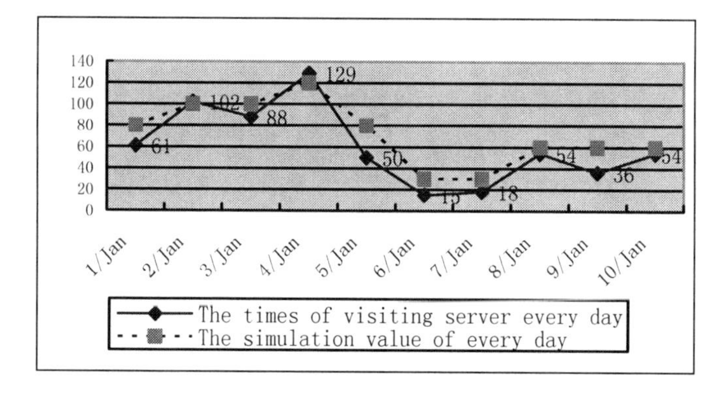

Fig. 5. The times of visiting the server every day map

3.2 Create the Table

Put the button "table" which is from the Log Check. From choice the condition we can get equal table. The stat mode is aim at server address (210.45.245.120).The time is one month, which is July 2010 year. The Particular thing is as shown in Fig. 4. We can find that during the month: 10.1.11.123 appear 377 times, 10.1.110.12 appear 320 times, 10.220.12.123 appear 258 times, 10.234.210.223 appear 219 times, 10.234.34.215 appear 177 times, 10.12.32.212 appear 121 times, 10.21.12.123 appear 83 times, 10.45.231.12 appear 60 times, 10.12.123.1 appear 53 times, 10.23.123.213 appear 51 times.

From this map we can clearly find which IP had ever visited server frequently, then manager can catch these IP to find the suspicious actions. By setting parameters in check model, we can see all relevant IP, and then make a detail analysis integrating the net background. The system will give the rational ideas when it finds the suspicious things. The most important to manager is solving the problem as quick as possible.

At the same time, another important standard to analyze the net is the time of visiting the server every day. Especially for the machine from one address which visit server frequently. The Particular example is as shown in Fig. 5. The time is the days, which is the first ten days of July 2010 year.

From this map we can find that, at the beginning of month, there are more people visit the server. Commonly, it's a natural visiting when the times under the simulation value. The value is based on analysis of visiting server for long time, and change along with the replace of the net and the equipment. If we find more times than simulation value, the log produced on this part will make a stress disposal. The aim is to avoid the dangerous of net, and protect the net work well.

References

1. Cao, J.-G., Zheng, G.-P.: Research on distributed intrusion detection system based on mobile agent. In: Machine Learning and Cybernetics, July 12-15 (2008)
2. Xing, H.Y., Shi, j.: A Log Analysis Audit Model Based on Optimized Clustering Algorithm. In: Network and Parallel Computing Workshops, pp. 841–848 (2007)
3. Ka, I.-m., Zhu, H.-b., Keie, et al.: A novel intelligent intrusion detection, decision, response system. IEICE Trans. on Fundamentals of Electronics, Communications and Computer Sciences E89-A (6), 1630–1637 (2006)
4. Jagannathan, G., Wright, W.R.N.: Seventh IEEE International Conference on Data Mining Workshops – Title. Data Mining Workshops, ICDM Workshops, pp. 1–3 (2007)
5. Qu, X., Liu, Z., Xie, X.: Research on distributed intrusion detection system based on Protocol analysis. In: Anti-Counterfeiting, Security, and Identification in Communication, pp. 421–424 (2009)
6. Yawl: The analysis of restructuring IP pieces and the common attack on fragmentation, http://www.nsfocus1.com
7. Dharmapurikar, S., Lockwood, J.: Fast and Scalable Pattern Matching for Content. In: Architecture for Networking and Communications Systems, Princeton,NJ, pp. 183–192 (2008)
8. Jang, C., Kim, J., Jang, H., Park, S., Jang, B., Kim, B., Choi, E.: Rule-based auditing system for software security assurance. In: Ubiquitous and Future Networks, pp. 198–202 (2009)

An Improved Traitors Tracing Scheme against Convex Combination Attack

Ya-Li Qi

The Computer Department,
Beijing Institute of Graphic Communication,
102600, Beijing, China
qyl@bigc.edu.cn

Abstract. The data supplier requires that traitor tracing schemes can reveal the identities of the subscribers who illegally receives the digital content. The flaw of traditional ElGamal scheme is that cannot resist convex combination attack. On this point this paper raises an improved ElGamal scheme, which shorten the length of the private key and use the users private key and the header h(s, r, k) to computer session keys. It is worth mentioning that the user's private key has relation with the header h(s, r, k), so the improved scheme has good revocability. The two random numbers help to resist the random number attack. The results show that the security of the improved scheme is enhanced in comparison with the traditional traitor tracing scheme.

Keywords: ELGamal, traitor tracing scheme, convex combination attack.

1 Introduction

The goal of a traitor tracing scheme is to provide a method so that the data-suppliers, given a pirate decoder, are able to recover the identity of the traitors [1,2,3,4, 5,6,7,8,9,10,11].

In[1], an ElGamal traitor tracing scheme is given. ElGamal has been raised in 1985[2], the security of which is based on the computation difficulty of discrete logarithm over finite domain. For its cryptographic properties it applies to data encryption, digital signature, etc. A recognized flaw of ElGamal encryption scheme is that the cipertext length is two times of plaintext length. So the traitor tracing scheme usually use two level encryption to improve the efficiency of the broadcast system.
For the traditional ElGamal traitor tracing scheme cannot resist the convex combination attack, many researchers study on it. In [3,4],by increasing the degree of f from k to 2k-1, k-inllusion resistance can be achieved.

This paper raises a new method based on ELGamal encryption algorithm. Data Supplier (DS) send the users with private key and the header h(s, r) to computer session keys. The private key and the header h(s, r) has close relation based on ELGamal. So the improvement has good revocation because DS can find the traitor from private decoder directly.

In Section 2, we describe the traditional ElGamal traitor tracing scheme and its flaws. Our construction scheme is described in Section 3. We analyze the performance of our scheme in Section4. Finally, conclusions are given in Section 5.

G. Zhiguo et al. (Eds.): WISM 2011, CCIS 238, pp. 233–237, 2011.

2 The ElGamal Traitor Tracing Scheme

The ElGamal traitor tracing scheme is two level encryption which can improve the efficiency of the broadcast system.. That is to say, information use symmetric encryption and is broadcasted to users. At the same time DS use the encryption algorithm in traitor tracing scheme to produce a data header which send to users too. For users, firstly use the key which is produced by traitor tracing scheme to get to get decryption key, then use the decryption key to decrypt information.

Let p be a prime power, q be a prime such that q|p-1, q>n, (where n is the number of users) and g be a qth root of unity over GF(p).

A Initialization

DS chooses a random polynomial $f(x)=a_0+a_1x+\cdots+a_kx$ over Zq and computes $y_0=g^{a_0}$, $y_1=g^{a_1}$,......, $y_k=g^{a_k}$., Then let $e_T=(p,g,y_0,y_1,\cdots y_k)$ be a public key. DS gives f(i) to authorized user u_i as personal decryption key e_i, $1\le i\le n$.

B Distributing a Session Key

For a session key s, DS computes a header as $h(s,r)=(g^r,sy_0^r,y_1^r,\cdots,y_k^r)$ where r is a random number. Then DS broadcasts h(s, r). Each user u_i computes s from h(s, r) and e_i as follows:

$$\left\{sy_0^r\times(y_1^r)^i\times\cdots\times^i(y_k^r)^{i^k}\right\}/(g^r)^{f(i)}=s(g^r)^{f(i)}/(g^r)^{f(i)}=s\,. \tag{1}$$

C Broadcasting Encrypted Data

To send actual plaintext data M, DS broadcasts $C=E_s(M)$, where E is a symmetric key encryption function. Every authorized user can recover s, and then decrypts C to obtain M .

D Detection of Traitor

When a pirate decoder is confiscated, the pirate key e_p is exposed. If e_p contains (j, f(j)), then DS decides that user u_j is a traitor.

E Convex attack

Let $f(i_1),f(i_2),\cdots,f(i_k)$ be k personal decryption key, $w=[u,w_0,w_1,\cdots w_k]$ be any convex combination of the vectors $v_1,v_2,\cdots v_k$, defined by:

$$\begin{aligned} v_1&=[f[i_1],1,i_1,i_1^2,i_1^3,\cdots i_1^k]\\ &\vdots\\ v_k&=[f[i_k],1,i_k,i_k^2,i_k^3,\cdots i_k^k] \end{aligned}\quad. \tag{2}$$

If there is $\sum_{i=1}^k a_i=1$, then we get a vector $w=\sum_{i=1}^k a_iv_i$. The w is not a legitimate personal decryption key, but it can be used to decrypt any ciphertext like $C=[a,b_0,b_1,\cdots b_k]$

since $b_0^{w0}\times b_1^{w1}\cdots b_k^{wk}/a^u=s$.

3 Improved Scheme with Two Random Numbers

The improvement scheme is two levels encryption too. This paper improves the h(r, s) with two random numbers. The algorithm describes as follow:

A Initialization

DS computes $\beta = g^{x} \bmod p$, x is random number as a secret. Then let $e_T = (p, g, \beta)$ be a public key. DS gives authorized user u_i as personal decryption key with e_i, $1 \le i \le n$.

B Distributing a Session Key

For a session key x, DS computes a header as $h(s, r, k)$ where r and k are random numbers. The three parameters meet the conditions as follows:

$$\gamma = g^{r} \bmod(p). \tag{3}$$

$$\lambda = g^{k} \bmod(p). \tag{4}$$

$$e_i = (x\gamma + r\lambda + ks) \bmod(p-1). \tag{5}$$

Then

$$s = (e_i - x\gamma - r\lambda)k^{-1} \bmod(p-1) \tag{6}$$

Because β is public key produced by x, and γ is produced by r, λ is produced by k, Now takes x as random number , r and k as private key. Then formula(6) becomes to:

$$s = (e_i - k\beta - r\lambda)x^{-1} \bmod(p-1) \tag{7}$$

Then DS broadcasts h(s,r,k). Each user u_i computes x from h(s,r,k) and e_i as follows:

$$x = (e_i - k\beta - r\lambda)s^{-1} \bmod(p-1) \tag{8}$$

The two random numbers have closer relation with the private key, so the structure gets stronger than one random number with private key.

C Broadcasting Encrypted Data

To send actual plaintext data M, DS broadcasts $C = Ex(M)$, where E is a symmetric key encryption function. Every authorized user can recover s, and then decrypts C to obtain M .

D Detection of Traitor

When a pirate decoder is confiscated, the pirate key e_p is exposed. The DS can compare exiting s and r with formula

$$e_i = (x\gamma + r\lambda + ks) \bmod(p-1) \tag{9}$$

then DS decides that user u_j is a traitor.

4 Performance Analysis

From the description of the tradition scheme based on ELGamal and the improvement scheme, the important distinguish is length of the private key. The private keys in improvement scheme are based on the ELGamal encryption algorithm. DS sends user i with ei and h(s, r,k) as private key of user i.

In fact e_i has relation with h(s, r, k). So the DS sends different user with different private key e_i and header h(s, r, k). Because every user' keys are based on random number r, to get session key x in improvement scheme every user uses his e_i and h(s, r, k) only. Then if the communication is security, the probability of combination attack gets lower. The improvement scheme can resistant convex combination attack. At the same time, the two random numbers have closer relation with the private key, so the structure gets stronger against attack aimed to private key .

Another advantage of improvement scheme is revocability. It is the revocability that the improved scheme requires the n headers for n users. Although the scheme is based on the public key system, the h(s, r, k). for every users has directly relation with his private key e_i . So every user has different private key. Once a private decoder has been confiscated, from the e_i we can location the user we can cancel his right to use the information. Of cause it based on there is a credible DS.

5 Conclusion

In this paper, we raise an improvement scheme based on ELGamal traitor tracing. By the construct relation between e_i and h(s, r, k) , every user use his only private key to get session key x.

The improvement scheme can resistant the convex combination attack which can attack the tradition scheme based ELGaml successfully. It has good revocability. In the other face, because the random r and k is very important in producing h(s,r,k), so the attack to random number get weaker.

Acknowledgments. The research is supported by The Foucs Project of Beijing Institute of Graphic Communication Ea2011005 and Funding Project for Academic Human Resources Development in Institutions of Highter Learning Under the Jurisdiction of Beijing Municipality PXM2010_014223_095557.

References

1. Komaki, H., Watanabe, Y., Hanaoka, G., Imai, H.: Efficient asymmetric self-enforcement scheme with public traceability. In: Kim, K.-c. (ed.) PKC 2001. LNCS, vol. 1992, pp. 225–239. Springer, Heidelberg (2001)
2. ElGamal, T.: A public key cryptosystem and a signature scheme based on discret elogarithms. IEEE Transactions on Information Theory 31(4), 469–472 (1985)
3. Kim, H.-J., Lee, D.-H., Yung, M.: Privacy against piracy: Protecting two-level revocable P-K traitor tracing. In: Batten, L.M., Seberry, J. (eds.) ACISP 2002. LNCS, vol. 2384, pp. 482–496. Springer, Heidelberg (2002)

4. Tzeng, W.G., Tzeng, Z.J.: A public-Key traitor tracing scheme with revocation using dynamic shares. In: Kim, K.-c. (ed.) PKC 2001. LNCS, vol. 1992, pp. 207–224. Springer, Heidelberg (2001)
5. Boneh, D., Franklin, M.K.: An efficient public key traitor scheme (Extended abstract). In: Wiener, M. (ed.) CRYPTO 1999. LNCS, vol. 1666, pp. 338–353. Springer, Heidelberg (1999)
6. Chor, B., Fiat, A., Naor, M.: Tracing traitors. In: Desmedt, Y.G. (ed.) CRYPTO 1994. LNCS, vol. 839, pp. 257–270. Springer, Heidelberg (1994)
7. Kiayias, A., Yung, M.: Breaking and repairing asymmetric public-key traitor tracing. In: Feigenbaum, J. (ed.) DRM 2002. LNCS, vol. 2696, pp. 32–50. Springer, Heidelberg (2003)
8. Kurosawa, K., Desmedt, Y.G.: Optimum traitor tracing and asymmetric schemes. In: Nyberg, K. (ed.) EUROCRYPT 1998. LNCS, vol. 1403, pp. 145–157. Springer, Heidelberg (1998)
9. Naor, M., Pinkas, B.: Efficient trace and revoke schemes. In: Frankel, Y. (ed.) FC 2000. LNCS, vol. 1962, pp. 1–20. Springer, Heidelberg (2001)
10. Yang, B., Ma, H., Zhu, S.: A Traitor Tracing Scheme Based on the RSA System. International Journal of Network Security 5(2), 182–186 (2007)
11. Li, X.f., Zhao, H., Wang, J.l., et al.: Improving ElGamal Digital Signature Algorithm by Adding a Random Number. Journal of Nort heastern U niversity(Natural Science) 31(8), 1102–1105 (2010)

Model of Trusted Dynamic Measurement Based on System Calls

Rui Hao[1,2], Xinguang Peng[1], and Lei Xiu[3]

[1] Colledge of Computer Science and Technology, Taiyuan University of Technology
[2] Department of Computer, Education Institute of Taiyuan University
[3] Shanxi branch, Agricultural Bank of China
Taiyuan, China
{Monica_haorui,Xiulei_ty}@163.com,
sxgrant@126.com

Abstract. The model of establishing the chain of trust presented by TCG is only the static integrity measure to system resources,it can not guarantee the dynamic trust after system operating. Therefore, this paper presents the trusted dynamic measurement model to platform application based on system calls for guaranteeing trust of the platform after computer system operating.

Keywords: system calls, dynamic measurement, trusted computing.

1 Introduction

In distributed computer network environment, how to validate security of remote computer system is the basic problem. we need to ensure data truthfulness, integrity confidentiality, effectiveness and refused denying, and not just increase to some security levels between pc and network such as the firewall. Therefore, in 1999, IBM HP, Intel,Microsoft and so on that are famous IT enterprises set up trusted computing platform alliance (TCPA). In 2003, TCPA reorganized to trusted computing group(TCG). it powerfully pushes the development of trusted computing.

TCG defines the trusted platform module [1] (TPM)that is a new embedded security subsystem in the computer. The core of trusted platform is TPM safety chip. it is a underlying core firmware of technology of trusted computing. TPM is called the trust origin of security pc industrial chain. In practical application, TPM is embedded to PC's mainboard. it provides measures and verification of data integrity, and services of security and identity certification. It changes the traditional concept of safety technology that software protects the computer system. it is that the underlying embedded systems protect the computer system.

The following briefly introduces TPM as core how to establish platform chain of trust. The trusted computing platform will guide the BIOS module as the root of trust for measurement integrity, and TPM acts as the root of trust for integrity. With electricity to start from platform, BIOS boot block measures the integrity of BIOS and storesthe value to TPM; Next BIOS measures OS loader and stores the value to TPM; Next OS loader measures OS and application and to the entire network. The

G. Zhiguo et al. (Eds.): WISM 2011, CCIS 238, pp. 238–241, 2011.

reporting of trust by continuous measurement of trust achieves an a full match of the entity and the expected value. Trust extends the whole computer system. Figure 1. shows establishing the chain of trust.

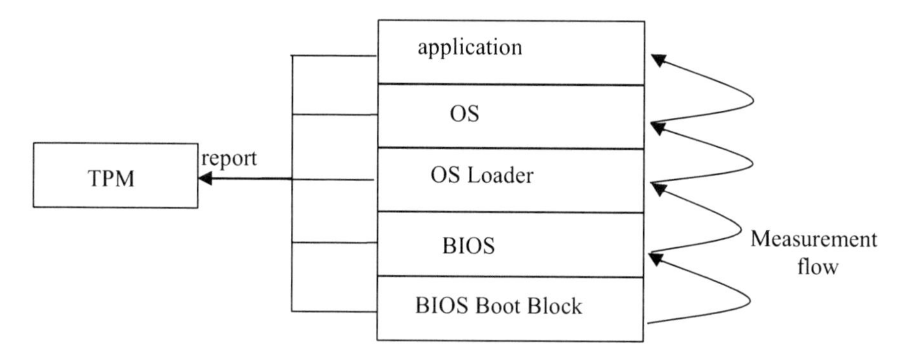

Fig. 1. Establishing the chain of trust

2 Measurement Model Based on System Calls

The model of establishing the chain of trust presented by TCG stays in the static [2] integrity measure to system resources, cannot guarantee the dynamic trust after system running. it is hardly enough just by judging the values of measurement beginning to run the system. So this paper present the corresponding dynamic [3] measurement of trust based on the system calls, for dynamic trust after running system.

As the interface of OS providing application to access system resources, the system calls is able to reflect the behavior [4] of the software in certain extent. Each process uses system calls correspond a track in the time from executing to end. Although the number of track of the application is probably a lot in different executing circumstances, the track is stable partially. The local patterns of the track (short sequences) show a certain consistency. So the normal behaviour of a program can be represented by the local patterns of the track. If the pattern is the departure of the local patterns of the track, it can be considered not trusted.

Based on the system calls, the basic idea of the dynamic measurement is: Getting the normal system call track of the executing programs, and in a way to form the normal behavior features of programme at the level of the system calls,and accordingly to the detection of the abnormity in actual operating process.

3 Dynamic Trusted Measurement Model Based on System Calls

TCG achieves the function of storage for trust, integrity measure,and the report for trust by adding the chip TPM to the hardware.Trusted platform begins from establishing the root of trust,from the root of trust to hardware platform, to OS, and toapplication. Higher level, By low level attesting higher level, the trust is extend the trust to the entire computer system. Figure 2. shows establishing the chain of trustIn the trusted computing environment, Trustworthy computing environment in the chart 2 shows the dynamic trusted measurement model [5] based on system calls in the trusted computing enviroment.

We add the behavior measurement modules to measure the behavior of executing codes.Behavior measurement module consist of two parts:obtaining system call sequences module and platform configuration register(PCR) extractor. The system calls sequences module is to obtain the system call numbers. In the Linux system, modified strace programme can be used to obtain system call numbers. PCR extractor is used to extract the value of measurement in TPM.Because operating of any software is not usually isolated, but inseparable in operating environment [6], the measurement of trust is consider of environment factors. Behavior measurement module records the measurement information in the behavior track table,and reports to TPM.When the attestation response module respond the attestation request by the time triggered, it extracts the corresponding record reports in behavior track table to attestation module, and attestation module matches patternly according to normal sequences of system calls in mode library for attesting the trust of platform. If you find out abnormal behavior, give an alarm. Here the data in the mode library of normal system calls is obtained by training massive data of normal system calls sequences.

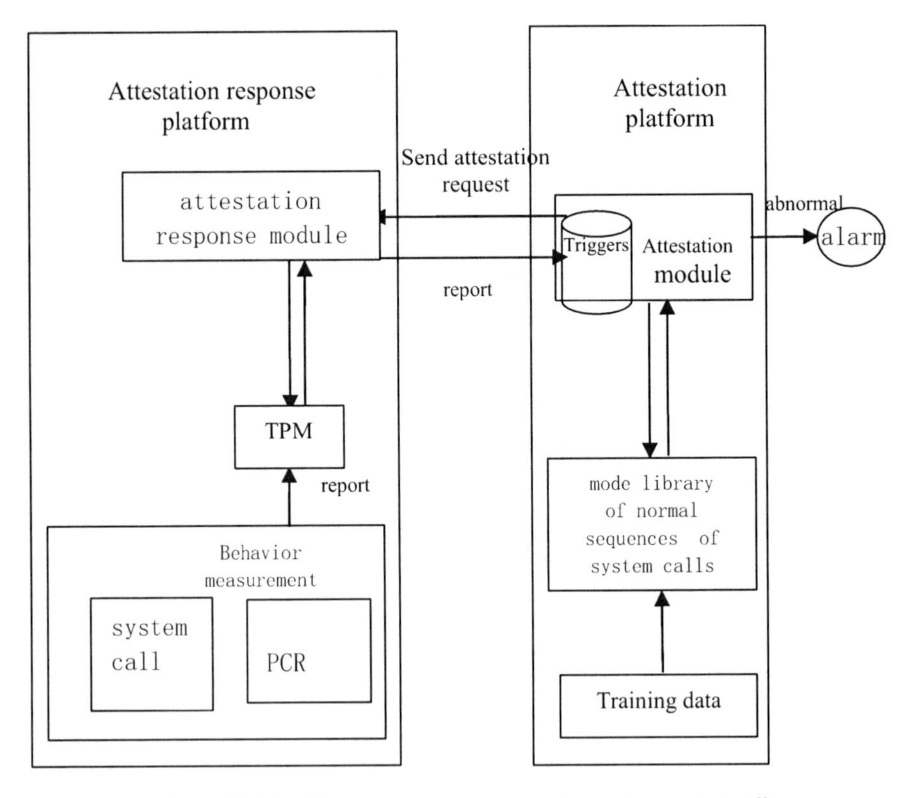

Fig. 2. Model of dynamic trust measurement based on system calls

4 Conclusion

This paper studies the existing problems in the attestation of trusted computing,and proposes the dynamic trusted computing attestation model. It can prove the software behavior running in the trusted computing platform, not static binary codes. Our solution to the problem is dynamic, real-time getting the sequences of system calls. If it can be proved that extracted characteristic value of system calls are consistent of existing normal sequences of system calls,the application running in the platform is guaranteed trust, So it is to achieve dynamic trust measurement in the entire platform. but how to get more accurate and effective normal sequences of system calls is the problems of the further research and debate.

Acknowledgments. This paper is supported by the Natural Science Foundation of Shanxi Province under Grant No. 2009011022-2 and Shanxi Scholarship Council of China Grant No. 2009-28.

References

1. Trusted Computing Group.TCG specification architecture overview (2007-08-02), `https://www.trustedcomputinggroup.org`
2. May, P., Ehrlich, H.-C., Steinke, T.: ZIB Structure Prediction Pipeline: Composing a Complex Biological Workflow Through Web Services. In: Nagel, W.E., Walter, W.V., Lehner, W. (eds.) Euro-Par 2006. LNCS, vol. 4128, pp. 1148–1158. Springer, Heidelberg (2006)
3. Foster, I., Kesselman, C.: The Grid: Blueprint for a New Computing Infrastructure. Morgan Kaufmann, San Francisco (1999)
4. Czajkowski, K., Fitzgerald, S., Foster, I., Kesselman, C.: Grid Information Services for Distributed Resource Sharing. In: 10th IEEE International Symposium on High Performance Distributed Computing, pp. 181–184. IEEE Press, New York (2001)
5. Foster, I., Kesselman, C., Nick, J., Tuecke, S.: The Physiology of the Grid: an Open Grid Services Architecture for Distributed Systems Integration. Technical report, Global Grid Forum (2002)
6. National Center for Biotechnology Information, `http://www.ncbi.nlm.nih.gov`

MSF-Based Automatic Emotional Computing for Speech Signal

Yuqiang Qin[1,2] and Xueying Zhang[1]

[1] Taiyuan University of Technology, Taiyuan, China
[2] Taiyuan University of Science and Technology, Taiyuan, China
aqinyuqiang@126.com, btyzhangxy@163.com

Abstract. In this paper, modulation spectral features (MSFs) are proposed for the automatic emotional recognition for speech signal. The features are extracted from an auditory-inspired long-term spectro-temporal(ST) representation. On an experiment assessing classification of 4 emotion categories, the MSFs show promising performance in comparison with features that are based on mel-frequency cepstral coefficients and perceptual linear prediction coefficients, two commonly used short-term spectral representations. The MSFs further express a substantial improvement in recognition performance when used to augment prosodic features, which have been extensively used for speech emotion recognition. Using both types of features, an overall recognition rate of 91.55 % is obtained for classifying 4 emotion categories.

Keywords: Speech emotional recognition, Modulation spectral features (MSFs), Spectro-temporal(ST), Automatic emotional computing, Speech emotion analysis.

1 Introduction

Speech emotional computing, an active interdisciplinary research field, is concerned with the automatic recognition, interpretation, and synthesis of human emotions. Within its areas of interest, speech emotion recognition (SER) aims at recognizing the underlying emotional state of a speaker from the speech signal. The paralinguistic information conveyed by speech emotions has been found to be useful in multiple ways in speech processing, especially serving as an important ingredient of “emotional intelligence” of machines and contributing to human–machine interaction [1][2].

In line with these findings, long-term modulation spectral features (MSFs) are proposed in this paper for emotion recognition. The proposed features are applied to two different SER tasks: (1) classification of discrete emotions (e.g. joy, neutral) under the categorical framework which characterizes speech emotions using categorical descriptors and (2) estimation of continuous emotions (e.g. valence, activation) under the dimensional framework which describes speech emotions as points in an emotion space [3].

2 Modulation Spectral Features for Emotional Speech

2.1 ST Representation of Emotional Speech

The auditory-inspired spectro-temporal (ST) representation of emotional speech is obtained via the steps depicted in Figure 1. The initial pre-processing module

G. Zhiguo et al. (Eds.): WISM 2011, CCIS 238, pp. 242–247, 2011.

resamples the emotional speech signal to 8 kHz and normalizes its active speech level to-26 dBov using the P.56 speech voltmeter. The preprocessed speech signal $s(n)$ is framed into long-term segments $s_k(n)$ by multiplying a 256 ms Hamming window with 64 ms frame shift, where k denotes the frame index[4][5].

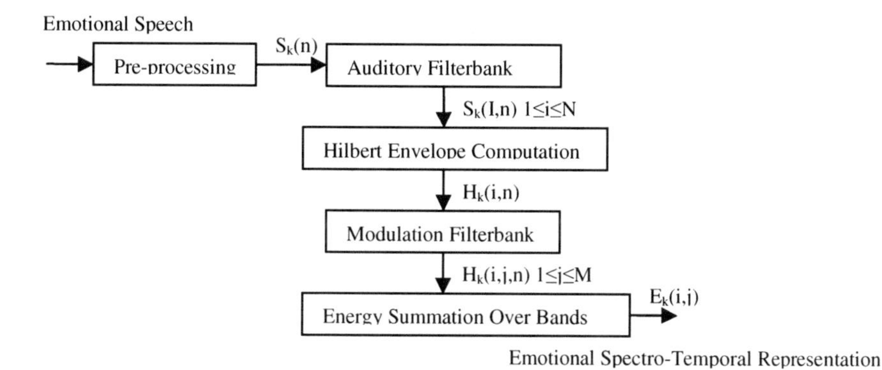

Fig. 1. Flowchart for deriving the ST representation

It is well-known that the human auditory system can be modeled as a series of over-lapping band-pass frequency channels, namely auditory filters with critical bandwidths that increase with filter center frequencies. The output signal of the i th critical-band filter at frame k is given by:

$$s_k(i,n) = s_k(n) * h(i,n) \tag{1}$$

where $h(i,n)$ denotes the impulse response of the i th channel, and * denotes convolution. The center frequencies of these filters (namely acoustic frequency, to distinguish from modulation frequency of the modulation filterbank) are proportional to their bandwidths, which in turn, are characterized by the equivalent rectangular bandwidth:

$$ERB_i = \frac{F_i}{Q_{ear}} + B_{\min} \tag{2}$$

where F_i is the center frequency (in Hz) of the i th criticalband filter, and Q_{ear} and $B_{\min}$ are constants set to 9.26449 and 24.7, respectively. The output of this early processing is further interpreted by the auditory cortex to extract spectro-temporal modulation patterns. An M -band modulation filterbank is employed in addition to the gammatone filterbank to model such functionality of the auditory cortex. By applying the modulation filterbank to each $H_k(i,n)$; M outputs $H_k(i,j,n)$ are generated where j denotes the j th modulation filter,

$1 \leq j \leq M$. The filters in the modulation filterbank are second-order bandpass with quality factor set to 2. In this work we use an $M = 5$ filterbank whose filter center frequencies are equally spaced on logarithm scale from 4 to 64 Hz. The filterbank was shown in preliminary experiments to strike a good balance between performance and model complexity[6] [7].

Lastly, the ST representation $E_k(i, j)$ of the K th frame is obtained by measuring the energy of $H_k(i, j, n)$, given by:

$$E_k(i, j) = \sum_{n=1}^{L} \left| H_k(i, j, n) \right|^2 \tag{3}$$

where $1 \leq k \leq T$ with L and T representing the number of samples in one frame and the total number of frames, respectively. For a fixed $j = j^*$, $E_k(i, j^*)$ relates the auditory spectral samples of modulation channel j^* after criticalband grouping. By incorporating the auditory filterbank and the modulation filterbank, a richer two-dimensional frequency representation is produced and allows for analysis of modulation frequency content across different acoustic frequency channels[8].

2.2 Modulation Spectral Features Extraction

Two types of MSFs are calculated from the ST representation, by means of spectral measures and linear prediction parameters. For each frame k, the ST representation $E_k(i, j)$ is scaled to unit energy before further computation, i.e. $\sum_{i,j} E_k(i, j) = 1$. Six spectral measures $\Phi_1 - \Phi_6$ are then calculated on a per-frame basis. For frame k , $\Phi_{1,k}(j)$ is defined as the mean of the energy samples belonging to the j th modulation channel ($1 \leq j \leq 5$):

$$\Phi_{1,k}(j) = \frac{\sum_{i=1}^{N} E_k(i, j)}{N} \tag{4}$$

Parameter Φ_1 characterizes the energy distribution of speech along the modulation frequency. The second spectral measure is the spectral flatness which is defined as the ratio of the geometric mean of a spectral energy measure to the arithmetic mean. In our calculation, $E_k(i, j)$ is used as the spectral energy measure at frame k for modulation band j and Φ_2 is thus defined as:

$$\Phi_{2,k}(j) = \frac{\sqrt[N]{\prod_{i=1}^{N} E_k(i, j)}}{\Phi_{1,k}(j)} \tag{5}$$

A spectral flatness value close to 1 indicates a flat spectrum, while a value close to 0 suggests a spectrum with widely different spectral amplitudes. The third measure employed is the spectral centroid which provides a measure of the "center of mass" of the spectrum in each modulation channel. Parameter Φ_3 for the j th modulation channel is computed as:

$$\Phi_{3,k}(j) = \frac{\sum_{i=1}^{N} f(i) E_k(i, j)}{\sum_{i=1}^{N} E_k(i, j)} \tag{6}$$

Two types of frequency measure $f(i)$ have been experimented: (1) $f(i)$ being the center frequency (in Hz) of the ith critical-band filter of the auditory filterbank and (2) $f(i)$ being the index of the i th criticalband filter, i.e., $f(i) = i$. In order to alleviate such information redundancy, $\Phi_{2,k}(j)$ and $\Phi_{3,k}(j)$ are only computed for $j \in \{1,3,5\}$ [9].

3 Emotional Corpus Database

An emotion database is specifically designed and set up for text-independent emotion classification studies. The database includes short utterances covering the 4 archetypal emotions, namely anger, joy, sadness and neutral. Non-professional speakers are selected to avoid exaggerated expression. A total of six native English language speakers (three males and three females), six native Mandarin language speakers (three males and three females) are employed to generate 720 utterances. Sixty different utterances, ten each for each emotional mode, are recorded for each speaker. The recording is done in a quiet environment using a mouthpiece microphone [10].

4 Experiment and Analysis of Results

4.1 ST -Based Emotional Speech MSF

Figure 2 shows the ST representation $E(i, j)$ for the 4 emotions in the speech corpus database, where every $E(i, j)$ shown is the average over all the frames and speakers available in the database for an emotion. As illustrated in the figure, the average ST energy distribution over the joint acoustic-modulation frequency plane is similar for some emotions (e.g. anger vs. joy), suggesting they could become confusion pairs, while very distinct for some others (e.g. anger vs. sadness), suggesting they could be well discriminated from each other. As reasonably expected, the less expressive emotions such as neutral and sadness have significantly more low acoustic frequency energy than anger and joy.

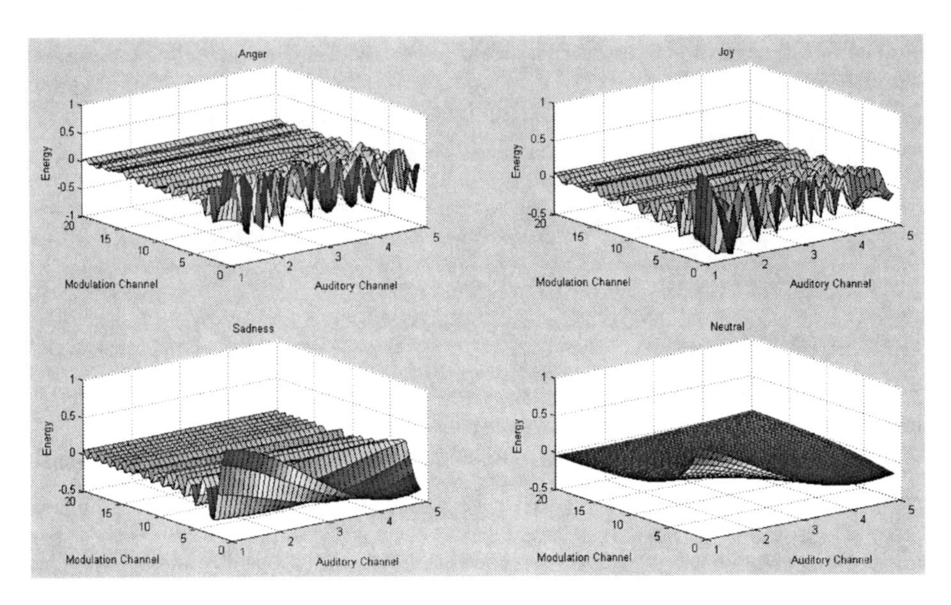

Fig. 2. Average E(i,j) for 4 emotion categories

4.2 Evaluation

Confusion matrix are shown in Tables 1 (left-most column being the true emotions), for the best recognition performance achieved by prosodic features alone and combined prosodic and proposed features (LDA + SN), respectively. The rate column lists per class recognition rates, and precision for a class is the number of samples correctly classified divided by the total number of samples classified to the class. We can see from the confusion matrix that adding MSFs contributes to improving the recognition and precision rates of all emotion categories. It is also shown that most emotions can be correctly recognized with above 89% accuracy, with the exception of joy, which forms the most notable confusion pair with anger, though they are of opposite valence in the activation–valence emotion space.

Table 1. MSF-base confusion matrix for automatic emotional recognition

Emotion	anger	joy	sadness	neutral	rate(%)
anger	119	8	0	0	93.7
joy	12	52	0	3	77.6
sadness	0	0	62	0	100
neutral	2	2	0	75	94.9
precision(%)	89.5	83.9	100	96.2	

5 Conclusion

This work presents novel MSFs for the recognition of human emotions in speech. An auditory-inspired ST representation is acquired by deploying an auditory filterbank as well as a modulation filterbank, to perform spectral decomposition in the conventional acoustic frequency domain and in the modulation frequency domain, respectively. The proposed features are then extracted from this ST representation by means of spectral measures and linear prediction parameters.

The MSFs are evaluated first on the database to emotional corpus database to classify 4 discrete emotions. Simulation results show that the MSFs serve as powerful additions to prosodic features, as substantial improvement in recognition accuracy is achieved once prosodic features are combined with the MSFs, with up to 91.6% overall recognition accuracy attained.

Acknowledgements. This work is sponsored by Research of Emotional Speech Recognition and HMM Synthesis Algorithm (China-Belgium cooperation) (Shanxi International Cooperation Foundation of Science and Technology of China, NO.2011081047), Research of HMM-Based Emotional Speech Synthesis Algorithm (Shanxi Natural Science Foundation of China, NO. 2010011020-1).

References

1. Ishi, C., Ishiguro, H., Hagita, N.: Analysis of the roles and the dynamics of breathy and whispery voice qualities in dialogue speech. EURASIP J. Audio Speech Music Process, article ID 528193, 12 pages (2010)
2. Falk, T.H., Chan, W.-Y.: A non-intrusive quality measure of dereverberated speech. In: Proc. Internat. Workshop for Acoustic Echo and Noise Control, pp. 155–189 (2008)
3. Falk, T.H., Chan, W.-Y.: Modulation spectral features for robust far-field speaker identification. IEEE Trans. Audio Speech Language Process. 18, 90–100 (2010)
4. Falk, T.H., Chan, W.-Y.: Temporal dynamics for blind measurement of room acoustical parameters. IEEE Trans. Instrum. Meas. 59, 978–989 (2010)
5. Giannakopoulos, T., Pikrakis, A., Theodoridis, S.: A dimensional approach to emotion recognition of speech from movies. In: Proc. Internat. Conf. on Acoustics, Speech and Signal Processing, pp. 65–68 (2009)
6. Wu, S., Falk, T., Chan, W.-Y.: Automatic recognition of speech emotion using long-term spectro-temporal features. In: Proc. Internat. Conf. on Digital Signal Processing, pp. 1–6 (2009)
7. Wollmer, M., Eyben, F., Reiter, S., Schuller, B., Cox, C., Douglas-Cowie, E., Cowie, R.: Abandoning emotion classes – Towards continuous emotion recognition with modelling of long-range dependencies. In: Proc. Interspeech, pp. 597–600 (2008)
8. Sun, R., Moore, E., Torres, J.: Investigating, glottal parameters for differentiating emotional categories with similar prosodics. In: Proc. Internat. Conf. on Acoustics, Speech and Signal Processing, pp. 4509–4512 (2009)
9. Lugger, M., Yang, B.: Cascaded emotion classification via psychological emotion dimensions using a large set of voice quality parameters. In: Proc. Internat. Conf. on Acoustics, Speech and Signal Processing, vol. 4, pp. 4945–4948 (2008)
10. Grimm, M., Kroschel, K., Narayanan, S.: The Vera am Mittag German audio-visual emotional speech database. In: Proc. Internat. Conf. on Multimedia & Expo., pp. 865–868 (2008)

Research on Social Trust of Internet Services

Anliang Ning[1], Xiaojing Li[1], Chunxian Wang[1], PingWang[2], and Pengfei Song[3]

[1] Computing Center, Tianjin Polytechnic University, 300160, Tianjin, China
[2] School of Electronic Engineering & Automation, Tianjin Polytechnic University
[3] School of Computer and Information Science, Southwest University, Chongqing, China
{yiran_ning,wchunxian}@163.com

Abstract. As the Web has shifted to an interactive environment where ever-growing set of applications is provided and much content is created by users, the question of whom to trust and what information to trust has become both more important and more difficult to answer. Social trust can be used to sort, filter, or aggregate information, or to visibly score content for the benefit of the users. Blending social computing with service-oriented computing leads to SWSs that "know" with whom they've worked in the past and with whom they would potentially like to work in the future. A social network-based strategy for Web service trust management intends clearly to reinforce Web services' performance capabilities through a fine-grained consideration of analysis and reasoning and consideration of "extra" information such as past experiences rather than just information related to Web services. For social trust comprise all sorts of immeasurable components, particularly challenging to working with be presented in the paper.

Keywords: Services Trust Management, Web Services, Social trust, social computing.

1 Introduction: Trust and Internet

The Web has dramatically changed the way the connected world operates. As the Web has shifted to an interactive environment where ever-growing set of applications is provided and much content is created by users, the question of whom to trust and what information to trust has become both more important and more difficult to answer. Unlike traditional computing environments that treat reliability, responsiveness, or accuracy as measures of trust, the fuzzier social trust is particularly well suited to these new Web settings. Computing with social trust—whether modeling it, inferring it, or integrating it into applications—is becoming an important area of research that holds great promise for improving the way people access information.

We frequently trust email systems in the sense that we believe we're receiving email from the asserted source. One reason we're comfortable doing this is that the context and content of the received messages can often establish a basis for trusting that the source is accurate. Email from strangers should be treated with more suspicion unless, perhaps, introduced by a trusted source. It's worth observing that trust doesn't always scale well. We can establish trust among a small group of people known to us, but it's harder to achieve trusting relationships on a larger scale.

G. Zhiguo et al. (Eds.): WISM 2011, CCIS 238, pp. 248–254, 2011.

Going back to the credit-card analogy, we're able to trust merchants we don't otherwise know because we know that the credit-card companies can withdraw payment for a fraudulent transaction. This is a clear example that risk mitigation enhances our willingness to trust an otherwise unknown party.

There isn't an argument that we should get off the Internet or stop using it. It's too convenient and increasingly too necessary. Instead, it's an argument for improving our security practices so as to mitigate risk and engender justifiable trust in the system. This preamble takes us to the question of trust in the use of the Internet. Its nearly 2 billion users probably trust the Internet's applications more than they should but don't engage in risk mitigation practices as much as they should. As with all things Internet, we must think through various means of compromising the system (such as "man-in-the-middle" attacks) so as to reduce the risk of trusting the authentication exchanges and subsequent transactions.

2 Web Users' Need for Social Trust

There are two important entities on the web: people and information. The numbers of both are in the billions. There are tens of billions of Web pages and billions of people with Web access. Increasingly, average Web users are responsible for content on the web. User-generated content is everywhere: discussion boards in online communities, social network profiles, auction listings and items for sale, video and photo feeds, reviews and ratings of movies, hotels, restaurants, plumbers, and more. While this content can be very helpful to users, the volume can be overwhelming.

With so much user-interaction and user-generated content, trust becomes a critical issue. When interacting with one another, users face privacy risks through personal information being revealed to the wrong people, financial risks by doing business with unreliable parties, and certainly inconvenience when spam or other unwanted content is sent their way. Similarly, information provided by users can be overwhelming because there is so much of it and because it is often contradictory. In some cases, it is simply frustrating for users to try to find useful movie reviews among thousands or to find a news report that matches the user's perspective. In other cases, such as when considering health information, finding insight from trusted sources is especially important. While users could formerly judge the trustworthiness of content based on how "professional" a website looked [1, 2], most of this new content is on third party sites, effectively eliminating visual clues as helpful factors.

There are, of course, plenty of cases where popular opinion or a central authority can help establish trust. A website with VeriSign credentials, for example, can be given some added degree of trust that it will protect the user's sensitive information. Peer-to-peer networks use measures of reliability for each peer as a measure of trust. However, these instances say more about the broad definition of "trust" than a solution to the problems described earlier. In the cases of user interaction on the Web or user-generated content, trust is often not an issue of security or reliability, but a matter of opinion and perspective. When looking at book reviews, for example, titles will often have a wide range of positive and negative reviews. Even dismissing the most misguided ones will leave a large set of valid opinions. Establishing trust, in this case, means finding people whose opinions are like the user's, that is, social trust.

In interactions with users or user-generated content, social trust can aide the user. Approximations of how much to trust unknown people can become a factor in interactions with them. In the case of user-generated content, trust in the users can serve as an estimate of how much to trust that information. Thus, content from more trusted people should be weighed more or considered more strongly than content from less trusted people. Trust can be used to sort, filter, or aggregate information, or to visibly score content for the benefit of the users [3].

3 Web Services Need Social Trust

Service-oriented architecture (SOA) and its flagship implementation technology known as Web services have changed the way software engineers design and develop today's enterprise applications. Web services help organizations maintain an operative presence on the Internet. Acting as building blocks that can provide and transform data, Web services connect together to create new on-demand value-added composite services. Services are treated as isolated components despite their previous interactions with other peers when complex services are built. Capturing service interactions using, for example, social networks could be beneficial for software engineers who can capitalize on the known successful interactions as needs arise [4].

3.1 Social Networks of Web Services

When enterprises discover and engage Web services for business needs, they're included in service compositions based on both the functionality they offer and the quality of service (QoS) they can guarantee, which implies the need for contracts. However, when consumers engage and compose services, it's much more informal and dynamic, much like how people download iPhone apps. But unlike iPhone apps, which are monolithic and operate independently of each other, Web services are intended to be composed, and their functionality and QoS are interdependent with other services. Moreover, they execute remotely and with some degree of autonomy. Their discovery and subsequent engagement thus become social activities, much like the collaboration and competition supported in social networks. [5,6]

Social networks exemplify the tremendous popularity of Web 2.0 applications, which help users become proactive; Prosumers, providers and consumers at the same time post definitions on wikis, establish groups of interest, and share tips and advice. These various operations illustrate the principles of "I offer services that somebody else might need" and "I require services that somebody else might offer" upon which SOA is built. [7] Service offerings and requests demonstrate perfectly how people behave in today's society, imposing a social dimension on how Web services must be handled in terms of description, discovery, binding, and composition.[8] What if this social dimension is the missing link? It could serve as an additional ingredient to the formal methods that support SOA needs, namely, service description, discovery, binding, and composition.

Weaving social elements into Web service operation means new social Web services (SWSs) that will [11]

• establish and maintain networks of contacts;
• put users either explicitly or implicitly in the heart of their life cycle, enabling additional functionalities through collaboration;
• rely on privileged contacts when needed;
• form with other peers strong and long-lasting collaborative groups; and
• know with whom to partner to minimalize ontology reconciliation.

SWSs is regarded as the result of blending social computing with service-oriented computing. On one hand, social computing is the computational facilitation of social studies and human social dynamics as well as the design and use of information and communication technologies that consider social context. [5] Social computing is also about collective actions, content sharing, and information dissemination in general. On the other hand, service-oriented computing builds applications on the principles of service offer and request, loose coupling, and cross-organization fow.[6]

Blending social computing with service-oriented computing leads to SWSs that "know" with whom they've worked in the past and with whom they would potentially like to work in the future. These two time-stamped elements constitute the "memory" of actions that SWSs can accumulate over time and apply in the future. In addition, they show the collective action of a group of SWSs that share respective experiences in response to requests for developing complex value-added composite services. SWSs are expected to take the initiative in advising users how to develop and reuse value-added services.[12]

3.2 Social Network-Based Strategy for Web Service Trust Management

A social-based connection between Web services offers a wider view by stressing the interactions that occur between users, between Web services, and between users and Web services. Table 1 highlights Web service management by comparing basic strategies to social-based strategies. The comparison criteria include user profiling, Web service description, Web service discovery, Web service composition, Web service advertisement, and trust between Web services and users.

A social network-based strategy for Web service management, as shown in this table, intends clearly to reinforce Web services' performance capabilities through a fine-grained consideration of analysis and reasoning [10] and consideration of "extra" information such as past experiences rather than just information related to Web services. A social network-based strategy offers better exposure, use, and follow-up of Web services compared to basic and community-based strategies.[13] As an example, at the composition level, social networks can include recommendations based on particular users' interests (as well as their immediate social relatives) instead of considering "general" and static behaviors of the services' composition. Another example of the social network-based strategy's strength is exemplified at the enterprise level by leveraging the diffusion property of a network for a better advertisement of Web services.[16]

Web services have progressed significantly from their inception for addressing business problems to their subsequent democratization to their anticipated socialization. Social networks, with their underlying principles and metrics, can offer innovative solutions to some of the issues Web services face today. The growing number of initiatives reflecting the blend of social computing with service-oriented computing is certainly a positive sign of this area's growing importance. [8,10,11]

Table 1. Web Service Trust Management Strategies

Comparative elements	Basic	Social Networks
		User Level
Profile	Built following regular use of Web service	Built following regular use of Web service and social relations that users maintain with others; relations are either explicit or implicit
Description	Developed by providers and then made available to all users	Made available subject to possible enrichment through annotations by members of the same social network, increasing the enriched description acceptance by the rest of this social network.
Discovery	Collective discovery after registry screening	Collective discovery after registry screening driven by the needs of each social network's members
Composition	Driven by individual users familiar with composition techniques and constraints.	Driven by the needs and previous experiences of each social network's members
Trust	Directly established between user and web service provider	Mainly related to the strength of the social relations that users have on top of their experiences of web service use.
		Enterprise Level
Advertising	By its provider	By users via their social contacts; better use of web services because of trust in these contacts.

4 Challenges to Social Trust of Internet Services

Each user knows only a tiny fraction of all the people they encounter online, either directly or through the information they produce. If we are to use trust as a way to evaluate people and information, we must be able to compute trust between people who do not know one another and translate the result into a form that is helpful to users. However, social trust is a fuzzy concept. [13]Unlike probabilities or trust derived as a measure of successful interactions between agents, social trust comprises all sorts of immeasurable components. Because of this imprecision, computing with social trust is particularly challenging. There are three broad computational challenges to working with trust: models of trust, propagation of trust, and applications of trust.

Modeling trust for use in computation is difficult, particularly when working in a Web context. Trust is a complex relationship based on a wide range of factors. Between two individuals, trust may be affected by their history of interactions, similarity in preferences, similarity in background and demographics, information from third parties about the reputation of one another, and each individual's separate life experiences that may impact their propensity to trust are only a few factors. Furthermore, trust depends on context; someone may trust a friend to recommend a restaurant, but not to perform heart surgery. And finally, trust changes over time for a variety of reasons. If that weren't enough, these factors are compounded on the web. Relationships online often have different features than relationships in real life; for example, people tend not to drop friendships in web-based social networks when those relationships are often long dead in reality [4]. The anonymous nature of the Web makes identity an important factor.

Propagation of trust: There are good models that respect these aspects of trust (e.g. [6]), but often, the information necessary to implement these models is missing. At the same time, web-based social networks offer a rich source of data that may be used. The benefits are clear: there are hundreds of social networks with over a billion user accounts among them [4]. Many of these networks have indications of trust between the users, either explicitly or implicitly. However, these networks generally provide only a simple trust rating between users; there is none of the information on history of interactions or other background data that a more complex trust model would require. Still, if trust can be propagated through the network, so two people who are not directly connected can be given estimates of how much to trust one another, and then this ample source of social data can be exploited to provide trust information for later use. Computing trust accurately, avoiding attacks by untrustworthy users, and understanding the differences between trust inference algorithms, all are important research problems that must be addressed if these values are to be useful.

Applications of trust: [17] Trust derived from simple or complex models becomes important when used in applications. As described earlier, there are many scenarios where an estimate of trust in a user can be important. Specifically, trust can help in judging the quality of user-generated content or determining whether or not to enter into a transaction with a person online. Exactly, how to apply trust estimates to these tasks is challenging. Depending on the context, trust should be indicated in different ways (as in feedback systems for online auctions) or hidden from the user all together (as in recommender systems). How to use estimates of social trust to build actual trust in people, content, and systems is also important.

Evaluation of Trust: Evaluation is very difficult when working with social trust, particularly in modeling, propagation, and other estimation methods. Because trust is sensitive, users do not publicly share these values. Nearly all social networks and other applications with trust ratings from their users keep this data hidden. On one hand, this is good for the users, because they are more likely to provide honest information when they know that it is private. From a research perspective, this is challenging because there are no open sources of data available to work with. For researchers who are fortunate enough to have access to their own social networks (e.g. the FilmTrust system [3] used in the chapter by Golbeck and Kuter, or the Advogato network [5] used in the chapter by Levien), they are still testing on only one network. Any analysis using a single network cannot make claims about robustness or scalability. In fact, this lack of data means that frequently research on trust models and inference algorithms is published with very little empirical analysis.

5 Conclusion

For work to progress and have significant impact in large Web environments, we must develop new and better ways of evaluating our work. A possible solution would be a trust network generator, which could generate simulated networks of varying size, structural dynamics, and with realistic trust properties. Another approach would be to gather as many social networks as possible with trust data, and generate anonymous versions for use in benchmark data sets. There are undoubtedly more solutions, and additional research in this work would be a great benefit to the community. More applications are also necessary to understand when and how trust is useful.

It is still an open question as to which applications see the best improvement for the user from these values and where trust is not as reliable as it needs to be to add significant value. This understanding can only come from further development and evaluation of a wide range of applications.

References

1. Maamar, Z., et al.: Using Social Networks for Web Services Discovery. IEEE Internet Computing (2011)
2. Wang, F.Y., et al.: Social Computing: From Social Informatics to Social Intelligence. IEEE Intelligent Systems 22(2), 79–83 (2007)
3. Maamar, Z., et al.: LinkedWS: A Novel Web Services Discovery Model Based on the Metaphor of Social Networks. Simulation Modeling Practice and Theory 19(10), 121–132 (2011)
4. Maaradji, A., et al.: Towards a Social Network-based Approach for Services Composition. In: Proc. 2010 IEEE Int'l. Conf. Communications (ICC 2010), pp. 1–5. IEEE Press, Los Alamitos (2010)
5. Nam Ko, M., et al.: Social-Networks Connect Services. Computer 43(8), 37–43 (2010)
6. Nguyen, N., et al.: Model for Trust Dynamics in Service Oriented Information Systems, in Computational Collective Intelligence. In: Semantic Web, Social Networks and Multi-agent Systems, pp. 573–583. Springer, Berlin (2009)
7. Kuter, U., Golbeck, J.: Semantic Web Service Composition in Social Environments. In: Bernstein, A., Karger, D.R., Heath, T., Feigenbaum, L., Maynard, D., Motta, E., Thirunarayan, K. (eds.) ISWC 2009. LNCS, vol. 5823, pp. 344–358. Springer, Heidelberg (2009)
8. Li, J., et al.: Denial of service attacks and defenses in decentralized trust management. International Journal of Information Security 8(2), 89–101 (2009)
9. Trcek, D.: Trust management for pervasive computing environments. Proceedings of World Academy of Science, Engineering and Technology 61, 185–190 (2010)
10. Rettinger, A., Nickles, M., Tresp, V.: Statistical relational learning of trust. Machine Learning 82(2), 191–209 (2010)
11. Tatham, P., Kovacs, G.: The application of "swift trust" to humanitarian logistics. International Journal of Production Economics 126(1), 35–45 (2010)
12. Yuan, W., et al.: Improved trust-aware recommender system using small-worldness of trust networks. Knowledge-Based Systems 23(3), 232–238 (2011)
13. Kovac, D., Trcek, D.: Qualitative trust modeling in SOA. Journal of Systems Architecture 55(4), 255–263 (2009)
14. Huang, J., Nicol, D.: A formal-semantics-based calculus of trust. IEEE Internet Computing 14(5), 38–46 (2011)
15. Liu, L., Shi, W.: Trust and reputation management. IEEE Internet Computing 14(5), 10–13 (2011)
16. Miller, K.W., Voas, J.: The metaphysics of software trust. IT Professional 11(2), 52–55 (2009)
17. Liu, B.: Efficient trust negotiation based on trust evaluations and adaptive policies. Journal of Computers 6(2), 240–245 (2011)

The Mobile Intelligent Network Platform for Pre-processing Based on Layered Architecture

Yong bin Li

Shanghai University of Electric Power,
Shanghai 200090, China
lybin4000@163.com

Abstract. Today there is a variety of pre-paid mobile service，particularly in relation to the properties of real-time billing extension. How to build the mobile intelligent network platform for pre-processing to resolve the problem between then External Platform and the Service Control Point is proposed, and some attempt to realize it has been made in this paper.

Keywords: Mobile intelligent network, External Platform, Service Control Point, Preprocessor Platform.

1 Introduction

Today with mobile communication technology's development and 3g communication technology's good performance in the market, the users' demand towards new mobile services are increasing. Thus, the competition among the three operators is becoming increasingly fierce. In this competition, rapid expansion of pre-paid subscribers plays a decisive role. The three operators have launched many prepaid services and value-added services to improve user experience. For the three operators, a flexible, fast and effective provision of new business on the internet is now an important factor to win the competition.

The diversity of applications brings a complex variety of business platforms. Though each provider develops their platforms in accordance with certain standards, the way business rules are set is always decided by the developers. According to conventional work pattern, when an external platform (referred to as EP) directly communicates with a service control point (referred to as SCP), every business change will arouse rules changes of SCP, thus hinder other work that SCP controls and further affect the service quality. As long as the business rules processing logic of SCP is not changed, users will not be able to deal with new services via self-help platform. This problem will not only result in artificial platform burden, but affect the promotion of new services.

An intelligent network platform for pre-processing is needed to facilitate self-help services and to better meet the operators' needs in market expansion. Such an intelligent network platform should be able to complete the analysis of EP data and send the processed data to SCP, as well as feedback to EP in terms of information that SCP returns.

G. Zhiguo et al. (Eds.): WISM 2011, CCIS 238, pp. 255–260, 2011.

2 The Design of IPP Communication

SMS platform is an operation system based on client / server model. Client-side and server-side exchange information through the Short Message Point-to-point Protocol (referred to as the SMPP) [1]. Currently the architecture is shown in Figure 1. In this architecture, user attributes from various external channels are packaged through EP at the first place, further, requests such as modification; setting and queries are made to the SCP[2]. Though this architecture requires communication in line with format prescribed, interact inflexibility may happen due to different EP firms and historical reasons of SCP.

In this paper, we consider to add an interlayer IPP to solve this problem.

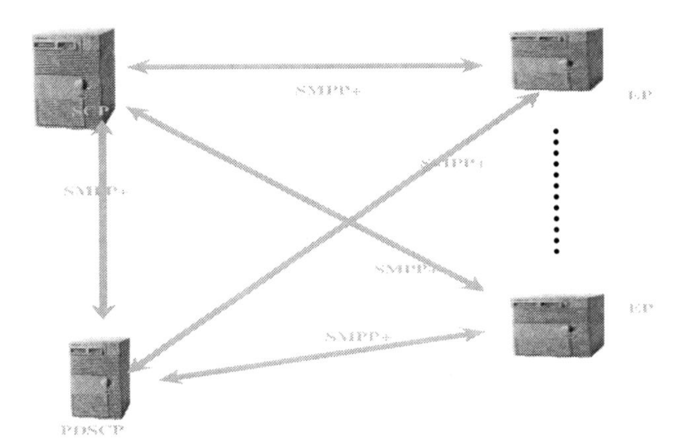

Fig. 1. The Architecture of SCP and EP

2.1 Function Structure

IPP system receives and sends messages on the basis of SMPP protocol and works mainly as a preprocessing unit between the EP and SCP. It receives bonus messages, which are later split into single bonus atoms, from EP in line with SMPP, and send the reply from SCP after having them processed.

As to partial activation, the system obtains records according to FTP protocol and recognizes the account to activate, and send the bonus from this account to SCP.

Each effective bonus is saved in the way of bonus bills and sent to a specified catalog under SCP. The IPP Scope of Work is as shown in Fig. 2.

According to its functions, IPP system can be divided into four functional modules:

SMPP Receiving & Sending Module. This module as an SMPP protocol encapsulation program is responsible for connecting and receiving connections, and translates strings that meet SMPP protocol into String of XML format, or string of XML format into SMPP strings that is consistent to SMPP protocol.

Data Translating Module. Data Translating Module is an socket server that intercepts specific ports all the time, which receives data from EP (SCP) and converts them into EP (SCP) ones.

System Configuration Module. This module configures original data the system involves and converting rules of EP and SCP.

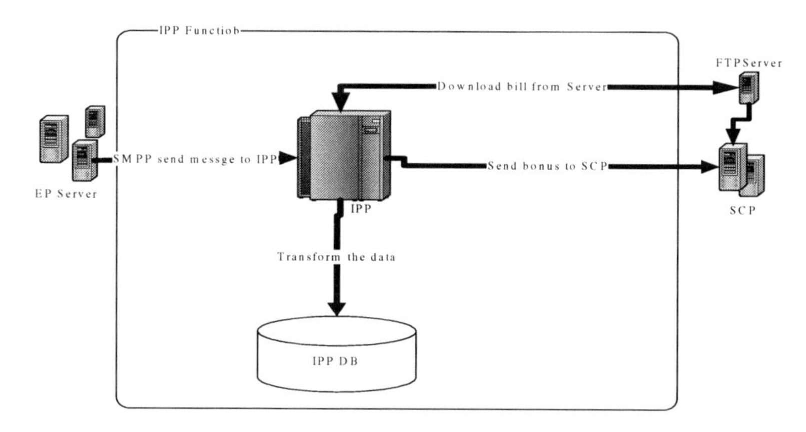

Fig. 2. IPP Scope of Work

Log Module. Log module records original data and those key operations received or sent.

Through these four modules, data conversion between EP and SCP is completed.

2.2 Software Architecture

Considering expansibility of the system and low coupling between the modules,the IPP communication mode is designed as shown in Figure 3. In this mode, software architecture is divided into three categories:

Business layer: This layer split and repackages specific business information (described in XML) from the network layer in accordance with certain rules and send the re-packaged information SCP or EP at a proper time. It works mainly as an adapter that executes defined logic operations to achieve specific business functions.

Network layer: This layer send packaged business information to SCP or EP in accordance with SMPP[2].

FTP layer: The major function of his layer is to check activated bills.

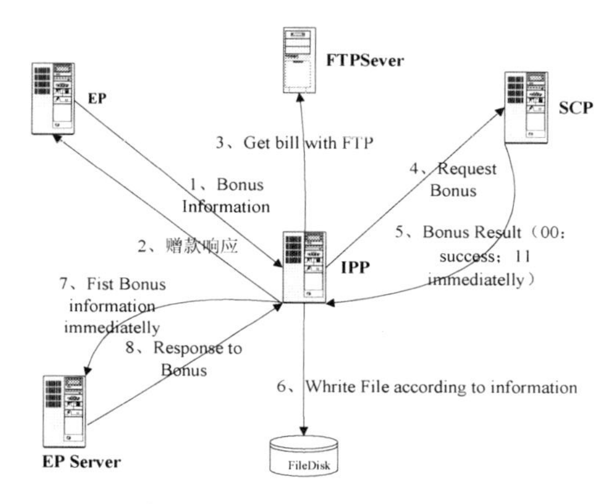

Fig. 3. The Communication of IPP

3 Implementation of Communication System

After the analysis above, the following part is going to describe from the key implementing points of network layer, FTP layer and business layer. Since the server platforms that operators apply are mostly UNIX (LINUX), the description will be described in JAVA.

3.1 Implementation of the Network Layer

The network layer establishes Socket communications ports and sets intercepting program to decide whether there is a connection or information transfer between EP and SCP. The communication applies to SMPP that runs on the basis of TCP protocol and connects with the SMS center through Socket. There is a basic progress in SMPP communication, in which each step represents a specific meaning called communication primitive.

The communication network layer is quite similar. Now we use EP connection and protocol analysis as examples:

ConnectionHandler

```
Public abstract class ConnectionHandler {
   Abstract void writeResponse(byte[] response) throws
IOException;
}// Abstract class, acts as the super-class[3] of other
connected classes.
```

EPConnectionHandler

```
Public class EpConnectionHandler extends ConnectionHandler
implements Runnable{
...}
```

This class, as client-side, establishes connections with EP, used in step 7 and step 8 in Fig 3. It succeeds ConnectionHandler and realizes runnable run function. Afterwards, the established connections are forwarded to EpProtocolHandler. [4].

EpProtocolHandler. EpProtocolHandler mainly analyses binary stream obtained in connections between client-side and Epsever, and then invokes related business layer to process.

3.2 FTP Layer Implementation

The main function of FTP layer is to connect with FTP server and to download activated bills.

FTPInterface. FTPInterface is mainly responsible for deleting downloaded files, as well as connecting with FTP and downloading.

```
Public class FtpInterface implements Runnable {}
```

FTP. FTP mainly picks needed information from downloaded files and save them into activebonus for later uses.

3.3 Implementation of the Business Layer

Business layer is set to make thread calls to business operating layers at regular time. It provides processXML call method, which invokes private doRequestTask or private doResponsetTask according to incoming information type; therefore invoke different business operating classes.

Since this platform is to complete the information transfer between EP and SCP, business layer can be blocked for the timeliness of operation. In this condition, each block runs by multiple threads and coordinate with each other in a event-driven or a time-triggered way in line with business need. Hence, the software architecture of this platform can be summarized as shown in Fig. 4. The overall progress is as followed:

First, the information obtained from the EP side (XML format) are saved directly to the EPBonus;

Messages from EP is analysed and split according to defined rules and then saved in the AtomBonus;

Since bonus are of different types, requests such as bonus be executed immediately can be realized through immediate operation of SCP processing thread, which enables the layer to package and send messages to SCP through network layer and to preserve them to the SCPBonus.

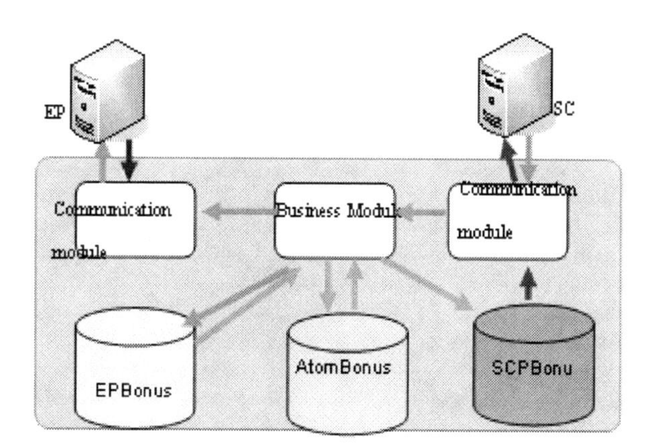

Fig. 4. The software Architecture of IPP

4 Conclusion

IPP platform plays as a buffer between business platforms and SCP platforms. It translates comparing complicated business requests into relatively simple requests that can be recognized and executed by SCP platforms.

Customers have put forward high demands to the configuration flexibility as well as response time. This paper gives a detailed description from the aspect of response time, while deep research about configuration flexibility is still needed.

References

1. ETSI GSM specification 03.40(4.9.1) European digital cellular telecommunications system: Point to Point Short Message Service technology (1999)
2. Guangdong Unicom Intelligent Network Interface Specification(1.2.1), China Unicom, China (2005)
3. Sierra, K., Bates, B.: Head First Java. Posts & Telecom Press, Beijing (2004)
4. Peng, B., Sun, Y.-l.: Java network programming examples. Tsinghua University Press, Beijing (2003)

The Design and Implementation of a Mobile Educational Administration System Based on WAP

Xue Bai, Shi Liu, and Jiaqi Jiang

College of Computer Science, Inner Mongolia University, Huhhot, China
kikibx_1@sohu.com, {ndzs2009,csjjq}@imu.edu.cn

Abstract. The Mobile Educational Administration System based on WAP is an educational information query platform, which is developed with network technology, mobile communication technology and information management technology, providing an immediate, convenient educational information access platform without any time and any regional restriction, for those who have a mobile phone, PDA or other mobile communication equipments. After a brief introducing WAP, this paper started from the framework of the system; subsequently, each functional module was introduced. Finally, the system has been tested on the WAP-based mobile phone, and the results indicates that the whole functions could be realized. The system has extended the function of educational administration system, which has provided users with convenient, anywhere, anytime, portable information services. It not only guaranteed the safety and efficiency of the whole system, but also expanded methods of information service, improved the administerial efficiency of whole system, and further strengthened the development of modernization and informationization on construction of educational administration.

Keywords: Mobile educational administration, Management information system, WAP, mobile terminal.

1 Introduction

Nowadays, the Educational Administration System based on Web services has a wide application in every school. With the demand of establishing digital campus, the Educational Administration System has played an important role in school management, performance management, program management, teacher management, order of teaching materials and tuition management. Educational Administration is the centre and foundation in college and it plays a dominant role in the allocation of resources for teaching management [1]. And its level of scientalization and modernization will influence the whole college's management standard and the quality of training talents.

However, the Educational Administration System based on Web services needs a support of internet and computer. As to the inquirement with few informations, such as testing scores, teachers' information, courses' information, academic notification, which will be complex by using a computer to realize our goal, especially where we lived has a restriction on internet. Thus, with the widely used of mobile phone, PDA or any other mobile equipment, we focus on developing a Mobile Educational

G. Zhiguo et al. (Eds.): WISM 2011, CCIS 238, pp. 261–268, 2011.

Administration System that can operate on the mobile phone, which provide students with the convenience to inquire information anywhere; on the other hand, it can also reduce the tedious daily affair of the teaching staff.

The Mobile Educational Administration System as an important supplement and expansion of the present one has its own unparalleled advantages and features [2], mainly reflect on the following aspects:

(1) Access flexibility: the used equipment of Mobile Educational Administration System broke the shackles of a wired connection; they can inquire information through the GSM mobile network no matter when and where, only if users had a WAP phone.

(2) Easy to find: user can find information intuitively via WAP, thus they can eliminate the trouble of need to connect to PC based on wired networks.

(3) Simple to operate: the Mobile Educational Administration System supports an open platform. Just having a cell phone or other mobile devices, parents can be very easy to master the learning of children in school.

In a word, the Mobile Educational Administration System has provided users with a convenient way. It not only guarantees the safety and efficiency of the whole system, but also expands methods of information services, and improved the administerial efficiency of whole system.

2 Introduction of WAP Technology

WAP is the abbreviation of Wireless Application Protocol and it is a global standard for all wireless systems. It provides data-oriented (non-voice) services to the mass market and is capable of benefiting—any-where and at any time—far more end-users than the personal computer [3]. The wireless application protocol provides end-users with new services in a wide range of applications, such as information retrieval on the Internet, mobile electronic commerce, and telephony applications [4].

As shown in figure 1, the network structure of WAP consists of three parts: the mobile terminal, the WAP gateway, the WAP content server. The content server of WAP is the widely used Web server, which adopted the construction of servlet technology(such as Java Server pages, PHP), and the language used to describe information is WML. Mobile terminal is the tool for users to access to a wap site, for example, mobile phone, digital secretary, etc. Wap gateway is a link bridge between mobile communications and internet, and it completes the protocol conversion function on the one hand, on the other hand completes transferring contents' encoding and decoding.

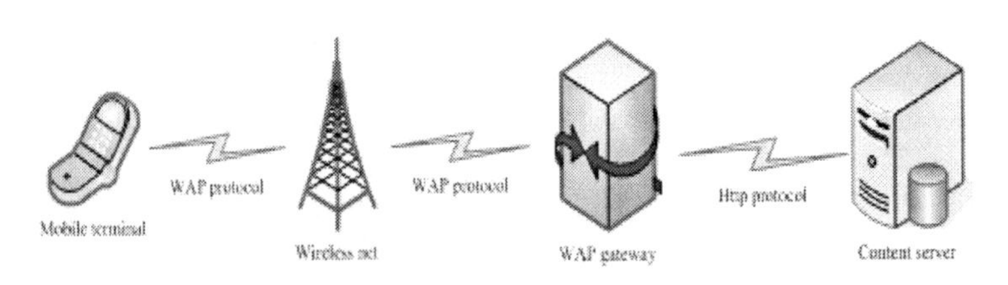

Fig. 1. The network structure of WAP

It contains four steps when users visit the WAP content server through mobile terminal:

(1) Mobile terminal establishes a connection to the WAP gateway and sends a request of obtaining information to the gateway;
(2) WAP gateway receives the request, then be sent to the content server of Internet after a series of treatment;
(3) The content server returns user request's information back to the WAP gateway when it receives a request message;
(4) The WAP gateway returns the information back to users after appropriately handling.

3 Mobile Educational Administration System Architecture and Functional Design

Software system architecture mainly contains of two models: C/S (Client /Server) and B/S (Browser/Server). B/S mode only is required to fix maintenance of the server, and the client uses web browser software. User sent a request to the server through the browser Distributed on the network, after the server processed the request, then it returned the needed information to the browser [5].

Mobile educational administration system mainly consists of WAP browser and the Web Administration server. According to actual needs combined with characteristics of those two models, we use the B/S architecture to design the WAP browser and use the MVC pattern to divide the application program into three core modules: model, view and controller. On the J2EE platform, we use a WAP enabled mobile phone as the browser terminal, and separate presentation layer, business logic layer and data layer based on the MVC design pattern which is called a three-tier architecture.

The Mobile Educational system consists of user terminals (mobile phones), WAP server (i.e. Mobile Information Server), application server, WAP gateway, the administrator (academic staff), database server, and Ethernet etc. System architecture shown in Figure 2:

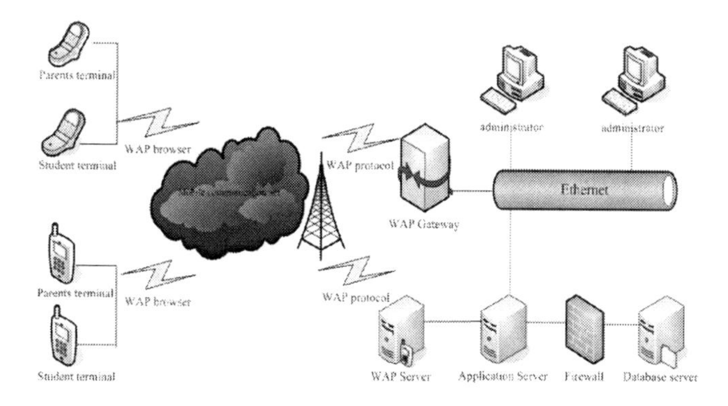

Fig. 2. System architecture

When users inquire information through Internet-enabled mobile devices (cell phone, PDA, notebook computers, etc.), the user firstly inputs a URL access to the WAP server from a WAP phone after the signal transmitting through the wireless network, and the request is sent as WAP agreement to the WAP gateway. Then, after a "translation" of URL specified by the WAP Gateway, it interacts with the WAP server in HTTP protocol mode. User requests to connect to the WAP server and the application server through the Internet. When the server accepts the request, it sends the request to the database server for data querying, and then the database server sends the query results to the mobile WAP server. Finally, the WAP gateway verifies the head of HTTP and WML content and returns the contents that has been compressed and processed into a binary stream which can be returned to the customer's WAP phone screen.

Figure 3 is a system flow chart; the system design of each module is completed in accordance with flow chart.

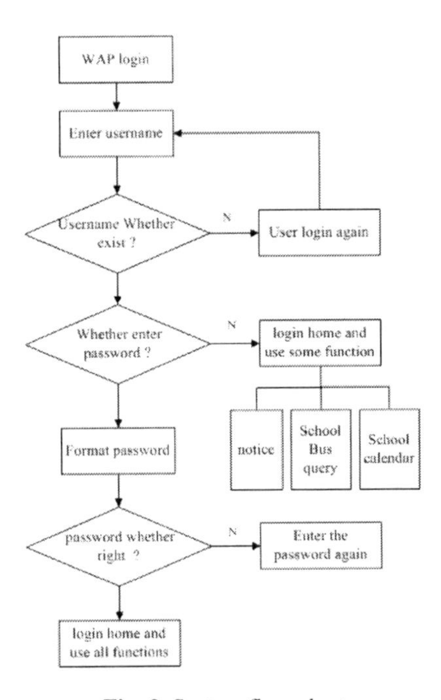

Fig. 3. System flow chart

WAP gateway as an important link in the WAP network is a primary part to contact the wireless network and the Internet. And its main function is to realize protocol stack conversion between the WSP / WTP / WTLS / WDP and HTTP / SSL / TCP / IP, so that the wireless Users can easily have access to WAP content on the Internet through operating a mobile terminal [3]. The construction of WAP gateway is generally built by operators (such as China Mobile, China Unicom, or some large ISP and other related enterprises) [6]. Because the design of system and achievement are supported by Inner Mongolia Branch of China Mobile, which combined with its own characteristics of Inner Mongolia university Educational administration System, the construction of WAP site use WAP gateway and public gateway, choosing China Mobile's WAP gateway as an open Public gateway.

The system is based on the existing educational system of Inner Mongolia University, using mobile communication technology, network technology and WAP technology, and according to the actual teaching management with users' needs, to design implement Mobile Educational Administration System using mobile communication devices as the user carrier. If students or parents would like to know the latest academic developments, or some personal information related to, they just need to hands in the WAP mobile phone, and log on to the school's Mobile Educational website, then you can easily and immediately to query information such as personal achievement, curriculum, examination arrangements, the latest Announcement, and school bus trips etc. System administrator can publish the latest announcement, and set the query methods through management interface.

The platform is divided into two parts: Web Admin and WAP content display. The WAP functions are shown in Figure 3:

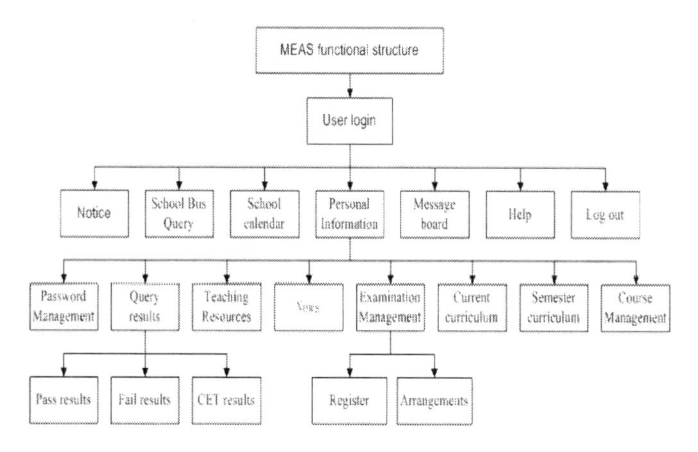

Fig. 4. System function architecture

The Mobile Educational Administration System includes the following modules: user login module, news module, school bus arrangements, personal information module, message board etc. In order to protect the privacy of student, the system needs a login account and key. Initial user's name and password is students' numbers and birth dates. After successful login, we can query information through the appropriate interface. Personal information module achieves main functions including: password management, results information, teaching resources, testing management, notice curriculum information, course management etc.

4 The Mobile Educational Administration System Implementation

The Information server system uses IIS6.0 which is able to make application process request to rout directly to the work process. It has built-in response, the request buffer and queue functions; thereby it improved the reliability and performance. MEAS is based on SQL Server 2000 as the database server, and its interface is developed using Visual Studio.Net [7]. SQL Server 2000 preferred software development has powerful database management capabilities and complete Web functionality, and closely linked with the internet.

This system is an expansion on the basis of educational system based on Web existed in Inner Mongolia University. It shares the same database with the existing Web-based Education Management Information System. By the analysis of the database entities and its relationships of WAP-based mobile Educational administration publishing platform and according to the relevant technical specifications of database and the SQL language, we designed these detailed database tables, including user information, student information form, Course information sheet, teacher information sheets, class information sheets, press release information form, test information form, performance information form, message information table and other database tables. We use ADO.NET technology to realize Database interface. And the database interface module is built mainly by the following steps to achieve:

1. The establishment of Web Services. First create a Web Services project, adding the methods to achieve system functionality in the project code files. The methods are different with other types of projects' methods because of its name preceded by the [Web Method]. And then DLL files after compiling are provided for external user-end to visit.

2. Deployment of XML Web Services. To deploy the XML Web Services, firstly is to create a virtual directory on Web server and put the XML Web Services .asmx in this directory. XML Web services discovery is the process of positioning intrusions of XML Web Services and querying them. This is a preliminary step to access the XML Web Services.

3. Set XML Web Services agency. If we want to visit Web Services from the client, we must firstly create a proxy class for the Web Services, or use the visual method to add "web references" provided by .NET. Defaultly, the proxy class uses SOAP to communicate with XML Web Services over the HTTP.

4. Client calls the Web Services. When creating a Web Services proxy, we can import the proxy to the client application, and then call the access interface provided by Web Services.

Part of the implementation code as follows:

```
Using System;
Using System_data;
Using wapjwServices;
Public class ListcourseINFO
{ Public static void Main ()
// create proxy object
wapjwServices course= new wapjwServices ():
// Call the Web method that returns a data set based on XML
```

```
Dataset Dscourse= course.GetAllcourseINFOBySPeciality();
```

// The following is the operation to return data sets

......}

Returned data set is a standard XML document, the following is part of course information XML documents.

```
<?XML version=''1.0'' encoding=''gb2312''>
<Listcourselnfo
xmlns='''http://localhost/courseinfo.xsd''>
<Course_info>
<courseid>140021</course_id>
<course_name>JAVA programming</course_name>
</graduate>
</course_info>
```

// Definition of the timer, and once every minute regulations

```
public void contextInitialized(ServletContextEvent
event) {
context = event.getServletContext();
         timer = new java.util.Timer(true);
         timer.schedule(new
MyTask(event.getServletContext()),0, 1 * 60 * 1000);
event.getServletContext().log("Task scheduling tables
have been added ");}
```

System test is according to “the Mobile Educational System’s design and achievement” of Inner Mongolia Nature and Science Fund project and“211 innovative talents”project. Research of the Educational Management Information System based on the mobile terminal” the system was tested. System uses a "black box" to test each functional module by entering the test data to detect whether each feature is normally available, and to review the module's source code. Figure 5 shows a screenshot of query results.

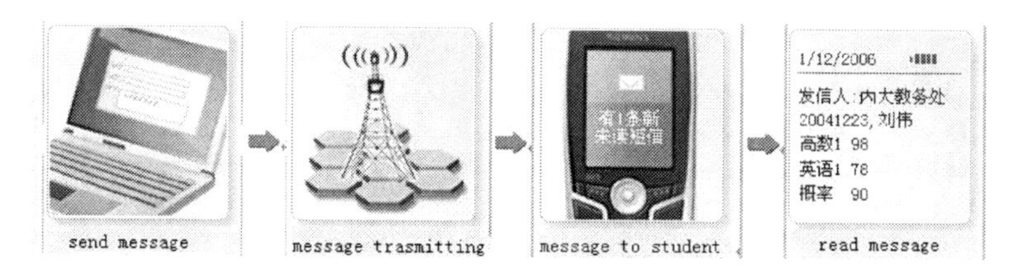

Fig. 5. Results show queried from mobile educational system

5 Conclusions

Mobile phones as educational management system terminal equipment are portable, strong movement without time and geographical constraints. In this paper, an implementation method of mobile educational administration systems has been proposed, which is an effective complement to the original Web Educational Administration System. On the basis of the WAP technology, querying information through mobile communication network really greatly facilitates the teachers and students. In addition, the system also has good scalability and versatility, further more models based on this system can also apply to mobile office, mobile commerce and other fields.

Acknowledgment. The authors thank referees of this paper for their constructive comments. This work was supported by the Inner Mongolia University 211 innovative talents project " Research of the Educational Management Information System based on the mobile terminal ", No. 2-1.2.1-046 and Inner Mongolia Nature and Science Fund "Research on Metadata Plan of Education Resources Based on Ontology", No. 2010MS1003.

References

1. hua, X., Jiang, D., Guo, D.: The innovation and development of Educational Administration technical support. Researches in Higher Education of Engineering (6), 45 (2005)
2. Ding, X., Feng, J.: The Research and Implementation of an Educational Administration System Based on the Mobile Terminal. master degree paper of SiChuan Normal University (2008)
3. WAP2.0 Technical White Paper, `http://www.wapforum.org`. WAP Forum
4. Yu, Q., Song, M.-n., Song, J.-d., Zhan, X.-s.: Research and design of wap service system based on MISC. J. the Journal of China Universities of Posts and Telecommunications 13(4), 34–38 (2006)
5. Kim, S.H., Mims, C., Holmes, K.P.: An Introduction to Current Trends and Benefits of Mobile Wireless Technology Use in Higher Education. J. Association for the Advancement of Computing In Education Journal 14(1), 77–100 (2006)
6. Wang, X., Zhang, J.: Detailed WAP technology and examples of the construction site M, pp. 1–295. Tsinghua University Press, Beijing (2001)
7. Microsoft Corporation. MSDN Library Visual Studio6.0,1998.J., `http://www.microsoft.com/support` (2006-05-12)

The Framework of Nuclear Management Information System*

Gangquan Cai, Shaomin Chen, Jianxiang Zheng, and Huifang Miao

Digital Instrument Control Center
School of Energy Research, Xiamen University
Xiamen, China
{cgq,shaominchen,zwu,hfmiao}@xmu.edu.cn

Abstract. As known to all, building nuclear power plants has the characteristics of large investment, long cycle, technical difficulties, too many interfaces and complex management and coordination, so the application of computer can help improving project management, improving work efficiency and quality, as well as cost reduction, wealthy information accumulation and enhancing the nuclear power enterprise market competitiveness. Nuclear power Management Information System (MIS) is a typical representative of this application. In this paper, taking the demand for document management, information search and real-time communication into consideration, we propose a solution and a framwork for development by using of SharePoint, RMS, Autonomy, exchange, Office Communication Server (OCS).

Keywords: Nuclear Power, Management Information System, Sharepoint, Autonomy.

1 Introduction

It is in great need to have a Management Information System to meet the nuclear power project management features for increasing nuclear power plant management level. The system can achieve demand management, engineering management, document management, records management, regulations management, efficiency management, knowledge management, and so on. Pre-construction experience can be accumulated in order for the latter part of the construction, be ensure quality and shorten the construction period by effective management, especially by knowledge management [1].

The International Atomic Energy Agency (IAEA) pointed out that the nuclear power Management Information System must have the following three main purposes:

- Integration tools. The system should provide a convenient, unified and comprehensive way to access its own resources.
- As other resources (internal and external) access tools. There are many useful information resources from outside those should be used by nuclear power staff.

* The project was supported by Science and Technology Planning Project of Fujian Province, China (Grant No. 2007H2002) and the Fundamental Research Funds for the Central Universities (Grant No. 2010121076).

G. Zhiguo et al. (Eds.): WISM 2011, CCIS 238, pp. 269–274, 2011.

- Communication tools. One key of Knowledge management is ensure that information can be shared, discussed and learned. A valid knowledge management should provide real-time formal and informal communication expediently for individual, the project team and various kinds of relevant personnel.

As mentioned above, we divide the function of nuclear power Management Information System into four core parts based on nuclear power users demand: document management, information search, real-time communication and rights management. Document management is implemented by using Sharepoint and redeveloping its GUI. Sharepoint is inlet and the core part of the Management Information System. By redeveloping Autonomy, the nuclear power users can search informations from internal or external. Exchange and OCS are used for building communication environment. In this environment everyone can communicate each other conveniently. And those above are controlled by rights management server which runs on Windows Server Active Directory[2].

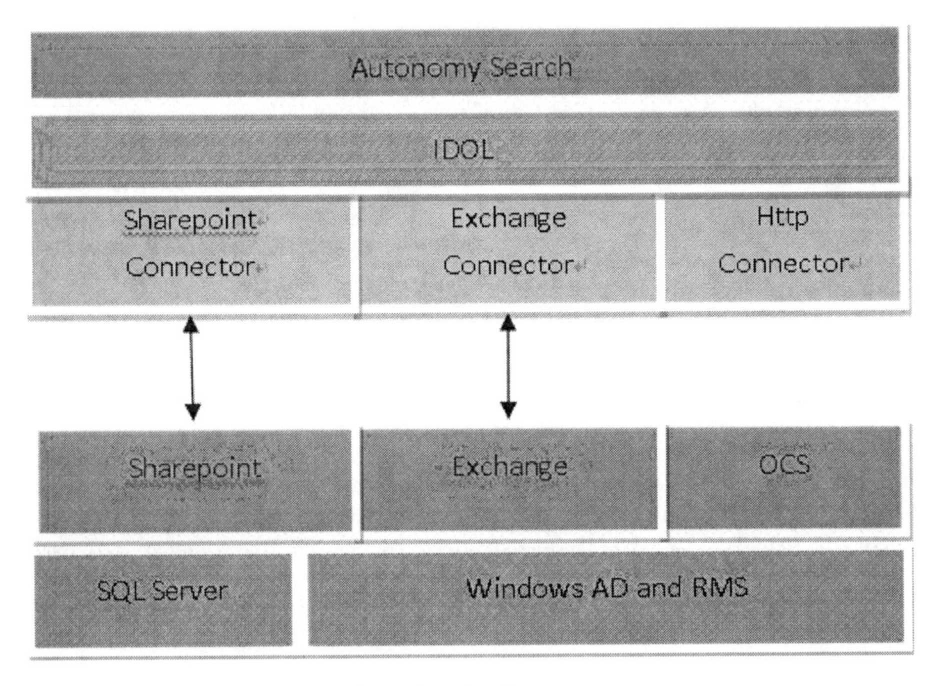

Fig. 1. Functional structure

2 File Management

2.1 Sharepoint and Its File Management

Office SharePoint Server 2007 is a new server program that is part of the 2007 Microsoft Office system. An organization can use Office SharePoint Server 2007 to facilitate collaboration, provide content management features, implement business processes, and supply access to information that is essential to organizational goals and processes[3].

In the framework of SharePoint, the content of document management includes documentation, workflow, and list, website, links, etc. The changing of documentation version state and workflow will be notified to related personnel through the ExChange, OCS, SMS, etc.

Directory management is based on the document libraries. All documents are classified as department or team, everyone has a specific rights to access files according to his appropriate permissions, which can be ensure the files security. All documents must be protected by rights, whatever they are on SharePoint or are on the hard disk download from Sharepoint. File modified version should be recorded, and users can execute restore, cancellation, delete operation. SharePoint also allows files to be modified by many people online simultaneously, that is working together[4].

2.2 Rights Manage

Sharepoint (Office) has its rights management calls IRM(Information Rights Manage), but IRM just controls all files on Sharepoint, and rights of files which out of the Sharepoint would be lost control. As a result, IRM should work with RMS together. What is the RMS? RMS is an abbreviation of Rights Manage Server, which is running on the windows server. We will elaborate about RMS subsequent chapters.

2.3 Communication

The nuclear power users can communicate each other by exchange mail or ocs. Exchange and ocs are Microsoft products.

Microsoft Exchange Server has a core function of e-mail, calendar and contacts system. The system runs on Windows Server 2008. From the perspective of the basic functions, Exchange Server provides the server-based message storage mechanism in a centralized way for handling e-mail messages, calendar appointments, contacts, and other users basic information. Users in the organization can connect to the Exchange server to access their e-mail and other information through Microsoft Outlook, Web browser, or other client systems.

Office Communications Server 2007 manages all real-time (synchronous) communications including: instant messaging, VoIP, audio and video conferencing. It works with existing tele-communications systems, so business can deploy advanced VoIP and conferencing without tearing out their legacy phone networks. Microsoft Office Communications Server 2007 also powers presence, a key benefit of Microsoft unified communications that unites all the contact information stored in Active Directory with the ways people communicate.

3 Nuclear Power Search

Autonomy is a recognized leader in the rapid development field of Meaning Based Computing (MBC). Autonomy is based on a unique combination of technologies by the Cambridge University, which was established in 1996.

A plethora of information exists throughout an enterprise, created and stored in different formats across multiple technologies, and exponentially growing by the moment. This information is the lifeblood of any organization as it is used to communicate to customers, power mission-critical operations, and drive strategic business decisions. As this information continues to grow, organizations require technology that is able to unify information scattered across an organization, make sense of the data, and instantly connect people with the relevant information that is required to drive business success[5].

Autonomy's Intelligent Data Operating Layer(IDOL) is a powerful infrastructure technology that surpasses keyword-matching to not only retrieve, but understand the meaning of all information that exists throughout an organization. Built on adaptive pattern recognition technology and probabilistic modeling, IDOL forms a conceptual and contextual understanding of digital content, and extracts meaning from the manner in which people interact with that content. Supported by the ability to recognize over 1,000 different file formats and securely connect to over 400 repositories, the technology provides advanced and accurate retrieval of the valuable knowledge and business intelligence that already exists within the organization. From a single platform, companies can access and process any piece data in any form, including unstructured data such as text, email, web, voice, or video files, regardless of its location or language.

IDOL can be considered as a database of Autonomy, which accepts data from the Connector and accepts the user's queries. When users search for data using the Query command, IDOL will produce the results in the form of XML and return it to the user[6].

There are several connectors Autonomy provides, we use HTTP Connector, Exchange Connector and Sharepoint Connector for this MIS. HTTP Connector is used for crawling external Web sites and sending these pages to IDOL. Sharepoint Connector is used for crawling Sharepoint informations which users can search for. Exchange Connector is used for capturing the contents of mails within the MIS. In addition to Http Connector, connectors are working with Windows AD (Active Directory), it is meaning that users can only search for messages which his permission allows[7].

Developers can use C# language to call APIs that Autonomy provides for completing a series of functions. The search function is integrated into the Sharepoint GUI, users can easily search the information and content that they want when they are using MIS[8].

3.1 Rights Management

Microsoft Windows Rights Management Services (referred to as RMS) is an information protection technology, it enables to help safeguard digital information from unauthorized use with the application which enable RMS - both online and offline, inside the firewall and outside.

All of documents from Management Information System are protected by sharepoint IRM, whatever they are created in the Sharepoint, or are uploaded from the local hard disk. When a user opens a document from Sharepoint, the document's application will combine with the IRM, sending certification to the RMS. if the authentication is valid, the application will open the document with the user's permission level. If a user is read-only access to a document library, when the user opens any documents under this library, the document will only be viewed, but not edited, copied and printed, let alone deleted. Supposing a user saves a document to a public computer, document is still protected by RMS, even if the document leaked out of the organization, but don't worry about exposing the core secrets.

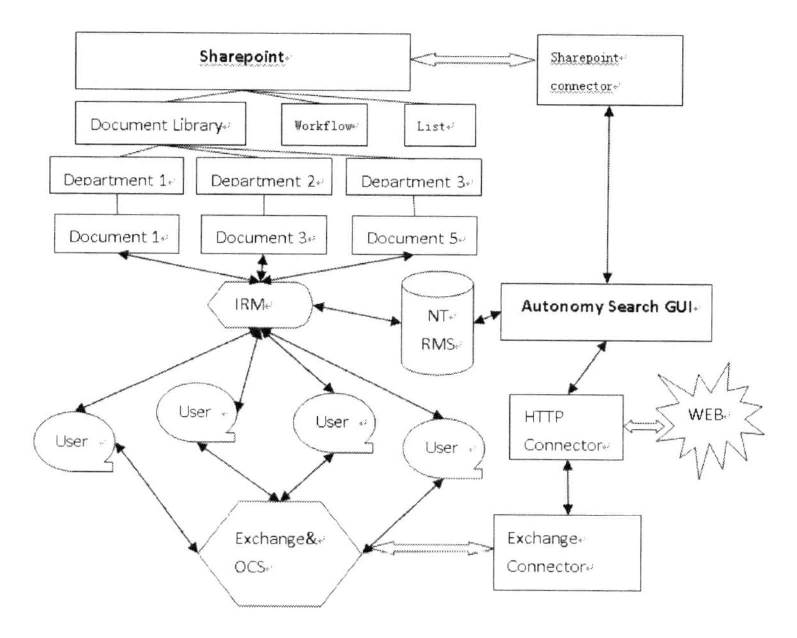

Fig. 2. Functional structure

4 Conclusion

We have completed the development of nuclear power Management Information System which meets the three requirements IAEA proposed, with the framwork of Sharepoint, RMS, Exchange (OCS) and Autonomy, and the MIS is actually put into use. There are too many document generated by nuclear power station in the daily management, which are well managed and controlled by MIS. In this Management Information System, Sharepoint redevelopment is the point, and the search is the characteristics. As long as a nuclear power enterprise proposes detailed requirements about Management Information System, this framework will be very useful to implement those requirements.

Acknowledgment. Thanks school of energy research of Xiamen university, thanks all staffs of digital instrument control center, especially thanks for prof.Ding, who is the director of the center. He is an expert in the field of nuclear power, under his leadership, the center is growing gradually, and this paper is accomplished so easily owing to a solid foundation of the environment and equipment that center provides.

References

1. Zou, L.-l., Zhang, P., Xiaozi, -y., Chun, Z.-j., Huang, F.-t.: Nuclear Power Project Management Information System. Nuclear Power Engineering 22(2), 184–191 (2001)
2. Chen, h.: Establishment and Development of Information System for Qinshan Phase III. Chinese Journal Of Nuclear Science And Engineering 21(2), 108–112 (2001)
3. Murphy, A., Perran, S.: Beginning Sharepoint 2007:Building Team Solutions with MOSS 2007. Wiley Publishing, Inc., Chichester (2007)
4. Tan, Z.: Implement of Dissertation Database System Based on Sharepoint File_Management. New Technology Of Library And Information Service (z1), 120–124 (2008)
5. Autonomy Technology Overview, Autonomy (2007)
6. HTTP Connector Administration Guide, Autonomy (2009)
7. Exchange Connector Administration Guide, Autonomy (2009)
8. Sharepoint Connector Administration Guide, Autonomy (2009)

Research on Online Public Opinions Analysis Based on Ontology for Network Community

Xin Jin and Wei Zhao

BeiYuan Road 178#9-4-718, ChaoYang District, BeiJing 100101, China
James.jin2009@gmail.com

Abstract. According to the requirement of online public opinion analysis, this paper builds an online public opinions extraction and analysis mechanism based on ontology. It builds public opinions' topics classification index system based ontology with which we can discuss and analyze how and what the netizen want to do. After analyzing the classification of the opinions, we can focus more on major opinions and response immediately. In this paper, we discuss how to building an ontology to classify the different opinions from lots of posters in network community, and we analyze the opinions about Yushu Earthquake information as an sample to discuss the classification result of public opinions further.

Keywords: Ontology, Online Public Opinion, Classification.

1 Introduction

With the rapid growth of Internet and World Wide Web(WWW), we have entered into a new information age so far. The Web has significant impacts on academic research, business, and ordinary everyday life. It revolutionizes the way in which information is gathered, stored, processed and used. The Web offers new opportunities and challenges for many areas, such as business, commerce, marketing, finance, publishing, education, research and development.

Since the Web is popular and easy to access, more and more people love to talk, publish/post, communicate their opinions each other through the network. The Network Community is one of important communication network platform, such as Tianya.cn, Sohu BBS. There are a lot of information of public opinions filled in the network community. Those public opinions should be focused, because they usually express the opinions or advices of some people[1-3].

Most people focus on significant or outbreak public events more, such as Yushu earthquake happened in QingHai Province last year. There were lots of public opinions posted online by netizen as soon as some big public event happened, those opinions maybe impacted the society. The government or related departments should focus and respond the event immediately to keep some problems clear out of normal schedule.

Due to the fullness of the online public opinions, it is important hard to capture and analyze the meanings of opinions information. In this paper, we present a mechanism based ontology to capture and analyze the opinions from major network community such as Tianyan.cn..

G. Zhiguo et al. (Eds.): WISM 2011, CCIS 238, pp. 275–280, 2011.

2 System Architecture

In Fig 1, it shows the system architecture. The site Crawler crawls the information of online public opinions from network community, then stores them into XML documents. The Wrapper tool extract the useful information from those XML document and detect and classify the topics for the extracted information. After the analyzed opinions, the opinions user needed would keep into one knowledge base. In the whole procedure from up to bottom, the ontology takes an crucial role to make the system goes well, and makes the system have the intelligent, semantic and automatic characteristics.

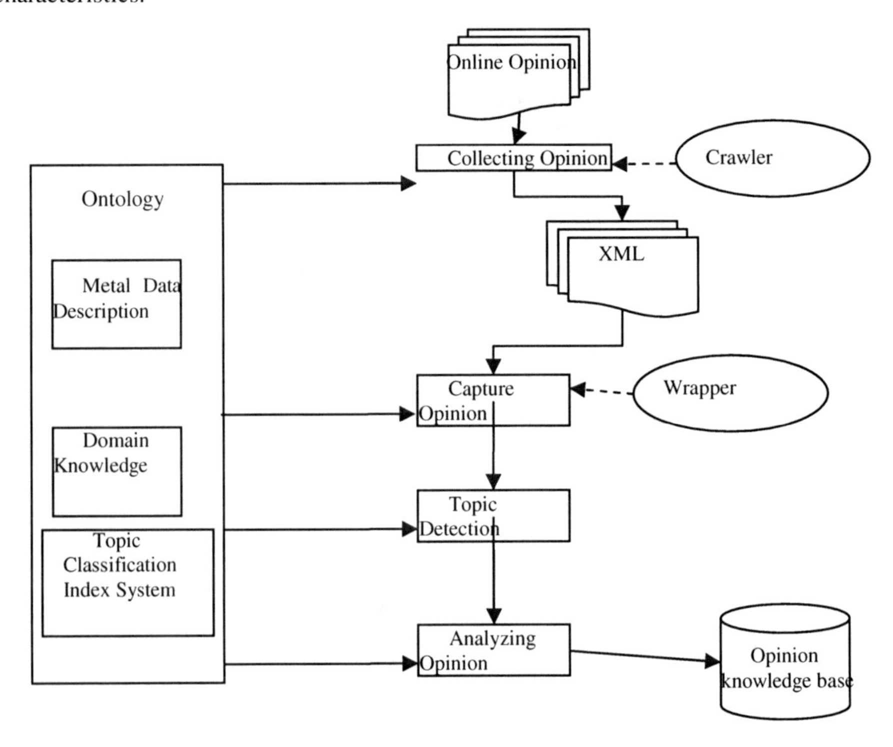

Fig. 1. System Architecture

The network communities provide such large amounts of data that accessing, finding or summarizing information remains a difficult task given the sheer amount of information to be found in each source and given the large number and variety of sources available through current technologies. The reasons underlying this problem are manifold, however one of the major causes lies in the large gap between the conceptualizations of information, such as envisioned by the user, and the actual storage and provision of information. By and large, the question remains how to bridge this gap at all and how to bridge it in a way that reduces the engineering task of providing information extraction methods for a large number and variety of information sources: free text, semi-structured information (e.g. XML), and relational database information all exhibit similar problems when it comes to providing a common conceptualization for underlying information. Ontologies describe shared conceptualizations for particular domains of interest on a high-level of technical abstraction. Rather then dealing with implementation issues, domain ontologies just describe the concepts relevant to this domain, their relationships, and rules about these relationships that enforce a well-defined semantics on the conceptualization. Regarding the information extraction and integration problem just mentioned, formal ontologies allow the precise description of a conceptualization common to varying information sources. Thus, ontologies offer themselves as (partial) solutions to the problems of extracting information and providing it to the user in a clear way.

3 Methods to Represent and Build Ontologies

3.1 Representation of Ontologies

Although many definitions of ontologies have been given in the last decade, the best one that characterizes the essence of an ontology is that an ontology is a formal explicit specification of a shared conceptualization. Here, conceptualization means modeling some phenomenon in real world to from an abstract model that indentifies the relevant concept of that phenomenon; formal refers to the fact that the ontology should be machine readable, that is an ontology provides a machine processable semantics of information sources that can be communicated between different agents; explicit means that the type of concepts used and the constraints on their use are explicitly defined. In other words, ontologies are content theories about the sorts of objects, properties of objects, and relations between objects that are possible in specified domain of knowledge. It provides a vocabulary of terms and relations to model the domain and specifies how you view the target world[4-6].

An ontology typically contains a network of concepts within a domain and describes each concepts' crucial properties through an attribute value mechanism. Such network is either directed or undirected one. It might also be a special type of network, that is , a concept hierarchy. Further relations between concepts might be described through additional logical sentences.

In this paper, we would adopt OWL language to represent ontologies. The Web Ontology Language (OWL) is a family of knowledge representation languages for authoring ontologies. The languages are characterised by formal semantics and RDF/XML-based serializations for the Semantic Web. OWL is endorsed by the World Wide Web Consortium (W3C)and has attracted academic, medical and commercial interest Units[7-9].

3.2 Building Ontology

Based on the methods of the text classification, the process of generation of ontology can be divided into the following two major atages.

The first stage is conceptual relationship analysis. We first compute the combined weights of terms in texts by Eqs1,2,, respectively,

$$D_i = \log d_i * t_i \tag{1}$$

$$D_{i,j} = \log d_{ij} * t_{ij} \tag{2}$$

Where d_i, d_{ij} are the text frequencies, which represent the numbers of texts in a collection of n texts in which term i occurs and both term i and term j occur, respectively, t_i 和 t_{ij} are the term frequencies, which represent the numbers of occurrence of the term i and both term i and term j, in a text, respectively

Then a network-like concept space is generated by using the following equations to compute their similarity relationships,

$$Rel(i, j) = \frac{D_{i,j}}{D_i} \tag{3}$$

$$Rel(j,i) = \frac{D_{i,j}}{Dj} \quad (4$$

Where Eqs 3 and 4 compute the relationship from term i to term j, and from term j to term i, respectively. We also use a threshold value to ensure that only the most relevant terms are remained.

4 Sample Analysis

4.1 Data Collection

Tianya is a powerful network community. It has about 20 million registered users by early 2011 and ranked 21 in the country's most visited sites (from Alexa.com). The Bulletin Board System (BBS) attracts more than 1.5 million page views per day. People go there to discuss a wide range of topics and issues. Anyone can start a discussion thread by posting a new message. The thread then grows when replies follow the original message. The Yushu earthquake caused a surge of online discussions on earthquake-related topics all over the Internet. Tianya's popularity and its swift reaction to the earthquake made it an ideal site for our study. Sample threads were extracted from one subforums on Tianya Earthquake. Immediately after the earthquake, people gathered on Earthquake to exchange information. We sampled threads from the Earthquake board from April 14-19 last year, which has a total of 43,110 new threads.

4.2 Classifying Sample and Building Ontology

After collect lots of texts of opinion posters(threads), we use text classifier and methods presented above to compute what and how many terms are related with term earthquake, then we building an ontology about Yushu earthquake, whose hierarchy of classes objects are showed in Fig 2. In that public opinion ontology, there are three supper classes: Organization, Individual and related Opinions. Here we mainly analyze and classify the opinions to tree-type hierarchical structure which includes five higher level classes.

Fig. 2. Opinions Classification Hierarchy in Ontology

All the sampled discussion opinion threads were categorized into 12 non-exclusive classes, which were then grouped into5 higher level categories (Fig 2). This classification of ontology revealed four major roles that Tianya played in this national disaster:

- Information-related. Effective and robust information exchange is extremely important for disaster response. The most important role of the forum is information sharing, seeking, gathering and integrating. Compared with other media such as news websites, the forum is particularly powerful in gathering and integrating information from users
- Opinion-related. Although seldom mentioned in disaster response literature, the expressing and exchange of opinions by community members may shape public opinions, provide feedback to authorities, and influence response policy. As a Bulletin Board System, Tianya is a natural platform for people to express and discuss different opinions.

- Action-related. Planning and coordinating actions are important functions of disaster response systems. Although there were not as many actionrelated threads as other types of threads on Tianya, we identified discussions on proposing, coordinating, and participating in rescue actions.
- Emotion-related. Emotional support is often needed in disaster recovery. Although most Tianya users are not direct victims of the earthquake, they need this online space to express their personal feelings and concerns and to provide comfort and sympathy to others. A considerable portion of the treads fell under this category.

The thread classification outlined the reaction of a large online forum to a national disaster. The four major top-level categories correspond to four ways the forum can help in disaster response. A closer examination of the categories and threads will shed more light on how to utilize such a public forum in disaster response.

5 Conclusion

In this paper, we present a system architecture which describes how the system goes well based on ontology, then discuss how to build ontology of public opinions to represent and classify thoses different and abundant opinions information. We use the data about Yushu Earthquake information from Tianyan network community as a sample, then discuss and build an ontology for the opinions about Yushu Earthquake. This paper provides a rational way to analyze online public opinions based on ontology, especially for opinions about huge disaster public event.

Acknowledgments. Thanks Dr. Yan Qu working at UMD give me a chance to understand her project which is related with this paper. This paper is partly supported by "211"project and Discipline Construction Foundation of Central University OF Finance and Economics.

References

1. Liu, B.: ACM SIGKDD Inaugural Webcast:Web Content Mining, November 29 (2006)
2. Wu, F., Huberman, B. A.: Social structure and opinion formation. HP Labs (2006)
3. Yi, J.: Sentiment analysis: Capturing favourability using Natural Language Processing. ACM 2003, pp. 70–77 (2003)
4. Wei, W., et al.: Opinion Mining for Web Public Sentiment Based on Dynamic Knowledge Base. In: ISEM 2009, pp. 199–203 (2009)
5. Ning, Z.: Representation and Construction of Ontologies for Web Intelligence. Internaional Journal of Fundations of Comouter Science (2002)
6. Bo, P., Lee, L.: A sentimental education: Sentiment analysis using subjectivity summarization based on minimum cuts. In: MACL 2004 (2004)
7. Qu, Y., et al.: Online Community Response to Major Disaster: A Study of Tianya Forum in the 2008 Sichuan Earthquake. In: HICISS 2009 (2009)
8. Wu, F., et al.: Social structure and opinion formation. HP Labs (2006)
9. Bar-Ilan, et al.: Method for Measuring the Evolution of a Topic on the Web: The Case of Informetrics. Journal of the American Society for Information Science and Technology 60(9), 1730–1740 (2009)

Research on Auto-regressive Load Balancing Model Based on Multi-agents

Yumin Dong and Shufen Xiao

QingdaoTtechnological University, Qingdao 266033, China
dym1188@163.com

Abstract. This paper proposes the auto-regressive load balancing model based on multi-agents. By the simulation experiments, we prove the load balancing mechanism can expand the server's "capacity" and improve the system throughput. The method overcomes the shortages of imbalance and instability of the server system. Therefore, the model can improve the system utilization factor of server system, achieve load balance.

Keywords: auto-regressive, load balance, multi-agents.

1 Introduction

Started from the 1990s, with the continuous development of network technology, regardless of the enterprise network, campus network or the Internet, all have a significant increase in system throughput. The traditional single-server mode can no longer meet the needs of fast-growing network. Therefore, the cluster which is composed of high performance servers becomes scalable and highly-efficient system.

Cluster technology is a group of linked computers, working together closely so that in many respects they form a single computer. To run parallel programs highly efficiently, service requests must be divided to each server, shorten the access time, and optimize the overall performance. Load balancing mechanism is the core issue of Cluster Technology.

Effective load balancing mechanism can expand the server's "capacity" and improve the system throughput. However, it is very difficult to achieve load balance of the whole system to use a simple scheduling scheme, because of the impact of performance parameters of the node server in cluster system, and the dynamicity and instability of the load. Whereupon, in order to overcome these shortcomings, this paper proposes the auto-regressive load balancing model based on multi-agents[1].

2 Auto-Regressive Model(AR)

The AR model is one kind of random time analysis model which developed from the 1920s Assume time-series $\{x_1, x_2 \ldots, x_p\}$, models which have the following structure, we called it p-order auto-regressive model, referred to as $AR(p)$ 。

G. Zhiguo et al. (Eds.): WISM 2011, CCIS 238, pp. 281–289, 2011.

$$\begin{cases} x_t = \varphi_0 + \varphi_1 x_{t-1} + \varphi_2 x_{t-2} + \ldots + \varphi_p x_{t-p} + \varepsilon_t \\ \varphi_p \neq 0 \\ E(\varepsilon_t) = 0, Var(\varepsilon_t) = \sigma_\varepsilon^{\ 2}, E(\varepsilon_t \varepsilon_s) = 0, s \neq t \\ Ex_s \varepsilon_t = 0, \forall s < t \end{cases} \tag{2-1}$$

In the above formula ε_t $(t = 0, \pm 1, \pm 2, \ldots)$ is white noise, p is a positive integer constant, we called it order. φ_i is a parameter. Especially when $\varphi_0 = 0$, we call it centralized $AR(p)$ model.

Decree $\mu = \dfrac{\varphi_0}{1 - \varphi_1 - \ldots - \varphi_p}$, $y_t = x_t - \mu$, $\{y_t\}$ is called the centralized series of $\{x_t\}$

Bring in the delay operator, centralized $AR(p)$ model can be short for $\alpha(B)Y_t = \varepsilon_t$, auto-regressive coefficient Polynomial is $\alpha(B) = 1 - \varphi_1 B - \varphi_2 B^2 - \ldots - \varphi_p B^p$.

AR model is one of the a commonly used fitting models of stationary series, but not all of the AR models are stable, here we have a brief introduction to the characteristics of AR models meeting the stability.

(1) Average value:

$$Ex_t = E(\varphi_0 + \varphi_1 x_{t-1} + \cdots + \varphi_p x_{t-p} + \varepsilon_t) \tag{2-2}$$

(2) Variance:

$$Var(x_t) = \sum_{j=0}^{\infty} G_j^2 \sigma_\varepsilon^2, G_j \text{ is the Green function} \tag{2-3}$$

(3) Covariance function:

$$\gamma_k = \varphi_1 \gamma_{k-1} + \varphi_2 \gamma_{k-2} + \cdots + \varphi_p \gamma_{k-p} \tag{2-4}$$

(4) Auto-correlation coefficient:

$$\rho_k = \varphi_1 \rho_{k-1} + \varphi_2 \rho_{k-2} + \cdots + \varphi_p \rho_{k-p}, \text{ among } \rho_k = \frac{\gamma_k}{\gamma_0} \tag{2-5}$$

(5) Tail of the auto-correlation coefficient:

$$\rho(k) = \sum_{i=1}^{p} c_i \lambda_i^k, \text{ among } c_1, c_2, \ldots, c_p \not\equiv 0 \tag{2-6}$$

Stable $|\lambda_i|<1$, $i=1,2,...p$, thereby ρ presents negative exponential decay, that is, $\rho(k)=\sum_{i=1}^{p} c_i \lambda_i^k \to 0$

(6) Partial autocorrelation coefficient:

$$\rho_{x_t, x_{t-k}|x_{t-1},\dots,x_{t-k+1}} = \frac{E[(x_t - \hat{E}x_t)(x_{t-k} - \hat{E}x_{t-k})]}{E[(x_{t-k} - \hat{E}x_{t-k})^2]} \tag{2-7}$$

So-called lag k partial autocorrelation coefficient refers to the related measure of the impact of x_{t-k} to x_t ,under the condition of giving middle $k-1$ random variables $x_{t-1}, x_{t-2}, \cdots, x_{t-k+1}$, or, after removing disturbances of middle $k-1$ random variables, Provable lag k partial autocorrelation coefficient is equal to the value of the k-th regression coefficient of the k -order autoregressive model $x_t = \varphi_{k1}x_{t-1} + \varphi_{k2}x_{t-2} + \cdots + \varphi_{kk}x_{t-k} + \varepsilon_t$ 。

(7) Partial autocorrelation coefficient p -order censored:

$$\varphi_{kk} = 0 \quad ,k > p \tag{2-8}$$

3 The Auto-regressive Load Balancing Model Based on Multi-Agents

In order to manage network resource,(in the paper, meanings server resource), we build resource agents, management agents, task agents. The server resources offer more better service to customers by the coordination of these agents[2].

In a general way, we illuminate the mathematic model and physical way by MAS distributing task programming and resource assign.

On the assumption that resource gather A={ A1, A2,..., An }, and task gather T={T1, T2,..., Tm}. Agent Ai has resource vector ri=(ri[1], ri[2],..., ri[h]). Task Tj demanding resource vector dj=(dj [1], dj [2],..., dj [h]). Figure 1 shows that the resource assigning status among n Agents and m tasks. Thereinto, aij=(aij [1], aij [2],..., aij [h]) is the resource vector of Agent Ai assigns task Tj . pij=(pij [1],pij [2],..., pij [h]) is the guerdon vector of task Tj defraying unit resource vector to Agent Ai. Muti-agents of participating in the same task form league. Considering that every kind of complicated social interactive behaviors and different autonomic characteristic and individuality (including different attention extent to MAS overall profit), Considering the reciprocity between the whole MAS and Agent individual. In this instance of all the tasks are finished to the best of our abilities, we receive the result of resource distribution and guerdon distribution to the best of our abilities. In order to utilizing composite-spring-net approach to parellel optimization and evaluating the approach, we define a synthesizing evaluating function. As income summation of all agents is the more, the residual capability of the agent of income being least is most, and the residual capability summation of all agents is the least. The resource vector of every task requirement matchs better to the resource vector summation of multi-agents, and the function value is the more.

MAS alignment and approach to distributed problem solving in composite-spring-net physics model as figure 2 .

Task / Agent	T_1	…	T_j	…	T_m
A_1	a_{11}, p_{11}	…	a_{1j}, p_{1j}	…	a_{1m}, p_{1m}
…					
A_i	A_{i1}, p_{i1}	…	A_{ij}, p_{ij}	…	a_{im}, p_{im}
…					
A_n	A_{n1}, p_{n1}	…	A_{nj}, p_{nj}	…	A_{nm}, p_{nm}

Fig. 1. The status of MAS alignment and resource distributing

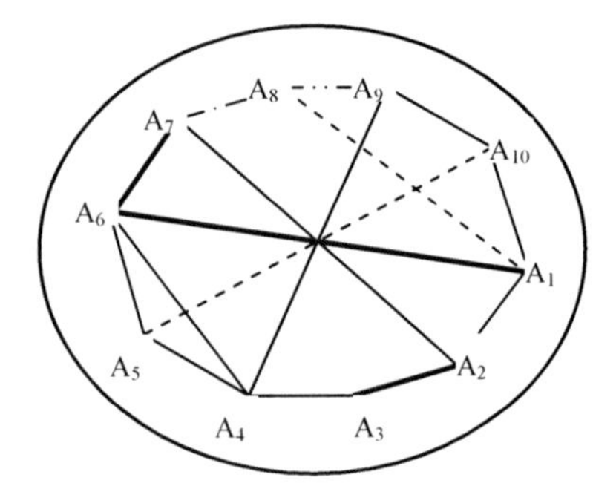

Fig. 2. MAS alignment and approach to distributed problem solving in composite-spring-net physics model

In figure 2, the nodes express Agent nodes, the circle express the gravitation circle to every Agent. The real lines, broken lines, dots-lines respectively express the pulls, press and composite-spring among the Agents. And the composite-spring is non-symmetrical. The thickness of the lines express different spring coefficient[3-8].

Every node is close-by the circle in same angles and optional orders. The distance of Agent Ai node to the center is direct proportion to Ai income in MAS currently situation. The length of the radius is greater than every income value of every Agent. Every Agent node lies on the circle gravitation field. In order to every Agent doing its best gains itself most income, the gravitation field drives every Agent node do its best moving along radial to the gravitation circle brink. Every Agent wants to take part in the league of higher income, and at the same time, it is restricted by the relation to the MAS whole in personality and autonomy. In addition, considering that every Agent resource wasting expenditure, and its correspond communicating expenditure of its multi-task league , the two kinds of expenditure are lower, the competing capability is stronger, and it gets the gravitation is bigger, its trend moving to the brink is more obvious.

The auto-regressive load balancing model based on multi-agents as follows fig.3

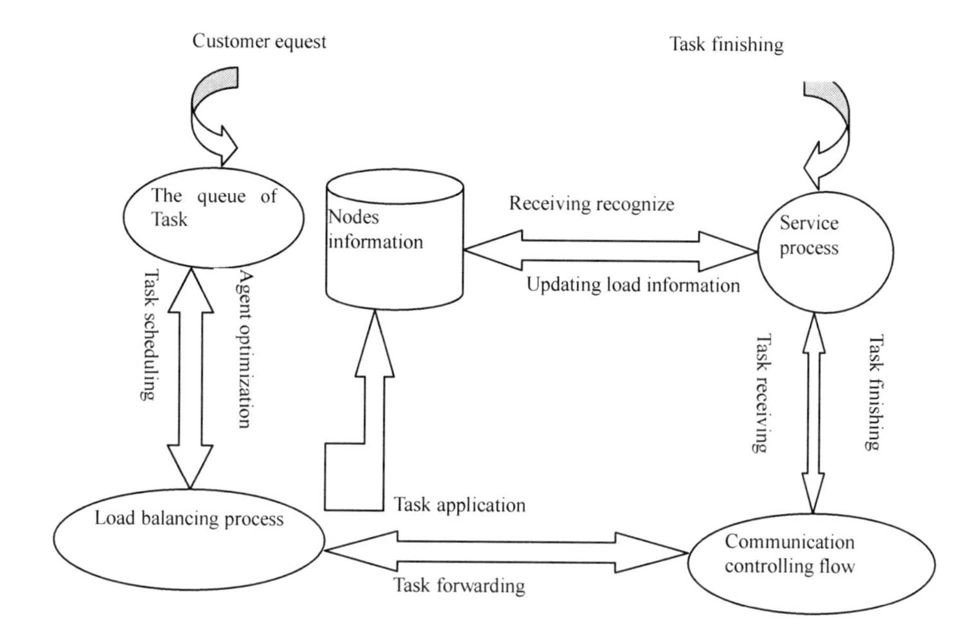

Fig. 3. The auto-regressive load balancing model based on multi-agents

Step 1: the customer sends access requests to application servers, simultaneously asks to access task queue waiting for scheduling.

Step 2: by multi-agents optimizing algorithm, the task accesses into load balancing process by scheduling, simultaneously the system starts its service process.

Step 3: by the auto-regressive load balancing algorithm based on multi-agents optimizing, the load balancing process distributes the customer requests into corresponding nodes servers according to the parameter information of the servers.

Step 4: the nodes servers confirm the tasks are received, simultaneously update itself load information.

Step 5: the service processes finished the requests of the customer, then close the service, waiting for the next task.

4 Simulation Experiments

4.1 Experiment Environment

We do our simulation experiments in the network center of one university. There are 10 servers with the operation systems of Solaris and Linux, more than 30 servers with the operation system of windows. These servers are built to one location network.

In order to do our simulation experiments, we chose 2 servers of Solaris system, 2 servers of Windows system and 1 server of Linux system. These servers structure is as follows fig.4.

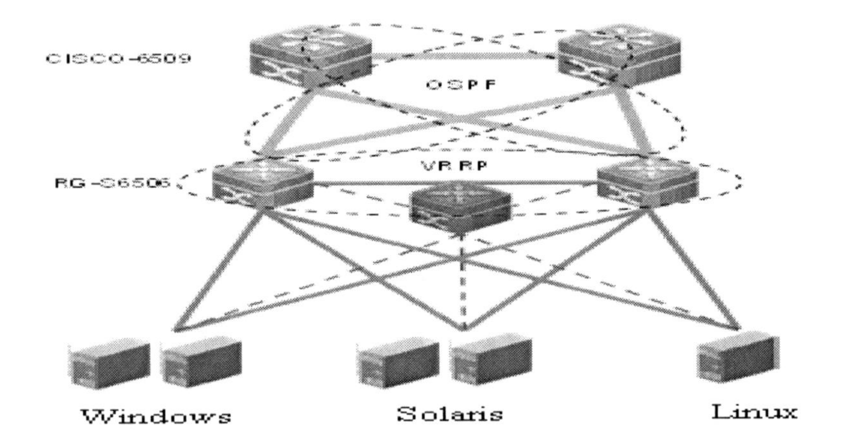

Fig. 4. These servers structure

These servers information is as follows fig.5.

Server parameter	CPU	main storage size	Network adapter	Operation system
Sun 880	AMD Opteron 2.4 GHz	4.0G	2*1000M	Solaris
Sun Fire	UltraSPARCIV 2.0GHz	4.0G	2*1000M	Solaris
IBM 366	Intel Xeon 3.0GHz	3.0G	2*1000M	Windows
IBM 346	Intel Xeon 3.0GHz	2.0G	2*1000M	Windows
IBM 346	Intel Xeon 3.4GHz	4.0G	2*1000M	Linux

Fig. 5.

4.2 Experiment Data

We extract data from 2011.05.01. am 7:00 to 2011.05.07. am 7:00, chose two servers, one is Sun 880, the other is IBM366. Their flow diagrams as follows:

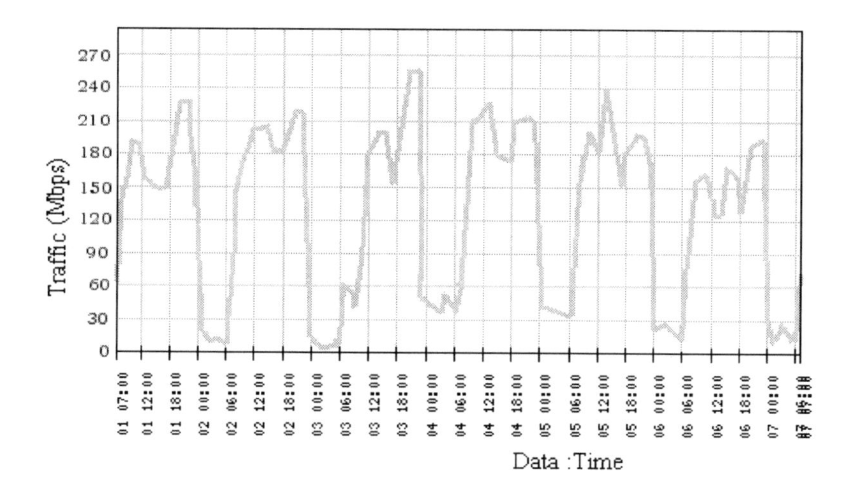

Fig. 6. Sun 880 after optimizing (load-balance)

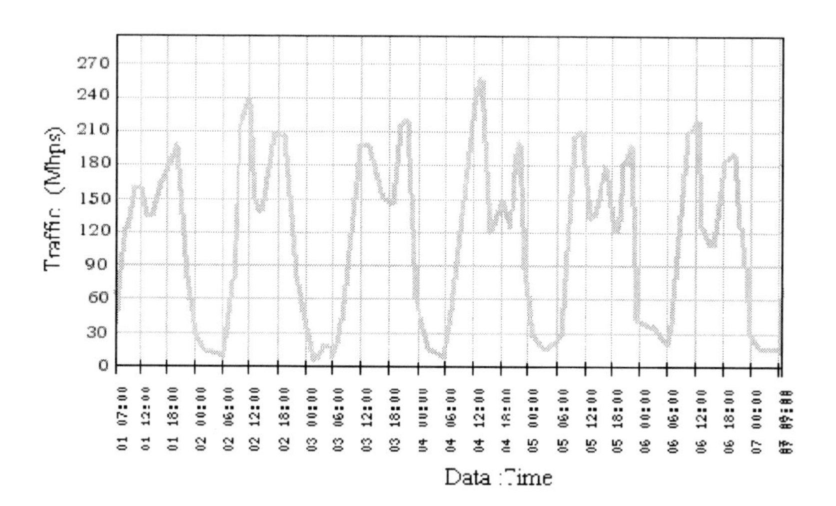

Fig. 7. IBM 366 after optimizing (load-balance)

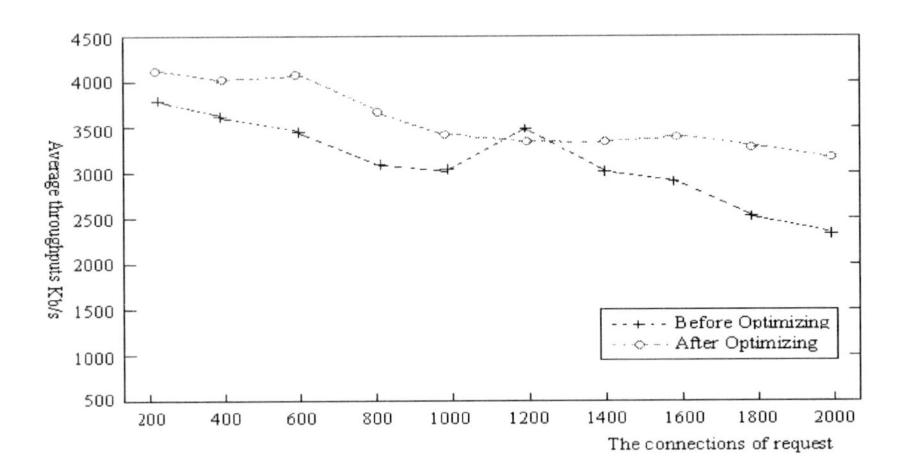

Fig. 8. The throughput of Sun 880 after optimizing compare with before optimizing

The fig.8 shows the average throughput of after optimizing the Sun 880 is more than that of before optimizing.

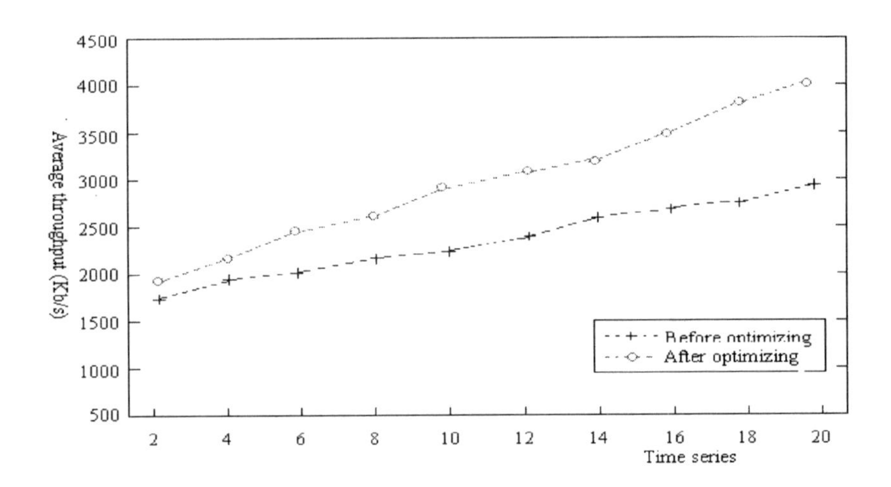

Fig. 9. The throughput of IBM366 after optimizing compare with before optimizing

The fig.9 shows the average throughput of after optimizing the IBM366 is more than that of before optimizing.

5 Conclusion

All this leads up to the Dynamic Load Balancing Method based on Load Prediction and combining to the advantage of multi-agents can increase the utilization ratio of servers.

References

[1] Chen, T., Lin, Y.-C.: A fuzzy-neural system incorporating unequally important expert opinions for semiconductor yield forecasting. International Journal of Uncertainty, Fuzziness and Knowledge-Based Systems 16(1), 35–58 (2008)

[2] Laszlo, M., Mukherjee, S.: A genetic algorithm that exchanges neighboring centers for k-means clustering. Pattern Recognition Letters 28, 2359–2366 (2007)

[3] Abraham, A., Das, S., Konar, A.: Document clustering using differential evolution. In: Proceedings of 2006 IEEE Congress on Evolutionary Computation, pp. 1784–1791 (2006)

[4] Easwaran, A., Shin, I., Lee, I.: Optimal virtual cluster based scheduling. In: Euromicro Conference on Real-Time Systems, vol. 43(1), pp. 25–59 (2009)

[5] Yan, K.Q., et al.: A hybrid load balancing policy underlying grid computing environment. Journal of Computer Standards & Interfaces, 161–173 (2007)

[6] Minh, T.N., Thoai, N., Son, N.T., Ky, D.X.: Project oriented scheduler for cluster system. In: Modeling, Simulation and Optimization of Complex Process, pp. 393–402. Springer, Heidelberg (2008)

[7] Ahmad, A., Dey, L.: A feature selection technique for classifieatory analysis. Pattern Recognition Letters 26, 43–56 (2005)

[8] Shuai, D., Shuai, Q., Dong, Y.: Particle model to optimize resource allocation and task assignment. Journal of Information Systems (32), 987–995 (2007)

Design and Implementation of Intelligent Monitoring and Diagnosis System Based on WSN and MAS

Bing Wu[1], Jian Lin[2], and Xiaoyan Xiong[1]

[1] Research Institute of Mechanic-electronic Engineering, Taiyuan University of Technology, Taiyuan, China
[2] Center of information management and development, Taiyuan University of Technology, Taiyuan, China
{wubing,xiongxiaoyan,linjian}@tyut.edu.cn

Abstract. Based on wireless sensor network (WSN), the process monitoring of mechanical equipment has some advantages like arranged flexibly, adaptable, maintenance cost. On the other hand, based on multi-agent system (MAS) fault diagnosis method has distributed, adaptive, and many other advantages. Combined the WSN and MAS, a fault monitoring and intelligent diagnosis system is proposed. The faults of equipment can be analyzed and fixed rapidly as well as improved equipment maintenance efficiency. As a key content, according to the requirements of high sampling rate, high precision, high speed and large amount of data transmission of mechanical equipment condition monitoring, the corresponding sensor node platform is design. Finally, through the production process in certain coal cleaning plant application equipment monitoring, the system validity and practicability are proved.

Keywords: Wireless Sensor Network (WSN), Multi-Agent System (MAS), Fault Diagnosis and Monitoring system.

1 Introduction

As the forefront of information acquisition and processing component of fault monitoring and diagnosis system, the sensor directly faces the targets. The measuring parameters are transformed into electrical signals, which is the important modern fault diagnosis technology foundation. Along with the digital circuit technology and large scale integrated circuit technology development, the sensor also gradually toward miniaturization and intelligent. The boundary of sensor and measurement system is also more and more vague. On the other hand, following the wireless communication technology, MEMs technology and embedded techniques become more mature, the wireless sensor network reliability gradually enhance and application range has been extended. More and more high reliability fields such as health condition monitoring [1], machinery manufacturing [2], mine safety monitoring requirements [3] are introducing the wireless sensor network. Due to the low cost, no maintenance convenience, convenient layout etc, the WSN is very suitable for bad conditions, large monitoring or man could not to reach of the complex equipment monitoring and diagnosis occasions, and it can avoid wiring trival, network maintenance difficulties, waste of resources, etc.

G. Zhiguo et al. (Eds.): WISM 2011, CCIS 238, pp. 290–297, 2011.

As the monitoring objects are complex mechanical and electrical equipment in the on-line monitoring system, the monitoring and diagnosis algorithms are based on the majority of machinery vibration signals in particular. The signal acquisition has some features such as long-term monitoring, high sampling rates, accuracy and so on. This leads the data generated by the sharp rise in the node. The node data storage and transmission brings many challenges. Therefore, the node hardware and software design trade-offs according to the Practical application, such as the power of data processing and energy consumption can be a secondary consideration (dual power supply can be resolved, etc.), and give priority to data storage and transmission accuracy.

Meanwhile, multi-agent technology over the past few years has come to be perceived as crucial technology not only for effectively exploiting the increasing availability of diverse, heterogeneous and distributed on-line information sources, but also as a framework for building large, complex and robust distributed information processing systems which exploit the efficiencies of organized behavior [4].

It has many advantages for multi-agent system (MAS) applied in fault diagnosis field, such as distributed, adaptive, and many other features. It not only can improve diagnostic accuracy and enhance the environmental adaptability, but also can be discovery and mining knowledge, improve learning ability to achieve self-diagnostic system performance improvement and improvement of intelligence for solving large-scale diagnosis of the problem.

Wireless sensor networks and multi-Agent system has the following similar characteristics:

- Wireless sensor network nodes constitute a network of self-organization and environmental monitoring, and Agents have similar features such as autonomy, interactivity, motivation, reaction mechanism, etc.
- Each node in wireless sensor networks cooperates with each other to complete a specific task of co-monitoring; As well as the agents in MAS work together to complete problem solving.
- When the network size is very large, the wireless sensor network will use the network structure of layered clustering; the multi-agent system hierarchy structure in the system has a good binding with it.

Therefore, this paper presents a fault monitoring system model based on wireless sensor network and multi-agent system. The system can better adapt to the industrial field environments, and meet the requirements of long-term online fault monitoring and intelligent diagnostics.

2 The Fault Monitoring and Intelligent Diagnosis System Based on WSN

In an intelligent monitoring and diagnosis system, the functional structure of intelligent sensor networks can be divided into four levels [5], as shown in Fig. 1.

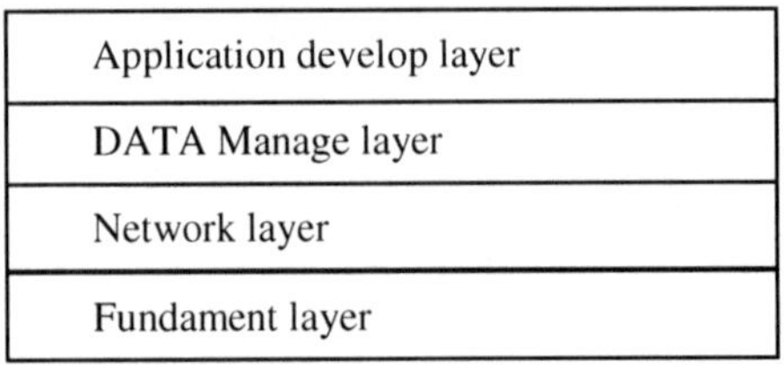

Application develop layer
DATA Manage layer
Network layer
Fundament layer

Fig. 1. Function structure of sensor networks

- The bottom is the foundation layer. It takes sensor node as the core. Including the each sensor hardware and software resources, such as perception device, embedded processor and memory, communication devices, embedded operating system, embedded database system, etc. This layer's features include monitoring sense objects, acquisition object information and transmission of information and released preliminary information processing.
- The second layer is the network layer. It is characterized by network as the core, realize sensor and sensor, sensors and data acquisition system, and support the communication between the multi sensor collaboratively large perception tasks. The network layer includes the communication network, the support of all kinds of agreement and network communication software and hardware resources.
- The third layer is the data manager layer. It is the sensor data management and processing software as the core, including support for sensory data collection, storage, retrieval, analysis, and mining and other data management and analysis software system to effectively support the sensor data storage, query, analysis, and mining.
- The top layer is the application develop layer. It provides effective software development environment and software tools for users in the basic layer, network layer and data management layer to develop various sensor network applications.

According to the above guidelines, this paper presents a monitoring and fault diagnosis system based on wireless sensor network. The overall architecture is shown in Fig. 2.

The architecture includes three parts: data acquisition agent (DAQ node), sink agent (sink node), gateway agent (gateway node) and the follow-up monitoring and intelligent fault diagnosis system (composed by fault analysis diagnosis agents, decision agents, remote access agent etc.). As the data collection layer, WSN nodes are connected with the follow-up monitoring and intelligent fault diagnosis system through the gateway and industrial Ethernet as the field bus.

As an agent, each sensor node in the system is independent of each other and to autonomy. Sensor nodes and the number of cluster (A cluster consists of a sink node and a number of DAQ nodes.) are determined by the application requirements, and can change dynamically as needed. The architecture is an open hierarchical network structure so that it has good scalability and manageability. It can be connected with enterprise management information system. When necessary, it can solve the difficult problem through the remote expert's help.

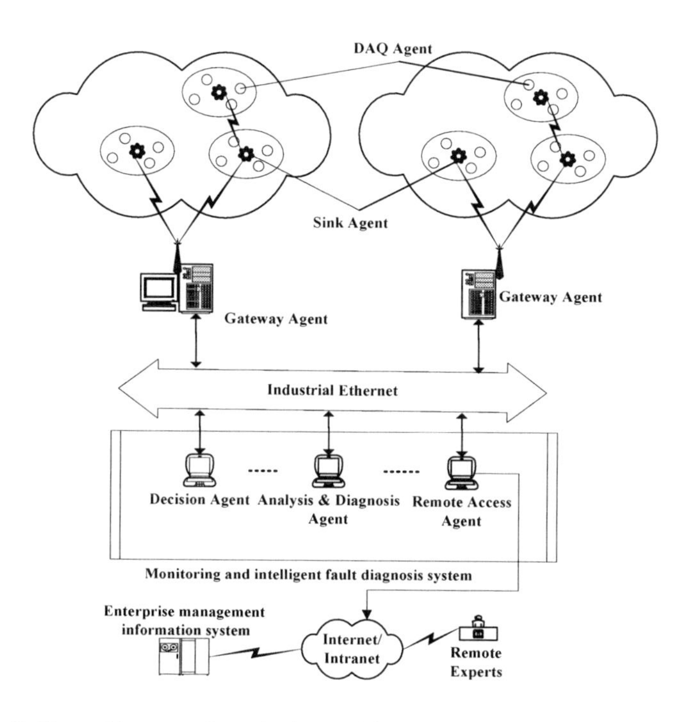

Fig. 2. The architecture of monitoring and fault diagnosis systembased on WSN

3 The Design of Agents

3.1 The Hardware Design of Wireless Sensor Node

As monitoring of mechanical equipment operation, the main consideration is the monitoring object's characteristics in the design of a wireless sensor network node (DAQ agent). For example, the relatively high acquisition frequency of the vibration signal (1 KHz and above), and the ability to detect high acceleration vibration signals (10g and above), the large amount of data (2Mbits and above) and so on [6].

Therefore, in the node hardware design, the chip ADXL345 is selected for digital accelerometer module and MSP430FG4618 MCU as the core of micro-controller module. The CC2420 is selected as the wireless transmission module. And M25P80 is selected to be used the expanded local storage. The power supply module is used USB, dc power supply, the battery power or other kind of ways. This design allows us to use ZigBee specification directly, regardless of its MAC layer and network layer. It can be developed for specific nodes in the hardware drivers and added the application at the application layer. The overall structure of the node platform is shown in Fig. 3.

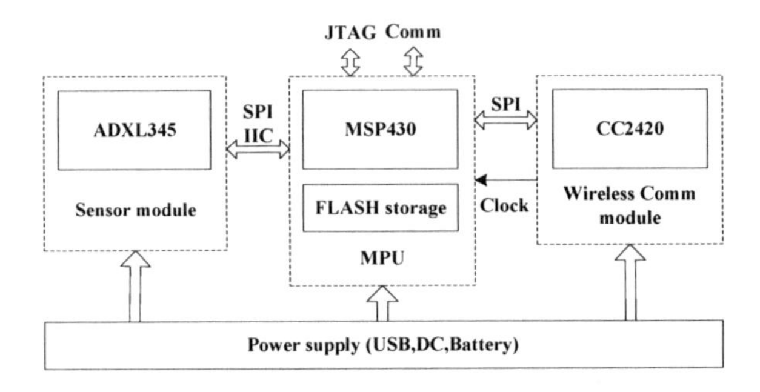

Fig. 3. Structure of the Node Platform

3.2 The Software Design of Wireless Sensor Node

For different objects and purposes of monitoring, the sensor network software may achieve very different processes and methods. But in the software design, the sensor network software is composed of the system layer and user layer. The system layer includes the file system, communication protocol, device management, distributed processing, low power, and memory management, process scheduling, interrupt management, etc. Application layer software are designed primarily the task collaboration. The application layer software is different based on specific applications. Whereas the DAQ node and sink node as well as gateway node has many similarities in software process. But they collected data and the existence of different approaches. But they are different in the way of data acquisition and processing.

Fig. 4 shows the process of a typical data acquisition node. According to the program, the first step of node is to initialize the hardware device and interfaces. The second step is the network initialization in order to establish communication channel with the subordinate sink node or gateway node. The third step is that the application needs to adjust the data acquisition parameters. And then the node can be into normal data acquisition and processing. In order to save the energy consumption and reduce the unnecessary data transmission, nodes are usually contained the hibernate function of the hardware itself provides. In certain trigger condition, node can return to normal working condition and acquire data. The main function of the information processing and control module is to respond various commands, may also include simple data processing. The data transport module is responsible for exchanging data with sink (or gateway) node and transmits instructions. After receiving instructions, the commands (including data acquisition, data processing and transit, node parameter and working status setting etc.) will be sent to information processing and control module.

However, for the system user, the core of the sensor network is sensory data rather than networking hardware and basic software. Users interested in the sensor data acquired rather than the sensor itself. Therefore, it is different from the transmission purpose of communication networks that the sensor network is a data-centric network. The management and operation of data into intelligent sensor network software, core technology. The core technology of intelligent sensor network software is data management and operation.

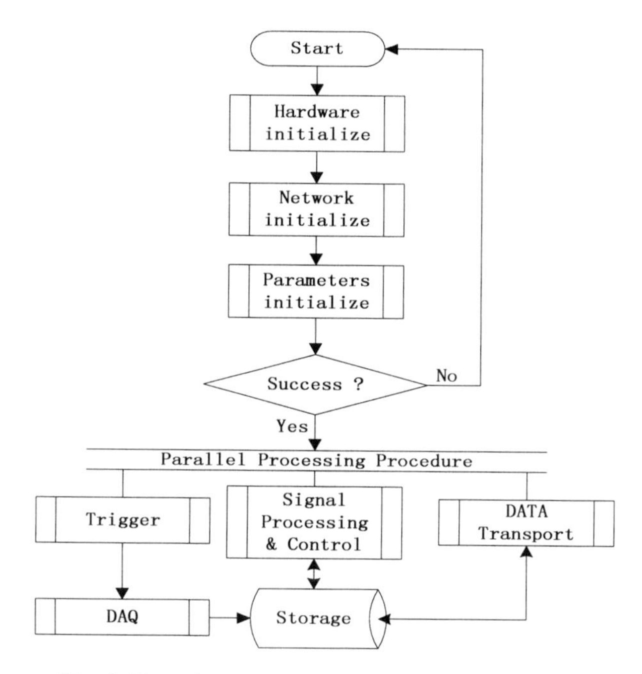

Fig. 4. The software flow chart of Data acquisition node

3.3 The Design of Other Agents

In addition to wireless acquisition node, other types of nodes are used general-purpose computers as their hardware platform. Just the gateway agent also includes a wireless transmission module for use with wireless DAQ nodes.

The design of these agents is concerned with how to write an algorithm to achieve the appropriate monitoring and diagnostic functions. As each equipment or its different parts' working principle and the different change of dynamic behavior of running of after local damage or failure occurred, in order to rapid detect the equipment failure from a large number of signals from the monitoring and make failure warning, it is needed to use appropriate signal processing methods to extract and identify the fault characteristics.

In the signal monitoring, fault automatic detection is the system can be given a certain part of the equipment failure warning when certain threshold to be reached. The signals including warning information will be added a corresponding identification signal and then together with the normal data into the historical database for subsequent further refined analysis, diagnosis.

4 The Application of the Monitoring and Intelligent Diagnosis System

The heavy-medium coal preparation technology has some characteristics such as numerous equipment long process and distribution and so on. The centralized intelligent diagnosis system does not meet requirements for monitoring and diagnosis obviously.

Therefore, the key equipment monitoring and fault diagnosis system of a certain coal preparation plant adopts the suggested fault monitoring and intelligent diagnosis system architecture. The monitoring objects includes key equipment in the coal preparing process like 3-product heavy-medium cyclones, vibrating screens, medium pump motor and so on. The system composition is shown in figure 5. All of the agents are networking via Industrial Ethernet except the wireless DAQ agents. As a bridge of DAQ agents and fault diagnosis agents, the gateway agent is also used as a sink node.

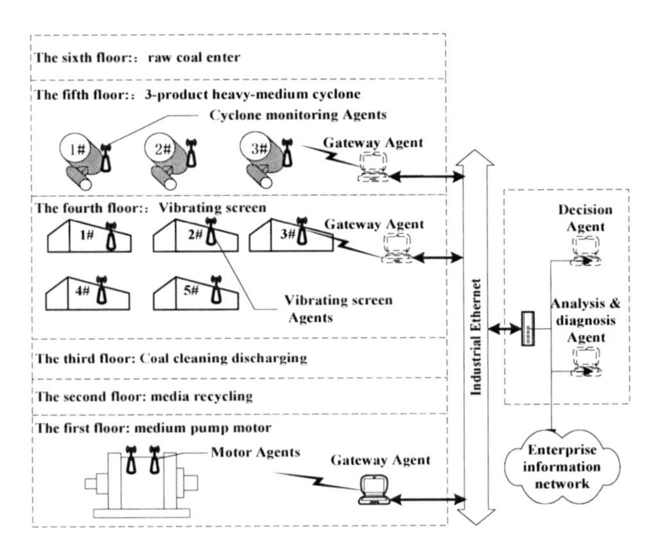

Fig. 5. The monitoring and diagnosis system topology of key-equipments of a coal preparing factory

Although the sampling rate of motor and vibrating screen agent is high, but the task does not require real-time of monitoring the state. Thus, the data can be transmitted at non-sampling time. But the cyclone agent need to continuous real-time monitoring. Figure 6 is one monitor interface of the group of cyclones fault diagnosis Agents. It shows the state of the system monitored when the fault occurred. The data is gained from the wireless data acquisition agent installed on cyclones.

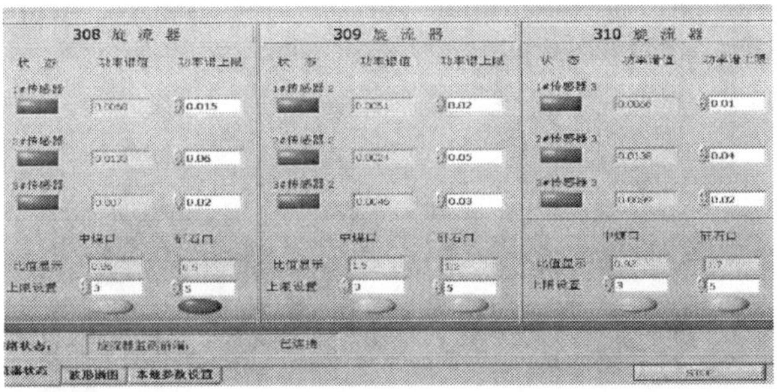

Fig. 6. The status monitor interface of a fault diagnosis Agent

And figure 7 is a set of waveform from a data acquisition agent. The left shows the time domain waveform of a monitoring point and the right is the corresponding RMS.

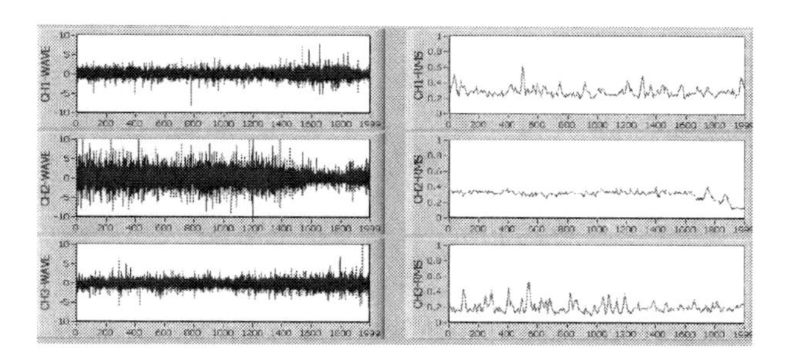

Fig. 7. The wave display of a cyclone monitoring Agent

5 Conclusions

Combining the wireless sensor networks and multi-Agent technology, an intelligent monitoring and diagnosis System based on WSN and MAS Model is proposed. First, the system architecture is designed. Second, key hardware and software design methods of the DAQ agent are described. Finally, the practical application example of a coal preparation plant proved that the system runs well.

Acknowledgments. This paper is supported by National Natural Science Foundation of China (50975188 & 51035007) and Key Science and Technology Program of Shanxi Province, China(20090322015).

References

1. Bai, H.: Wireless sensor network for aircraft health monitoring. In: First International Conference on Broad Nets, pp. 748–750 (2004)
2. Evans, J.J.: Wireless sensor networks in electrical manufacturing. In: Electrical Insulation Conference and Electrical Manufacturing Expo 2005, pp. 23–26 (2005)
3. Zhao, J., Liu, G., Zhu, L.: Multi-robot system for search and exploration in the underground mine disasters based on WSN. J. Journal of China Coal Society 34(7), 997–1002 (2009)
4. Li, J.-Z., Li, J.-B., Shi, S.-F.: Concepts, Issues and Advance of Sensor Networks and Data Management of Sensor Networks. J. Journal of Software 14(10), 1717–1727 (2003)
5. Wu, J., Yuan, S., Ji, S.: Multi-agent system design and evaluation for collaborative wireless sensor network in large structure health monitoring. J. Expert Systems with Applications 37(3), 2028–2036 (2010)
6. Zhao, M., Xu, K., Ni, W.: A Node Design of Wireless Sensor Network and Study on Communication Protocol. J. Chinese Journal of Scientific Instrument 26(8), 630–633 (2005)

Design of Dual-Shared DRAM Controller Based on Switch

Yifeng Li, Bo Zhang, Xiaoxia Han, and Gang Zhang

College of Information Engineering, Taiyuan University of Technology, Taiyuan, 030024, China
Yifeili163@163.com, zhangbo0351@163.com, hanxueeryang@163.com, tyzhgang@sohu.com

Abstract. According to the harsh desire to share memory of multi-processors system on chip, this paper presents a switch-based Dual-Shared DRAM Controller (DSMC). The DSMC is composed of center control module, two IP(interface to processor) module, two ID(interface to DRAM) module, CM(Clock Manager) module, and IR(initialize and refresh) module. It use two memories as shared memory and resolves conflicts which may occur when two processors access the same memory, and actualizes every control to DRAM, such as initializing, refreshing, reading and writing. At last, it makes two processors access the shared memories coinstantaneous at will.

Keywords: Switch-based, DRAM, Multi-processors, FPGA Sharing Memory.

1 Introduction

In the light-communication transfer equipment based on saving–sending, CPU has been replaced by memory, and memory has been the center of the system. Making use of multi-processors and buses to improve the whole index point of the system is the developing trend, and they are sources of conflict and competition for DMA. However, the dual-ported memory existing is not only expensive but also low capacity. The DRAM which is low capacity and cost are all single port, which cannot support the simultaneous access of multi-processors [1].

This paper presents an interconnected memory model based on switching and designs a memory framing which may support the dual-port access. The conflict-free access to the shared DRAM of multi-processors may be realized. Making use of FPGA, we supply the IP(Intellectual Property) meeting the design. The design is realized in VHDL and verified with the FPGA [2,3].

2 Principle of the System

The design philosophy and characteristics of the dual-ported DRAM control system based on switching are as follows.

Step 1: The conflict-free access of multi-processors can be realized through stitching. Two or more processors can access memory in parallel by using the control system, which upgrades the performance of the system.

G. Zhiguo et al. (Eds.): WISM 2011, CCIS 238, pp. 298–302, 2011.

Step 2: The shared memory used the DRAM parts which are high capacity and low cost. The initializing and refreshing DRAM can be realized.
Step 3: Two same DRAM are mirror images, which are the shared memory of multi-processors. At the cost of the capacity of one memory, we get the high performance of the system.
Step 4: Defining the memory n (n equals to 1, 2) as the local storage of the processor n, and defining memory m (m not equals to n) as the remote storage. In the ordinary way, processors visit their local storages respectively, so that reading and writing data is fast. The stationary problem between memories can be resolved by using writing back.
Step 5: Writing back is writing to the remote storage. When the remote storage is at leisure, the control system write data to remote storages by using writing back, which makes data synchronization between the two memories. The time of writing back is transparent for the processor, which is when the processor is not visiting to memory, writing back is executed.
Step 6: The processor can access the memory in the way of multi-sequential system. The sequence model of processor maybe the sequences of DRAM or remote memory. Asking no odds of DRAM in the processor makes the control system easy to use.
Step 7: The collision-free technique:

- A. Two processors access shared memory in different periods, which makes it conflict-free.
- B. Two processors read data at the same time. Two processors can access local memories simultaneously.
- C. Two processors read and write data respectively at the same time. If two processors own the same reading and writing address, the data wrote by one processor could be sent to the other processor which reads, and be wrote to the shared memory at the same time. If two processors own different addresses, the data is read from and wrote to the local storage, and then making two memories simultaneously through writing back.
- D. Two processors writing at the same time. If two processors own the same address, then one processor's writing data is discarded, and the other processor's writing data is wrote to the shared memory. If two processors own different addresses, data is wrote to the local memory first, and then making two memories simultaneously through writing back.

What is said above (B, C, D) is the detail of collision-free technique. Processors can access the shared memory fast, and the high performance of the control system will be recognized.

3 Design of the System

3.1 Design Methodology

Using two DDR SDRAM as shared memories, DSMC (DSMC, DUAL Shared Memory Controller) lies between two processors and two storages. Two processors access shared memories in a high speed by using the system. The internal structure of DSMC is as illustrated in Fig.1. The modules and their functions are as follows [4].

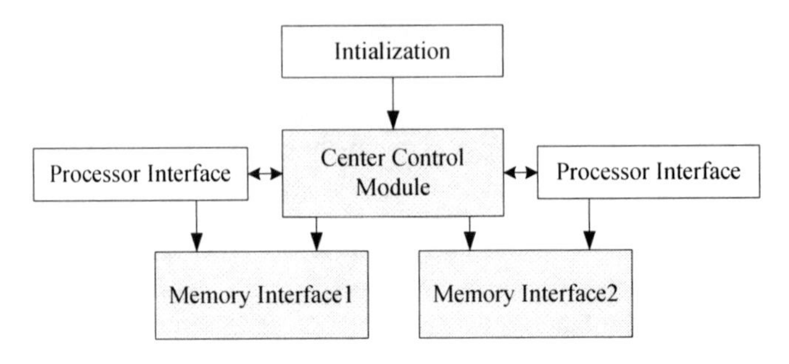

Fig. 1. DSMC Internal Structure

3.2 Modules Function

1) The Center Control Module

The center control module is the core of the DSMC, it is charge of switching working stating of the whole system, and coordinating the operating requests between modules. It is comprised of the major state module, the request-judgment module, the signal interface gating module, the control demand gating module, and the memory interface gating module.

The request-judgment receives requests of reading-writing, writing-back, initializing, and refreshing from two processors' interfaces. Initializing and refreshing have the highest priority, then is the reading-writing requests of processors' interfaces, and the last is the writing-back requests of processors' interfaces. According to the feedback information from the request-judgment, the major state module switches the whole system's working state, and controls other modules' work. The control command generates modules and control signals to other modules in the control system coordinating with the major state module. Reading-writing, writing-backing, initializing and refreshing signals are selected to two memory interface modules from the signal gating module and memory interface gating module.

2) The Center Control Module

There are two processor interface modules in the system, corresponding to two processors respectively. It contains 4 parts: control segment, processor's command interface, data buffer area, and fast track, which is shown in figure 2.

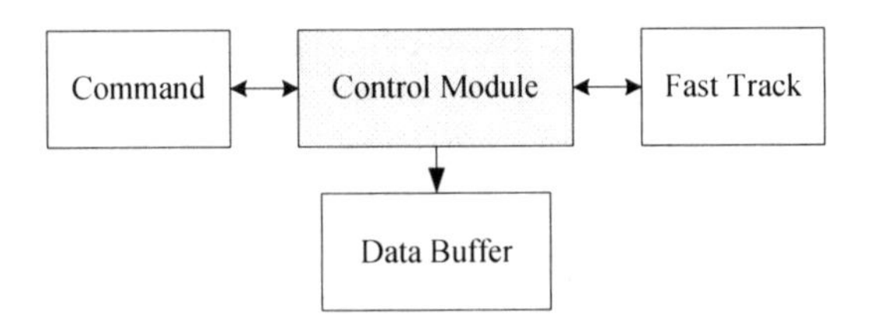

Fig. 2. Internal Structure of Processor Interface Module

Control segment contains: controller, feedback state, command generating. According to the commands of center control module received by feedback state module, and the state of other modules, control segment coordinates the running of four parts and five state modules, and control those communicating signals generated from control command for other modules.

Processor's Command Interface contains: processor's state machine, command analyzing, address switching, reading-writing state machine, writing-back state machine. According to the design of asynchronous memory sequence, the command translation and processor's address switching to memory's address is completed by command analyzing and address switching modules respectively. Reading-writing state machine reads from or writes to local memory, control read-write buffer in data buffer area, generated corresponding control commands. Writing-back state machine is designed for writing-back. It controls write buffer module remotely, and writes processor's writing data to remote memory simultaneously.

Data Buffer Module contains: data separating dating module, read buffer, local write buffer, remote write buffer. Data separating dating module divides shared signals into read passage and write passage. Read passage connects read buffer, and sends data from memory to processor. Write passage connects local and remote write buffers, and sends writing data from processor to the two buffers respectively. Read buffer handles data from memory to processor. The data processor needed and the data in the continuation address is read to the buffer one-time, then the data is sent to processor according to reading requests. Write buffer handles data from processor to memory. On one hand, it adjusts sequence differences between processor and memory, and on the other, the data wrote to memory could be merged in subsequent module, and could be wrote to memory one time. Remote write buffer module saves data from processor to remote memory, which is designed for the synchronous between two memories. Writing data is saved from remote memory to remote write buffer, and writing data in remote write buffer to remote memory when remote memory is at leisure. Coordinating with the fast track, data is sent to another processor immediately.

Fast Track Module contains: sending state machine, receiving state machine, address compare. Data could be switched between two processors and bring into correspondence with memory could be realized by using fast track. Sending state machine sends address command information, and receives feedback information. Receiving stating machine receives information, compares addresses and locates data by using address compare module. Particularly, if the data processor 2 needs to read is in remote writing buffer of processor 1, then processor interface 2 will send address information through sending state machine. Processor interface1 will compare addresses and locate data in the remote writing buffer through receiving state machine. So that processor2 need not read from storage, which not only saved time, but also made time for the memory. If data to be wrote by processor2 lies in the large buffer of processor2, then processor interface2 will compare information through sending state machine, processor interface1 will find data in the large buffer through receiving state machine, and make data be invalided, so that data in two memories could be the same. The function of fast track in processor1 is the same as processor2.

3) Memory Interface Module

The internal structure of memory interface is as figure 3. It is composed of control module, internal command analysis, BANK management, DDR command module, DDR data separating gating module [5].

The control module receives commands from the center control module, and feeds information which is charged by BANK management module back to processor interface module. At the same time, it controls DDR data separating module according to the command sent by command analysis module.

Command analysis module analyzes operating commands, and then feed it back to the control module. BANK management module records the state of lines in the memory, and feed it back to processor interface module. Data separating module separates read data from write data, so that the reading or writing for other modules is easier.

4) Memory Interface Module
The operating of initializing and refreshing for DDR SDRAM is generated by the module. DDR SDRAM comes to the working state by initializing. Moreover, because DDR SDRAM stores information depending on capacitance, every line of DDR SDRAM must be refreshed within a certain period of time. There is a counter in the refreshing module. It will make a request for refreshing within a certain period of time.

4 Conclusion

The above modules was described and designed in VHDL, and was simulated with MODELSIM. The simulating results show that the DSMC designed here makes two processors' access to the same DDR SDRAM without any confliction.

Acknowledgement. This work was supported by grants from the Natural Science Foundation of China (Project Number: 60772101), and by the Fundamental Research Funds for the Technology Development Project in Shanxi Province University (Project Number: 2010107).

References

1. Zhao, Q., Li, W.: Analyzing multiprocessor structure. Microcomputer Applications 26(1), 113–115 (2005)
2. Lee, K.-B., Lin, T.-C., Jen, C.-W.: An Efficient Quality-Aware Memory Controller for Multimedia Platform SoC. IEEE Transactions on Circuits and Systems for Video Technology 15(5), 620–633 (2005)
3. Acacio, M.E., Gonza lez, J.: An Architecture for High-Performance Scalable Shared-Memory Multiprocessors Exploiting On-Chip Integration. IEEE Transactions on Parallel and Distributed Systems 15(8), 755–768 (2004)
4. Li, w.: Control System Based on Switching of Shared Dynamic Memory with dual-processor. [Master thesis], Taiyuan University of Technology (2007)
5. Zhang, N., Zhang, P.-y., Liu, X.-c., Jiang, X.-y.: FPGA-based DDR SDRAM Controller Implemention. Computer Engineering and Application (24), 87–90 (2006)

The Formal Definitions of Semantic Web Services and Reasoning

Duan Yuexing

Division of Computer Foundational Teaching Course,
Taiyuan University of Technology, TaiYuan China
dyxty01@163.com

Abstract. Description logic is used as the formal theoretical foundation of semantic web services, the paper analyzed description language of Semantic Web Services OWL-S, WSDL-S, WSML and SWSL. Give a formal definition of description about semantic web services, further Web services is interpreted with the semantic of description logic. The satisfiability and subsumption in description logic can be applied to inferring semantic web services. For further research semantic annotation of web services, relation between web services and services, reasoning problems for semantic web services, our work was important and fundamental.

Keywords: Description logic, Semantic Web Services, Satisfiability.

1 Introduction

At present the research for semantics web services[1] mainly adopts two major technology: one is based on artificial intelligence (AI) Planning Technique, another is formal and automatic reasoning method. Both have their own advantages and disadvantages, and gained a series of Results, such as Article [2] by describing the logical axiomatic to depict the web service of IOPR (Inputs, Outputs, Preconditions and Results), puts forward the semantic web service combination problem into a description logic reasoning method. Article [3] is introducing a action formalism based on description logic, and analyzes on the different description logic reasoning tasks on standards for research, the influence of relevant action that reasoning laid solid foundation. The article [4] puts forward the SHOP2 HTN planning system based on the Web service automatic combination method, the author thinks in combination HTN planning task decomposition concept and DAML - S in the concept of decomposition of the composite process is consistent, and the semantic Web services adopted DAML -S description can translated into one of the domain SHOP, also SHOP2 output planning can be transformed into the format of the operation on DAML - S execution engine directly . The papers [1] used description logic to depict the Web service method we think overly complex, take into consideration on semantic Web service research only in OWL - S situation, also article[2] also lacked of variety of semantic Web service of consideration, if used the method on the paper[3] introduction of a Web services, Web services executing process produced a new individual when system cannot handle. This paper will present Web service various

G. Zhiguo et al. (Eds.): WISM 2011, CCIS 238, pp. 303–311, 2011.

semanteme abstracted as a simplified form, based on description logic ALC related the definition of the concept explains of the semantic on the web services concept, using knowledge on subsumption and satisfiability relationship under Knowledge base TBox, the research for semantic Web services formal and reasoning is discussed.

2 Description of Semantics Web Servics

On the semantic Web service study, Web services was added semantics annotation by ontology, and be translated into semantics web services. Such be ability to further study. Industry representative work has OWL-S, WSDL - S, WSML, SWSL etc. Below we give a brief account:

2.1 OWL-S

OWL-S (formerly DAML-S) is an ontology of services that makes these functionalities possible. It describe the overall structure of the ontology and its three main parts: the service profile for advertising and discovering services; the process model, which gives a detailed description of a service's operation; and the grounding, which provides details on how to interoperate with a service, via messages. The service profile tells "what the service does", in a way that is suitable for a service-seeking agent to determine whether the service meets its needs. This form of representation includes a description of what is accomplished by the service, limitations on service applicability and quality of service, and requirements that the service requester must satisfy to use the service successfully. The service model tells a client how to use the service, by detailing the semantic content of requests, the conditions under which particular outcomes will occur, and, where necessary, the step by step processes leading to those outcomes. That is, it describes how to ask for the service and what happens when the service is carried out. For complex services, this description may be used by a service-seeking agent in at least four different ways: (1) to perform a more in-depth analysis of whether the service meets its needs; (2) to compose service descriptions from multiple services to perform a specific task; (3) during the course of the service enactment, to coordinate the activities of the different participants; and (4) to monitor the execution of the service. A service grounding specifies the details of how an agent can access a service. Typically a grounding will specify a communication protocol, message formats, and other service-specific details such as port numbers used in contacting the service. In addition, the grounding must specify, for each semantic type of input or output specified in the ServiceModel, an unambiguous way of exchanging data elements of that type with the service.

2.2 WSDL-S

In 2005, W3C modified reports submitted by IBM and the LSDIS laboratory at the University of Georgia and developed Web Service Semantics---WSDL-S, This technical note prescribes a mechanism to annotate Web Services with Semantics. The advantage of this evolutionary approach to adding semantics to WSDL is multi-fold. First, users can, in an upwardly compatible way, describe both the semantics and operation level details in WSDL- a language that the developer community is familiar with. Second, by externalizing the semantic domain models, we take an agnostic approach to ontology representation languages. This allows Web service developers to annotate their Web services with their choice of modeling language (such as UML or OWL). This is significant since the ability to reuse existing domain models expressed in modeling languages like UML can greatly alleviate the need to separately model semantics. Moreover, this approach realizes the need for the existence of multiples ontologies, either from the same or different domains to annotate a single Web service and provides a mechanism to do so. Finally, it is relatively easy to update the existing tooling around WSDL specification to accommodate our incremental approach. WSDL-S is guided by the following.

Build on existing Web Services standards: they consider that any approach to adding semantics to Web Services should be specified in an upwardly compatible manner so as to not disrupt the existing install-base of Web Services.

The mechanism for annotating Web services with semantics should support user's choice of the semantic representation language: There are a number of potential languages for representing semantics such as OWL, WSML, and UML. Each language offers different levels of semantic expressivity and developer support. WSDL-S position is that it is not necessary to tie the Web services standards to a particular semantic representation language. By keeping the semantic annotation mechanism separate from the representation of the semantic descriptions, it offers flexibility to the developer community to select their favorite semantic representation language.

The mechanism for annotating Web services with semantics should allow the association of multiple annotations written in different semantic representation languages: Service providers may choose to annotate their services in multiple semantic representation languages to be discovered by multiple discovery engines. They believe that the mechanism for annotating Web Services with semantics should allow multiple annotations to be associated with Web Services.

Support semantic annotation of Web Services whose data types are described in XML schema: A common practice in Web services-based integration is to reuse interfaces that are described in XML.

Provide support for rich mapping mechanisms between Web Service schema types and ontologies: Given their position on the importance of annotating XML schemas in Web service description, they believe that attention should be given to the problem of mapping XML schema complex types to ontological concepts.

2.3 WSML

WSML is a language for modeling Web services, ontologies, and related aspects, and is based on the Web Service Modeling Ontology WSMO. The formal grounding of the language is based on a number of logical formalisms, namely, Description Logics, First-Order Logic and Logic Programming. Besides providing its own language for modeling ontologies, it allows the import and use of RDF Schema and OWL ontologies for Web service description. The WSML syntax consists of two major parts: the conceptual syntax and the logical expression syntax. The conceptual syntax is used for the modeling of ontologies, goals, web services and mediators; these are the elements of the WSMO conceptual model. Logical expressions are used to refine these definitions using a logical language. A WSML document has the following structure:

wsml = wsmlvariant?namespace?definition*

wsmlvariant shows the different variants of WSML: WSML-Core WSML-Flight WSML-Rule WSML-DL WSML-Full. The namespace keyword is followed by a number of namespace references. Each namespace reference, except for the default namespace, consists of the chosen prefix and the IRI which identifies the namespace. Definition element defines the statement with which wsml is related to. It has involved goal | ontology | webservice | mediator | capability | interface. A WSML Web service specification is identified by the webService keyword optionally followed by an IRI which serves as the identifier of the Web service. If no identifier is specified for the Web service, the locator of the Web service specification serves as identifier.

A Web service specification document in WSML consists of:

webservice = 'webService' iri ? header * nfp * capability ? interface *

2.4 SWSL

The Semantic Web Services Language (SWSL) is used to specify the Semantic Web Services Ontology (SWSO) as well as individual Web services. The language consists of two parts: SWSL-FOL, a full first-order logic language, and SWSL-Rules, as rule-based language. SWSL-FOL is primarily used for formal specification of the ontology and is intended to provide interoperability with other first-order based process models and service ontologies. In contrast, SWSL-Rules is designed to be an actual language for service specification.

The SWSL-Rules language is designed to provide support for a variety of tasks that range from service profile specification to service discovery, contracting, policy specification, and so on. The language is layered to make it easier to learn and to simplify the use of its various parts for specialized tasks that do not require the full expressive power of SWSL-Rules.

SWSL-FOL is used to specify the dynamic properties of services, namely, the processes that they are intended to carry out. SWSL-Rules have monotonic semantics and therefore can be extended to full first-order logic. Above each layer of SWSL-Rules, we know corresponding SWSL-FOL extension. The most basic extension is SWSL-FOL. The other three layers, SWSL-FOL+Equality, SWSL-FOL+HiLog, and SWSL-FOL+Frames extend SWSL-FOL both syntactically and semantically. Some of these extensions can be further combined into more powerful FOL languages.

3 Semantics Web Services on Description Logic

3.1 Description Logic

Description Logics are a family of knowledge representation languages which can be use to represent the terminological knowledge of an application domain in a structured and formally well-understood way. The main expressive means of description logic are so-called concept descriptions, which describe sets of individuals or objects. Formally, concept descriptions are inductively defined with the help of a set of concept constructors, starting with a set N_C of concept names and a set N_R of role names. The available constructors determine the expressive power of the DL in question. We only consider concept descriptions built from the constructors shown in Table1, where C,D stand for concept descriptions, r for a role name. In the description logic ALC, concept descriptions are formed using the constructors negation, conjunction, disjunction, value restriction, and existential restriction. The semantics of concept descriptions is defined in terms of an interpretation $I=(\Delta^I, \cdot^I)$. The domain Δ^I of I is a non-empty set of individuals and the interpretation function $\cdot^I$ maps each concept name $A \in N_C$ to a set $A^I \subseteq \Delta^I$ and each role name $r \in N_R$ to a binary relation $r^I \subseteq \Delta^I \times \Delta^I$. The extension of $\cdot^I$ to arbitrary concept descriptions is inductively defined, as shown in the third column of Table 1.

Table 1. Syntax and Semantics of concept descriptions

Constructor	Syntax	Semantics
negation	$\neg C$	$\Delta^I \backslash C^I$
conjunction	$C \cap D$	$C^I \cap D^I$
disjunction	$C \cup D$	$C^I \cup D^I$
existential restriction	$\exists R.C$	$\{x \in \Delta^I \mid \exists$ 一$y \in \Delta^I$, 使得：$(x, y) \in R^I \wedge y \in C^I\}$
value restriction	$\forall R.C$	$\{x \in \Delta^I \mid \forall$ 一$y \in \Delta^I$, 使得：$(x, y) \in R^I \rightarrow y \in C^I\}$

3.2 Formalism for Semantics Web Services

The description of Semantic Web services has shown a variety of forms, so that we can use different semantics description language to describe the same Semantic Web Services in a domain with OWL-S, WSDL-S,WSML, and SWSL. As described above, as each of the semantic Web service language has its own characteristics. However, their semantics are inseparable from the ontology-related. Any a services are formed in specific domain, however the service forming in different domain falls far short of its function. For example, on the field in the financial domain you can not get a service of ticketing about travel. Contrarily in the transport sector you can not carry out a service in financial management. Below we formalize the semantic Web services; we can put the set of all services generated by a semantic description language in a field to be discussed as the domain. Each individual on domain is the one related services in the field. Service is characterized by mass, and fair, and it involves two basic aspects: one is the output, one input.In front of the fairness we believe that the input can be anonymous and output have unique after a certain abstract. On precondition, Semantic Web services in a domain can be seen as an abstraction of similar services formed by using a semantic Web service description language. It is considered to be a concept described by description logic in a domain. The relationship between semantics web services was seen as the role of description logic. Below we will discuss in a domain. The domain is represented by ontology T. Being discussed, the domain is the set of all services generated by a semantic description language in a field.

Definition 1(Formalism services). We use the following expression to describe a Semantic Web services in domain: LetΓbe an acyclic TBox, an atomic service $SWS\equiv(I,O)_\Gamma$ for an acyclic TBox Γ,where $I=<I_1,I_2,...In>$,$O==<O_1,O_2,...Om>$, when SWS was executed, I was a group of concepts related with ontology T and need to be entered; When the SWS is executed, O is a group of concepts related with ontology T and is the output.

Definition 2(new services). Let a atomic denumerable set N_C of concept names and a atomic denumerable set N_R of role names. Service descriptions are inductively defined as follows:

(1) Any a atomic service A, If $A \in N_C$, then A is an ALC web service.
(2) If C and D are ALC services, R is atomic role, $R \in N_R$, then $\neg C$, $C \cap D$, $C \cup D$, $\exists R.C$, $\forall R.C$ is an ALC service.

We use the top services$\top$to correspond to the set of all the services the individual in the field, and use the bottom services $\bot$.to correspond to the empty set.

3.3 Description Logic Based Semantics Description for Web Services

We know that in the field T a semantic interpretation of the concept is a binary group and semantic interpretation are varied. In the field T, explanations of a concept are endless. For example, in terms of Web services semantic interpretation, OWL-S, WSDL-S, WSML and SWSL is a variety of expression of semantic interpretation. In other words we can use OWL-S, WSDL-S, WSML and SWSL for different semantic interpretation on the same web service. In the future there will be more semantic interpretation.

Definition 3(semantics for web services). In the field T a semantic interpretation of the concept is a binary group $I=(\Delta^I, \cdot^I)$, The domain Δ^I of I is a non-empty set of services and the interpretation function $\cdot^I$ (owsl-s|wsdl-s|wsml|swsl)maps each service name $C \in N_C$ to a set $C^I \subseteq \Delta^I$ and each role name $R \in N_R$ to a binary relation $R^I \subseteq \Delta^I \times \Delta^I$. The extension of $\cdot^I$ to arbitrary service descriptions is inductively defined, as shown in the third column of Table 2.

Table 2. Semantics for new web service

new services Semantics	atomic services Semantics
$(\neg C)^I$	$\Delta^I \backslash C^I$
$(C \cap D)^I$	$C^I \cap D^I$
$(C \cup D)^I$	$C^I \cup D^I$
$(\exists R.C)^I$	$\{x \in \Delta^I \mid \exists$ 一$y \in \Delta^I$, 使得：$(x,y) \in R^I \wedge y \in C^I \}$
$(\forall R.C)^I$	$\{x \in \Delta^I \mid \forall$ 一$y \in \Delta^I$, 使得：$(x,y) \in R^I \rightarrow y \in C^I \}$

4 The Formation of Semantic Web Services and Reasoning

Let semantics web services $U \equiv (I,O)_\Gamma$, $C \equiv (I^*,O^*)_\Gamma$, $D \equiv (I',O')_\Gamma$, I=<I_1,I_2,...In>, O==<O_1,O_2,...Om>, I^*=<I_1,I_2,...Ip>,O^*==<O_1,O_2,...Oq>, I'=<I_1,I_2,...Ir>, O'==<O_1, O_2,...Ow>, when SWS U,C,D was executed Respectively, I, I^*,I' was severally a group of concepts related with ontology T and need to be entered; When SWS U,C,D was executed Respectively, O, O^*,O' was severally a group of concepts related with ontology T and is the output. According to the existing service we can inductively form the number of new services. Atomic services include:

$U \equiv \neg C$ the input I of service U belong to a non-C service; the output of service U is corresponding to the output of the above non-C service.

$U \equiv C \cup D$ the input I of service U suffice the input I^* of service C or the input I' of service D; the output O of service U is the output O^* of service C or the output O' of service D. In other case, part of the input I of service U suffice the input I^*, another part of the input I of service U suffice the input I'; Part of the output O of service U is the output O^*, the another part of the output O of service U is the output O'.

$U \equiv C \cap D$ the input I of service U suffice the input I^* of service C and the input I' of service D at the same time; The output O of service U is the common part of output O^* of service C and the output O' of service D.

$U \equiv \exists R.C$ At least one service U with which the service C satisfies role R, the input I and output O of service U and it's of service C has a relationship.

$U \equiv \forall R.C$ Any one service U with which the service C satisfies role R, the input I and output O of service U and it's of service C has a relationship.

There, the relation R has many forms like as described in description logic. It possesses different characteristics, example for reversible, transitive, hierarchy Etc. With different characteristics between the input and output of service U and it's of service of V have a definite link. This is where we need to study further.

In a field, (no matter what kind of semantic web service language) generation services are limited. We can put these limited services as atomic services. However new services composed of atomic services are infinite and countable. How to determine whether a new service makes sense?

From a description logical point of view, we need to judge whether a new service is satisfiable with respect to T. We all know satisfiability checks of service are one way of reasoning mainly. Further, Article [10] has proved most reasoning of service can be reduced to satisfiability checks. We give four forms of reasoning for a TBox in the paper:

Satisfiability: A ALC service C is satisfiability with respect to T if there exits a model I of T such that C^I is nonempty. In this case we say also that I is a model of C. The individual x in domainΔ^I is an instance of service C with respect to ABox, iff $x^I \in C^I$.

Subsumption: A ALC service C is subsumed by a service D with respect to T if $C^I \sqsubseteq_T D^I$ for every model I of T. In this case we write $C \sqsubseteq D$.

Equivalence: Two services C and D are equivalent with respect to T if $C^I \equiv D^I$ for every model I of T. In this case we write $C \equiv_T D$.

Disjointness: Two services C and D are disjoint with respect to T if $C^I \cap D^I \equiv \emptyset$ for every model I of T.

By subsumption we obtained some inferences:

Proposition 1 For service C and D

(1) C is unsatisfiable iff C is subsumed by $\perp$;

(2) C and D are equivalent iff C is subsumed by D and D is subsumed by C;

(3) C and D are disjoint iff $C \cap D$ is subsumed by $\perp$.

Proof: (1) because of C is unsatisfiable, according to the definition of satisfiability, we know the interpretation C^I of service C is empty. In other words, then $C^I \equiv \emptyset$. However, the interpretation of $\perp$ service is empty service. So service $C \subseteq \perp$.

On the contrary, according to the definition of $\perp$ service, we know where is no any individual in the interpretation of service $\perp$. Then there is no any individual in a subset of $\perp$ service .If $C \subseteq \perp$, have $C^I \equiv \emptyset$, then service C is unsatisfiable.

Similarly according to the knowledge of set theory, proposition (2) and (3) can be proved.

There is a negation constructor in ALC description logic. By negation constructor, we achieved that a relations of service subsumption, equivalence and disjointness can be reduced to unsatisfiability. Then to inference satisfiability of service by unsatisfiability.

Proposition 2 For service C and D

(1) C is subsumed by D iff $C \cap \neg D$ is unsatisfiable ;

(2) C and D are equivalent iff both $(C \cap \neg D)$ and $(\neg C \cap D)$ are unsatisfiable;

(3) C and D are disjoint iff $C \cap D$ is unsatisfiable.

Proof: (1) assume $C \subseteq D$, then service C does not belong to $\neg D$. Then $C \cap \neg D$ is empty. According to the definition of satisfiability, so $C \cap \neg D$ is unsatisfiable.

On the contrary, if $C \cap \neg D$ is unsatisfiable, there is no any individual in interpretation of service $C \cap \neg D$. In other words, $C \cap \neg D$ is empty. Then service C belong to service D, $C \subseteq D$ established.

(2) if C and D are equivalent, have $C \subseteq D$ and $D \subseteq C$. According to (1), because $C \subseteq D$, then $C \cap \neg D$ is unsatisfiable; because $D \subseteq C$, then $\neg C \cap D$ is unsatisfiable. So $(C \cap \neg D)$ and $(\neg C \cap D)$ are unsatisfiable.

On the contrary, if $(C \cap \neg D)$ and $(\neg C \cap D)$ are unsatisfiable. According to (1), because $C \cap \neg D$ is unsatisfiable, then have $C \subseteq D$; because $\neg C \cap D$ is unsatisfiable, have $D \subseteq C$; So that we can get: $C \equiv_T D$.

Similarly according to the knowledge of set theory, proposition (3) can be proved.

5 Conclusion and Future Work

In this work, we have shown how to use description logic to formalism for semantic web services. We proposed a approach and considered that a semantic web services is the concept of a field. Give formal definition of semantic web services and explain web services by the semantics of describing logic. Explain why the semantics of

Semantic Web Services can be varied. In the future there will be a more appropriate the language for description semantics. Our ideas was proposed for the first time which the formal definition for the semantic web services is a abstract conception contained two aspects of input and output.

As a future work, we are working on the relationship between services conception example for reversible, transitive, hierarchy. We are also investigating the possibility of composing new service from this, and further discussed the satisfiability of formal definition of semantic web services. As DLs play an important role in the Semantic Web, we expect to find an increasing number of activities in semantics web services.

References

1. McIlraith Sheila, A., Cao, S.T., Honglei, Z.: Semantic Web Services. IEEE Intelligent Systems. Special Issue on the Semantic Web 16(2), 46–53 (2001)
2. 王杰生, 李舟军, 李梦君. 用描述逻辑进行语义Web服务组合. 19(4), 967–980 (2008)
3. Baader, F., Lutz, C., Milicic, M., Sattler, U., Wolter, F.: A description logic based approach to reasoning about Web services. In: Vasiliu, L. (ed.) Proc. of the WWW 2005 Workshop on Web Service Semantics: Towards Dynamic Business Integration, pp. 636–647. ACM Press, Chiba (2005)
4. Sirin, E., Parsia, B., Wu, D., Hendler, J.A.: HTN Planning for Web Service Composition Using SHOP2. Journal of Web Semantics 1(4), 377–396 (2004)
5. The, OWL Services Coalition. OWL-S: Semantic Markup for Web Services, OWL-S 1.0 Release (2003)
6. Akkiraju, R., Farell, J., Miller, J., Nagarajan, M., Sheth, A., Verma, K.: Web Service Semantics-WSDL-S Version1.0 (2005), http://www.w3.org/Submission/WSDL-S/
7. Fensel, D., Kifer, M., de Bruijn., J., et al.: D16.1 v1.0 WSML Language Reference, http://www.wsmo.org/TR/d16/d16.1/v1.0
8. Battle, S., Bernstein, A., Boley, H., Grosof, B., Gruninger, M., Hull, R., et al.: Semantic Web Service LanguageVersion1.0, http://www.w3.org/Submission/SWSF-SWSL/
9. Schmidt-Schauβ, M., Smolka, G.: Attributive concept descriptions with complements. Artificial Intelligence 48, 1–26 (1991)
10. Baader, F., Nutt, W.: Handbook of Description logic, ch.2. Cambridge University Press, Cambridge (2003)

Semantic Video Retrieval System Based on Ant Colony Algorithm and Relevant Feedback

Jianjun Liao[1,*], Jianhui Chen[1], Xiaoming Liu[2], and Xiaoning Li[3]

[1] Zheng Zhou Institute Of Aeronautical Industry Management, Zheng Zhou, China
{liaojj,chenjianhui}@zzia.edu.cn
[2] Henan University of Technology hut, Zheng Zhou, China
ming616@gmail.com
[3] Wanfang College of Science & Technology HPU, Zheng Zhou, China
25787040@qq.com

Abstract. How to effectively use the user's relevance feedback information to semantic-based video retrieval is an important and challenging problem.This paper presents a relevance feedback video retrieval method based on ant colony algorithm, which is the traditional relevance feedback methods for an improved video retrieval. With the idea of ant colony algorithm, using user feedback to establish the video key frame of the semantic web and the iterative method used to retrieve the video. Experiments show that this method is not only effective, but storage capacity is small, little calculation.

Keywords: video retrieval, relevance feedback, ant colony algorithm, semantic web.

1 Introduction

Content-based video retrieval, visual features are often difficult to correct the underlying expression of the user's understanding of video, so the use of semantic search to narrow the "semantic gap" has been the video retrieval of the current hot and difficult. [1] In order to solve the "semantic gap" problem, people try to user feedback in a retrieval system, established the correlation between lower video features and high-level semantics through Human Computer Interaction, which is Relevance Feedback. [2] Relevance Feedback from text retrieval, was later applied to the image and video retrieval. The technology is implemented in 1998, the first time in the MARS system.The results show that the Relevance Feedback can improve the efficiency and accuracy of image retrieval. [3]

This paper presents an improved relevance feedback methods of video keyframe retrieval. [4] Using the idea of ant colony algorithm to establish the semantic web of the keyframe. Semantic relevance of the keyframes are stored in a pheromone matrix. Users to retrieve the video process is considered as one ant foraging. Every time, the pheromone matrix according to the semantic information retrieve video. All data is

---

* Jianjun Liao, Lecturer, Master, the direction of research: Information Management and Information System.

G. Zhiguo et al. (Eds.): WISM 2011, CCIS 238, pp. 312–319, 2011.

stored in a matrix. When updating data, the matrix only needs to be updated. Compared to conventional video retrieval methods, the method stores a small amount, less calculated.

2 Video Retrieval Method Based on Ant Colony Algorithm and Relevance Feedback

2.1 Introduction to Ant Colony Algorithm

Ant colony algorithm is proposed by Italian scholars Dorigo M and others in 1991. Especially, after 1996, Dorigo M and others expounded the basic principles of ant colony algorithm and the mathematical model systematically. The algorithm gradually caught the attention of researchers from countries, and the fields of their applications had been rapidly expanded. [5]

Ant colony algorithm is a simulation which was faced on the foraging behavior of real ants in nature. When ants foraging, Their secretion of a chemical irritants - Pheromone will be left on their path. The more the ants passed the path, which caused pheromone increased intensity, the more attractive for the ants passed the path later; while there are fewer ants passed the path, the pheromone on the path will gradually evaporate over time. The algorithm seeks the optimal solution by the groups evolution of candidate solutions.

2.2 Relevance Feedback Retrieval

Over the recent years, there are more and more technology research about relevance feedback, what's more, there have been many relevance feedback algorithms. Generally speaking, that mainly divids into the following categories: Based on the method of inquiry points on movement, based on characteristics on the method of weight adjustment, based on the method about machine learning, based on the approach to memory-type. The first three approaches to categories are putting the user's current effects on feedback information to the retrieval about lower-level features which based on the image to optimize the search. However, the image similarity of lower-level features does not mean that they are similar to the semantics. Therefore, the effects of these three categories approachs are always lower than the expectations. The relevance feedback based on the memory style is through learning the feedback log from users to optimize the search. These methods are called the first relevance feedback based on long-term learning, and later Li M, Han J, Shyu M, who makes the methods for the further development, puts forward a method of the memory- type image retrieval. This method will transform the logs of the user's feedback into the semantic association between the images and then store them into the matrix associated with the semantics information. Along with the accumulation of the user's feedback information, the semantic association matrix is increasingly improved, and gradually forms a semantic network,thus image retrieval is conducted by the Semantic Web. Experimental results show that this approach works well. The implementation of this method needs to the definite three matrices: the semantic association matrix as sc, at the same time the co-positive-feedback frequency matrix as cop, also the co-feedback frequency matrix as cof. The image i and the image j of the semantics on relational degree sc (i, j) is defined as:

$$sc(i,j) = \begin{cases} 1 & if \quad i = j \\ \dfrac{cop(i,j)}{cof(i,j)} & if \quad i \neq j \, and \, cof(i,j) \neq 0 \\ 0 \quad i = j & if \quad cof(i,j) = 0 \end{cases} \tag{1}$$

Among the Formula, sc(i,j)=sc(j, i), and the $0 \leq sc(i,j) \leq 1$. In which, cop(i,j) of an image i and image j is the number of positive feedback at the same time, cof(i,j) means an image i and image j were with the same number of positive feedback and the number of positive feedback.

Advantage of this method is: The image retrieval based on the the semantic information come directly from users themselves which truly realized image retrieval based on semantic will have a good effect. Disadvantages of this method is: It requires a lot of the user's feedback and three similar matrix which are used to store the image information Every time you update data, matrix to be updated simultaneously on the three, which will lead to large amount of memory and calculations.

2.3 Semantic Web Based on Ant Colony Algorithm

Ant colony algorithm is based on informations. It seeks optimal solutions by the groups evolution of candidate solutions. The keyframe search based on the semantic, for candidates solution can be regarded as the group composed of (each related degree between the keyframes) evolution to seek optimal solutions.Each user is regarded as a ant, and the process of user retrieve video is regarded as a ant out to eat.The user can find the keyframe of video they want along the informations predecessors left, and then choose the most satisfactory video. At the same time, they left new informations, update the pheromone matrix. With the accumulate of user feedback, the video between the keyframes of semantic web also gradually formed.

This work needs to be based on an assumption that: the video made users satisfied is based on a certain relevant semanteme at the same query.[6] First of all, we provide the definition of the pheromone matrix which storage the semantic web of video keyframe.

Define 1: Pheromone Matrix.

Defining pheromone matrix is the matrix of pheromone,which are used for storing the relevant degree of semanteme between the keyframes of video.If there have N pieces of keyframe, then matrix size is N×N. The main diagonal of the matrix is initialized to a value of 1, other 0 ,it is called unit matrix.

When the user completed a video retrieval, the system will update pheromone matrix according to the user's choice. If the keyframe i and the keyframe j were choosed by the users finally, it is considered that they are similar and leave pheromone among them:

$$pheromone(i,j)_t = pheromone(i,j)_{t-1} + \frac{\mu \times (1 - pheromone(i,j)_{t-1})}{length} \tag{2}$$

Among them, the pheromone $(i, j)_t$ is the information between the keyframe i and the keyframe j at time t, namely their similarity degree. The higher the value, the similar the two keyframes. pheromone $(i, j)_t \in [0, 1]$, and when i = j, pheromone $(i, j)_t = 1$. μ is pheromone growth factor, and $0 < \mu < 1$. length is the number of keyframes the user choice of query the video to their own satisfaction.

With the time passed, the pheromone will volatile slowly, so updated pheromone matrix regularly:

$$pheromone(i, j)_t = \begin{cases} \eta \times pheromone(i, j)_{t-1} & i \neq j \\ pheromone(i, j)_{t-1} & i = j \end{cases} \tag{3}$$

Among them, η is the decay factor of pheromone, and $0<\eta<1$. Such, it can make pheromone matrix update efficiently, and the semantic information contains more close to the user's interests; At the same time,it also can gradually fade the user error feedback.

After many trainings, the pheromone matrix storage the semantic information associated between each keyframes of video, which constitutes a semantic web, keyframes in the semantic web were by fuzzy clustering. Below are the definition of fuzzy clustering center of the keyframes.

Definition 2: Fuzzy clustering center of keyframes.

if the key frame i is fuzzy clustering center, then meet the following conditions:
While pheromone(i,j)>0,
sum_pheromone(i)>sum_pheromone(j)

Among them, $sum\_pheromone(i) = \sum_{k=1}^{n} pheromone(i,k)$

Obviously, every keyframe is a cluster center before trained. After several trains, keyframes well be fuzzy clustered according to the similarity of semanteme. Cluster center have many semantemes meaning, the keyframe which is related to it must has one or several same semantemes with it. [7]

3 Video Retrieval System Based on Semantic Relevance Feedback

This video keyframe retrieval system design as follows: After intercepted and extracted the underlying characteristics of the frame successfully, exploit the methods of multivariate regression SVR to make semantic annotation, [8] and then used the feedback correction semantic weight to establish feedback retrieval based on ant colony algorithm, and improve it. [9] System framework shown in Figure 1.

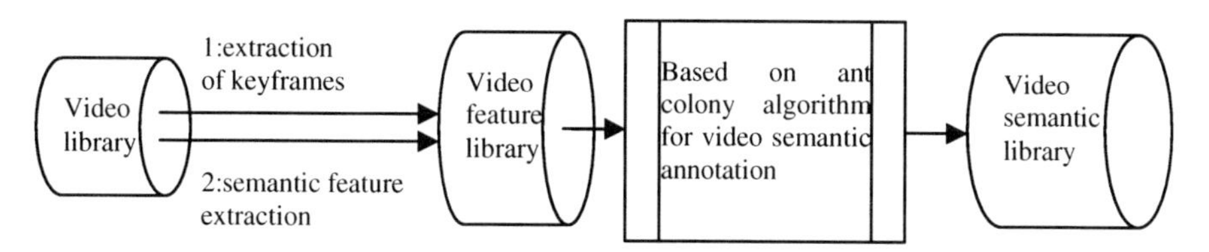

Fig. 1. Video retrieval system based on ant colony algorithm

Pheromone matrix is actually recorded by user's relevance feedback formed the formation of semantic web. Video retrieval is conducted based on the semantic web, and while use it ,learning the user's behavior at the same time.

When user begin to search the videos,the system will offer every cluster center in the Pheromone matrix to the users.User choose the keyframe of video (Pheromone matrix) which close to their target, then system will offer these keyframes and related to the same meaning of the keyframes to the user to choose.Then the user make a choice again and then submit to the system. The user do it again and again until they find their satisfied key frame. The steps of the video retrieval system based on ant colony algorithm shown in Figure 2.

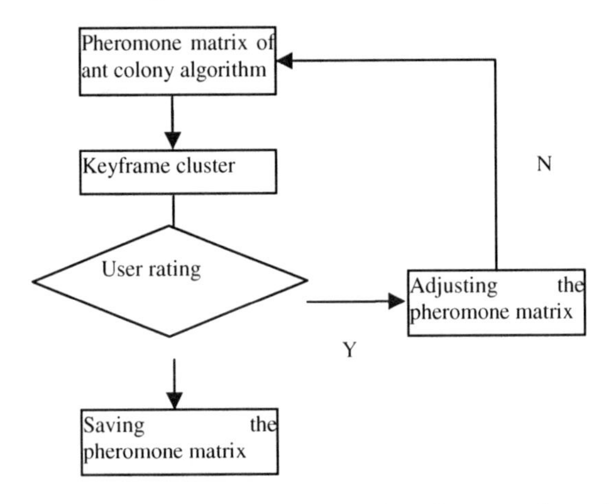

Fig. 2. The steps of the video retrieval system based on ant colony algorithm

Every iteration, the user is close to their target gradually.Until seeing their satisfied picture, the user will finally choose these keyframes which satisfied to their require. The system regard the final keyframe choosed by the user as semantic related, and renew the related part of the pheromone matrix.

4 Experimental Results and Analysis

In this paper, using visual studio 2008 implements a relevance feedback video retrieval system based on ant colony algorithm. The system's interface is shown on Figure 3 as follows.

Fig. 3. Retrieval interface

The video library includes one thousdands videos, its contents involved more than ten topics, such as airplanes, cars, animals, high mountains, constructions ,and so on. It adopts 1000×1000 matrix as the pheromone matrix.First of all, selecting 3 trains this system six times with different purposes respectively, and then count the system test results. The results showed that an average of 6 iterations can find satisfactory results. With the accumulation of user feedback, it needs average of 3-4 iterations finally that can find one saticfactory result,and this number tends to normality. Selecting 4 topics makes a test,calculate the initial search and the first 15 keyframe precision, recall, which is used the ant colony algoriyhm to return after 3-4 iterations.The results show in Figure 4 and Figure 5.

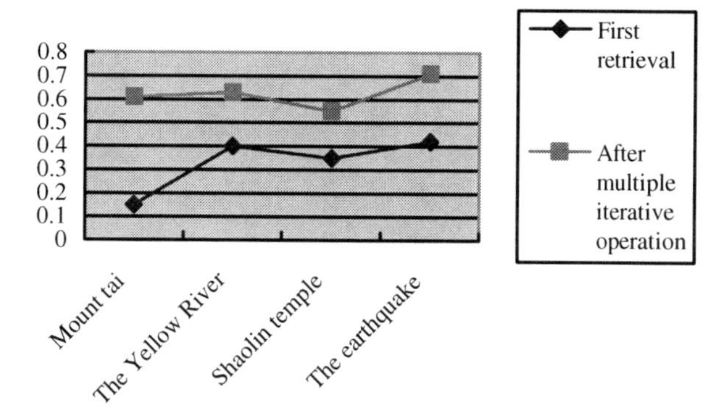

Fig. 4. The comparison of precision of the initial search and 3~4 iterations by ant colony algorithm

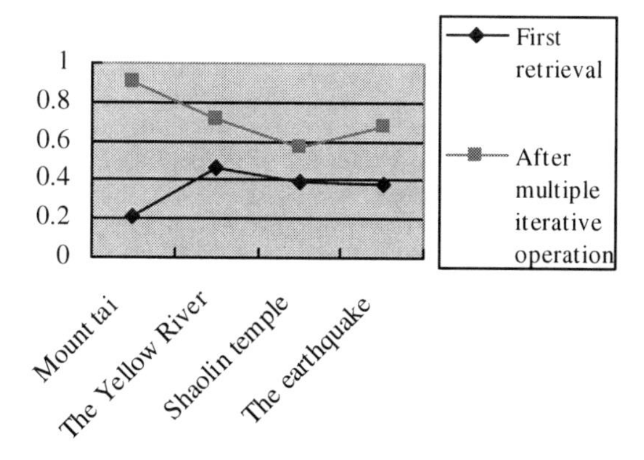

Fig. 5. The comparison of recall of the initial search and 3~4 iterations by ant colony algorithm

The three literation process that the user searchs for the earthquake in vedio library shown in Figure 6. (a) is an initial state, the system provides all kinds of clustering center in the vedio library to the users, and they page to look the keyframe that alike his own goal; (b) is a picture, after make a decide for the user, the system will appers a picture similai with it, the user will look for the more similar his idea once more. (c) The use serearched the satisfactory image, the system will remember the choice you have made, and then updating the pheromone matrix.

(a) The first search (b) The second iteration (c) The third iteration

Fig. 6. The process of three iterations for searching yellow sport cars in the image library

Beacause the basis of retrieval is from the users who feedback on the semantcs associated with the information between keyframes. So it is ture that this is entirely based on the semantic retrieval, which is very effective. The conventional method requiring three matrices to store the similar information coming from the keyframes.Every time updating data, you should update on the three matrices simultaneously, which has a large memory capacity and a caculation of valume. The method of this paper just need a matrix to store the video information from keyframes under the circumstance of guaranteing the efficiency, which geatly reduces the amount of storage and computation.

5 Conclusion

This paper is an improved relevance feedback method of video retrieval. It is not only introduced the idea of ant colony algorithm to create a video keyframe of the semantic web, but also proposed a semantic web-based video retrieval relevance feedback model, and the model is an iterative approach to gradually close to the user's goals. Compared with traditional methods, it only to use a matrix to store the video keyframe semantic information,and effectively record the user's feedback, what's more, it has the advantages of storage capacity is small and less computation. Of course, as the same with all the traditional video retrieval relevance feedback methods, the inadequacy of this method is requires a lot of feedback.

References

1. Datta, R., Joshi, D., Li, J., et al.: Image retrieval: Ideas, influences, and trends of the new age. In: ACM Computing Surveys (CSUR), pp. 17–60 (2008)
2. Shyu, M., Chen, S., Chen, M., et al.: Probabilistic semantic web-based image retrieval using MMM and relevance feedback. Springer Journal of Multimedia Tools and Applications, 50–59 (2006)

3. Qi, G.-J., Tang, J., Wang, M., et al.: Correlative multilabel video annota-tion with temporal kernels. ACM Trans. Multimedia Comput Commun., 27–30 (2008)
4. Zavesky, E., Chan, S.F.: Columbia University's Semantic Video Search Engine. In: CIVR 2008 (2008)
5. Duan, H.: Ant colony algorithms: Theory and applications, pp. 99–130. Science Press, Beijing (2005)
6. Ralph, E., Bernd, F.: Semi-supervised learning for semantic video retrieval, pp.154–161. CIVR, Amsterdam Netherlands (2007)
7. Amir, A., Argillander, J., Campbell, M., et al.: IBM research TREC-VID-2005 video retrieval system. In: Proceedings of the TREC-VID 2005 Workshop, pp. 73–75. NIST Special Publications (2005)
8. Chen, L.-h., Chin, K.-h., Liao, H.-y.: An integrated approach to video retrieval. In: Proc. 19th Australasian Database Conference(ADC 2008), pp. 49–56. Australian Comuter Society, Inc., Wollongong (2008)
9. Ngo, C.W., Pong, T.C., Zhang, H.J.: On clustering and retrieval of video shots through temporal slices analysis. IEEE Transactions on Multimedia, 446-459 (2002)

A Semantic-Based Mobile Publishing Framework with Copyright Protection

Chen Hejie and Hua Yuhong

Beijing Institute of Graphic Communication, School of Economics and Management,
Daxing Huangcun Xinhuabeilu 25, China
chenhejie81@sina.com, huayuhong@bigc.eud.cn

Abstract. With the rapid development of Internet and mobile telecommunication technology, more and more people use mobile terminal to reading. The rapid development of mobile reading promoted the mobile publishing industry development. But there is no authoritative definition of mobile publishing. This paper firstly presented a kind definition of mobile publishing. The definition emphasized on the differences between mobile publishing and mobile phone publishing. The differences included terminal, content classification, technology of implementing mobile publishing. In a word, the bound of mobile publishing is larger than mobile phone publishing. On the other hand, it is important for mobile publishing industry chain to copyright protect. So the paper emphasized on copyright protection adopted semantic web. The paper introduced copyright protection ontologies of mobile publishing. The paper also presented a semantic-based mobile publishing framework with copyright protection and copyright protection process.

Keywords: Mobile publishing framework, Copyright protection, Semantic web, Semantic Service System, Ontology.

1 Introduction

With the rapid development of Internet and mobile telecommunication technology, digital reading took on an explosive growth. The rate of mobile reading significantly increased. According eight national reading survey, the rate of digital reading was 32.8% in 2010, which it increased 8.2% than in 2009[1]. Further analysis of the digital reading survey found these conclusions.

- 23% residents used mobile phone to reading and mobile phone reading increased 8.1% in 2009.
- 18.1% residents used Internet to reading and Internet reading increased 1.4 % in 2009.
- 3.9% residents used electronic reader to reading and electronic reader reading increased 2.6%t in 2009.

On the other hand, the rapid development of the mobile Internet technology ensured wireless Internet reading by mobile terminal. More and more people use mobile terminal to reading.

G. Zhiguo et al. (Eds.): WISM 2011, CCIS 238, pp. 320–327, 2011.

The rapid development of mobile reading promoted the mobile publishing industry development. Up to the end of 2009, the active users of Chinese mobile publishing reached 155 million, mobile publishing market sized up to 3.0 billion[2]. Mobile publishing maintained high-speed development trends. Amazon, Sony, hanvon and telecommunications operators are entering the industry. The market competition is constantly intense.

As a young industry, there are many questions in the mobile publishing industry. For example, industry chain has not yet formed. Up enterprises, middle enterprises and forward enterprises and their functions haven't identified. Copyright reduce the industry development speed. The profit distribution isn't reasonable. These crucial questions constrained industry development. This paper presented a mobile publishing architecture to solve some difficulty questions. The system used of service oriented architecture and semantic web.

The paper is structured as follows. Section 2 provided mobile publishing concept. Section 3 outlined some ontologies specification of providing copyright protection. Section 4 introduced an ontology-based mobile publishing framework with copyright protection. Section 5 provided the conclusions and the future works.

2 Mobile Publishing Concept

So far, there is no authoritative definition of digital publishing. The fact recognized by the public is that digital publishing is a kind of the digital publishing. Before the paper presented mobile publishing concept based on our team research, the paper firstly introduced the digital publishing concept.

Digital publishing refers to apply binary technology in entire publishing process, including digital original creation content, digital editing, digital printing, digital copying, digital distribution and digital reading. Digital publishing emphasizes digital content, digital process of production mode and operation process, digital communication carriers and digital read consumer, and digital learning styles.

Definition of mobile publishing is currently no authoritative accepted by the publishing industry. We referred to the definition of the mobile phone publishing. According to the 'Standards Architecture of Mobile Publishing', mobile pone publishing is the concept that the service providers adopted the words, photos, audios, videos or other forms media to create contents for reading. The selected and edited contents which are provided by the contents providers or others authors are converted the digital publishing and translated through the wireless Internet, cable Internet, and embed-in phone. The users used mobile phones or similar mobile terminal to read or download [3]. Our team believed that the area of mobile publishing is larger than mobile phone publishing.

- First difference is the terminal. The device characters of mobile publishing are mobile, stored, download resource and upload data. And the screen size is limited. So, the device of the mobile publish included table PC, notebook PC of limited screen size, mobile phone, digital reader etc.
- Second difference is sorts of mobile publishing. By the content classification, mobile phone publishing can be divided into mobile literary works, mobile animation games, mobile music, mobile TV, etc. But the bounds of mobile

publishing sorts are broader than mobile phone publishing. Our team considered that interactive form of content distribution is a kind of mobile publishing. For example, micro-blog is a kind of mobile publishing (Micro-blog used 140 words to distribute information of the promulgator and instantly share in other person.).

- Third difference is that mobile publishing used more technologies. Mobile phone publishing can be divided into the Wireless Internet, Cable Internet, and embedded mobile phone according to technology. But mobile publishing used wireless Internet, Cable Internet, WAPI, WIFI, Bluetooth, GPS, CMMB.

3 Mobile Publishing Industry Chain and Copyright Protection

3.1 Mobile Publishing Industry Chain

Traditional publishing industry chain included authors, publishers, wholesales, retailers and readers.

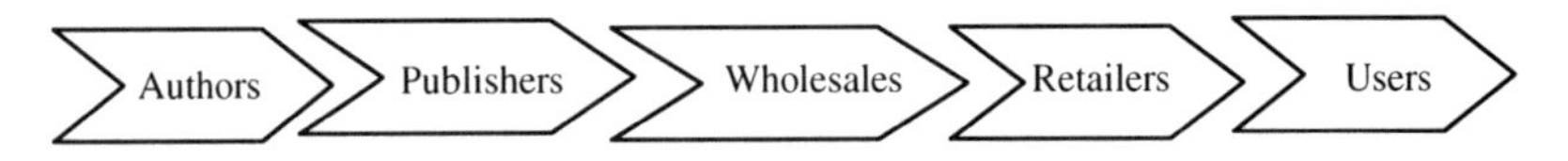

Fig. 1. Traditional publishing industry chain

Traditional mobile telecommunication industry chain included equipment providers, telecommunication operators, terminal manufacturers and users.

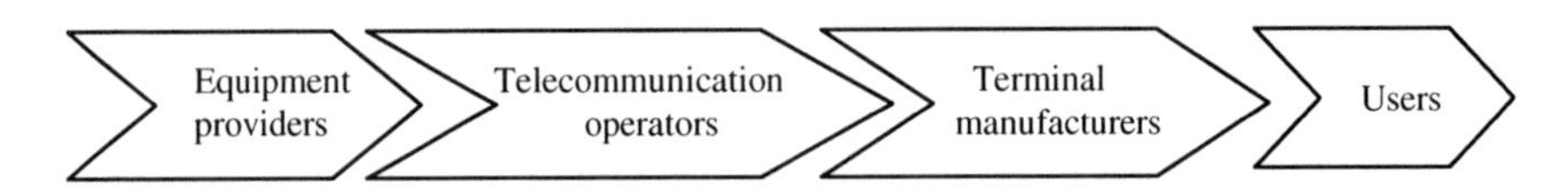

Fig. 2. Traditional mobile telecommunication industry chain

Mobile publishing industry chain is the combination of upper two chains. With the development of 3G technology, mobile publishing industry chain gradually divided and refined. Our team considered that mobile publishing industry chain included equipment providers, network providers, service providers, content integrators, terminal providers, users (Fig.3.).

Equipment providers and network providers owned or operated telecommunications network. They provided telecommunication network equipment, network access and etc. And they are basic condition of mobile publishing. Usually telecommunication operators acted as the network providers. For the telecommunication operators, mobile publishing became the most valuable new business of 3G and mobile Internet. After SMS and coloring ring back tone, mobile reading can gain good economic benefits.

Content providers usually didn't the authors. Service providers provided the available content to users through the menu. And users paid according to traffic.

Service providers didn't own basic network, but they charge users according to the protocol and billing system. Network providers and Service providers are same in many countries such as Japan, China. The content providers not only aggregated a large number of contents from independent authors, but also ensured the legitimacy and true of contents. When the contents are spreading, the content providers ensured that the contents are not being illegally copied in the end.

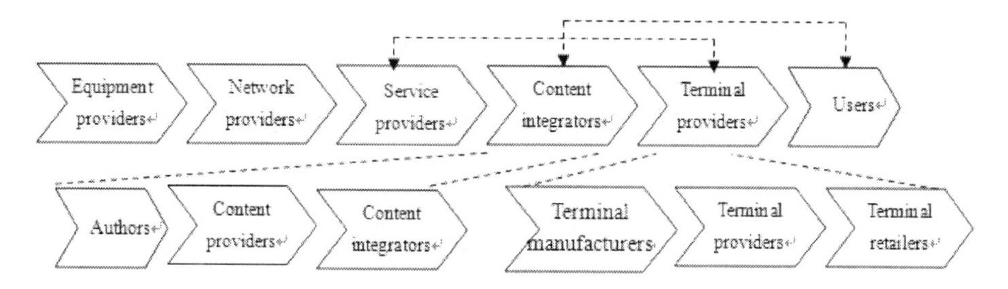

Fig. 3. Mobile publishing industry chain

Content providers provided the content to user through the website. Content providers usually didn't the authors. SP usually act as the content providers. Content integrators mainly collected different content from diverse authors or content providers to integrate into a specific website. Content integrators are known as portal site. In theory they focused on contents compounding and didn't created their own content. Telecommunication operators acted content integrators. But they wanted to do more things in many countries. China Mobile set up an R & D base to create contents in Zhejiang province.

Terminal manufacturers are responsible for the terminal production. Terminals usually are mass production. Terminal manufactures independently researched and produced terminals. Sometimes they produced terminals according to standards provided by telecommunication operators.

Terminal providers play the bridge between terminal manufactures and retailers. Terminal retailers sold terminals to the consumers. They are traditional retail stores or any roles legally sold terminals. Mobile terminals included mobile phones, electronic reader. They can provide on-line or off-line reading functions through the telecommunication network or wireless internet network.

The customers not only read all kind of contents from traditional books, but also from new media which included digital library, digital books, the networks and mobile devices. On the other hand, the customers wanted the products of reasonable price and better service quality. It is noteworthy that customer participation is becoming increasingly important.

Another issue of concern is that the mobile publishing industry chain has changed from the original single-chain into a reversed chain. For example, users not only read contents but also created contents. In addition, in order to ensure profitability of the mobile publishing chain, users' requirements are primary. Services providers, content providers and terminal retailers are directly contacted with users.

Mobile publishing maintained high-speed development trends. But there are many challenges in the Mobile publishing industry. For example, copyright protection, service satisfaction of customers, the press functions in the mobile publishing industry, and content integration. Copyright protection determined the development of mobile publishing, so the paper deeply discussed copyright protection. The semantic web and ontology technology provided a solution. The Semantic Web offered semantic annotations that explicitly described Web resource. These annotations are built on ontologies, representing domain through their concepts and semantic relations between them [4]. The following focused on the copyright protection.

3.2 Copyright Protection

Copyright protection and publishing industry is very close links. And copyright protection is the base to achieve rapid development of mobile publishing industry. But copyright protection of mobile publishing is the most complex than traditional publishing and digital publishing.

In fact, the General administration of Press and Publication of the P. R. C and the State Intellectual Property Office of the P. R. C are legislating relevant laws. Copyright Law in PRC (adopted with amendments on October 27th, 2001) has fit the right of communication of information on network into the context of copyright. Copyright law can be divided into personal rights and property rights, of which there are 4 personal rights and 13 property rights. Copyright owner may allow others to use the copyright in the property, or transfer all or part of the property rights for agreed remuneration.

But the mobile publishing has not yet formed an integrated and effective copyright protection system. Like the China Mobile acted as telecommunication operator and content provider, these companies can independently provide contents, they also used a vast number of content provided by the publishing house and website. Copyright protection needs some new management methods and information technologies.

4 Some Ontologies Specifications of Providing Copyright Protection

Ontology is the term used to refer to the shared understanding of some domain of interest. Or ontology is an explicit account or representation of a conceptualization. Ontology necessarily entails or embodies some sort of world view with respect to a given domain[5]. Ontology implements knowledge share and knowledge management.

Our team published some paper to describe some key ontologies of the mobile publishing [6].This paper mainly introduced Copyright protection ontology whose function is realized the copyright protection. Copyright protection ontology included copyright registration ontology, copyright marker ontology, and service composition ontology.

4.1 Copyright Registration Ontology

Some publishing works must register copyright at copyright registration agency. But copyright registration ontology included formal legal copyright registration and copyright registration which finished by mobile publishing service distribution platform provider. The important part of this ontology is copyright serial number. The copyright serial number must unique. Copyright registration ontology included publishing content serial number, copyright serial number, author, publishing time, publishing manner, provider of mobile publishing service distribution platform.

4.2 Copyright Marker Ontology

Some content can't require copyright protection, because distributing content is too simple or other reasons. But service management enforce service distribution platform to save origination of content. Copyright marker ontology is used in order to note origination information. The ontology included publishing content serial number, registered author, composition service time, service composition serial number, provider of mobile publishing service distribution platform.

4.3 Service Composition Ontology

This ontology described responding results according to the user needing. The ontology included service composition serial number, downloaded users serial number, file size, element-service identification. A multiplex service can include several element-services, so element-service identification appeared many times. Note that element-service identification is publishing content serial number of copyright registration ontology. Service composition serial number is service composition serial number of copyright notice ontology. The service composition serial and the element-service identification linked three ontologies.

5 A Semantic-Based Mobile Publishing Framework With Copyright Protection

Today many companies found that services are the fastest growing parts of their business and use some service systems. Service systems are value-co-creation configurations of people, technology, value propositions connecting internal and external service systems, and shared information. A Service Oriented Architecture (SOA) combined with Semantic Web technologies is assumed to provide interoperability between services, devices and networks [7]. Mobile network are based on an unreliable radio transmission, round-trip-delay and uneasily predictable costs for mobile usage. Mobile publishing also has dynamically changing contents for reading and the limitation resource of the mobile devices. So the ideal of the virtual mobile service integrations proposed by J. Noll etc is been adopted [8].

Our team presented a semantic-based service composition framework for mobile publishing [9]. This paper mainly introduced the semantic-based mobile publishing system with copyright protection system described figure 4. Copyright information process described figure 5 and figure 6.

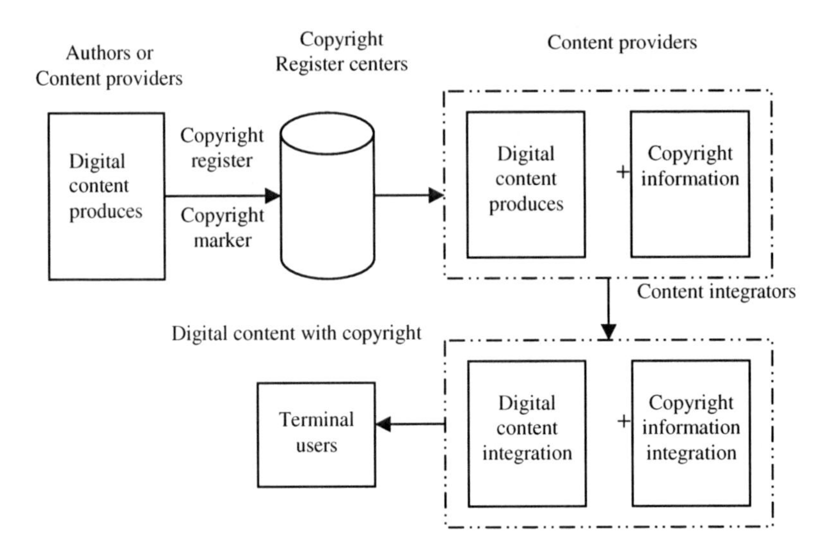

Fig. 4. Copyright protection of mobile publishing

Fig. 5. Copyright protections principle

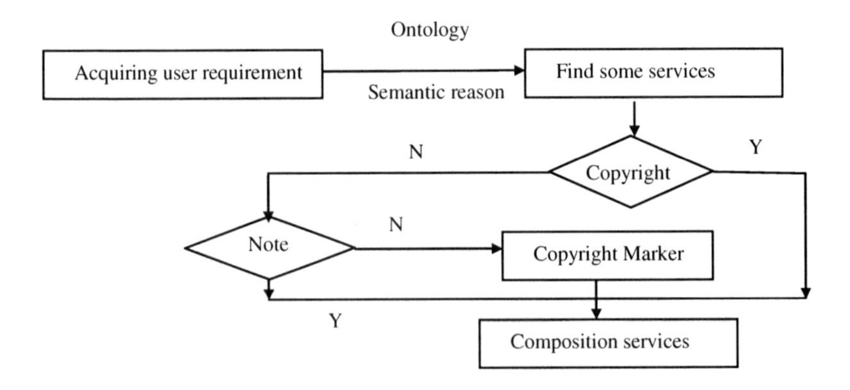

Fig. 6. Copyright protection process of mobile publishing

6 Conclusion and Future Work

Mobile publishing maintained high-speed development trends. And the market competition is constantly intense. Mobile publishing promoted several industries development, which included cultural and creative industry, telecommunication operators, software vendors and hardware vendors. Every industry wanted to control the mobile publishing. Mobile publishing depended on the telecommunication and wireless network. However, mobile network are based on an unreliable radio transmission, round-trip-delay and uneasily predictable costs for mobile usage. Consumers wanted to reduce costs and improved the quality of mobile communication. 3G provided a solution. Mobile publishing also had own characteristics, which included dynamically changing contents for reading and the limitation resource of the mobile devices. This paper firstly presented a kind definition of mobile publishing according to our research. The definition emphasized on the difference between mobile publishing and mobile phone publishing. Copyright protection is a key and difficulty question. And the paper introduced a mobile publishing industry chain.

It is important for mobile publishing industry chain to copyright protect, so the paper emphasized on copyright protection. A copyright protection framework and process are introduced.

Acknowledgement. This research was supported by China cultural and creative industries cluster and cluster management research.

References

1. Eight national reading survey, `http://www.xwcbj.gd.gov.cn/news/html/zxdt/article/1303522672922.html`
2. 2009-2010 China mobile Internet reading survey, `http://www.iimedia.com.cnortnew`
3. What is mobile phone publishing?, `http://labs.chinamobile.com/mblog/399321_73304?`
4. Corby, O., Dieng-Kunta, R., et al.: Searching the Semantic Web: Approximate Query Processing Based on Ontologies. IEEE Intelligent System, 20–27 (January/February 2006)
5. Uschold, M., Gruninger, M.: Ontologies: Principles, Methods and Applications. Knovledge Engineering Review 11(2) (June 1996)
6. Hejie, C., Dan, C.: Mobile publishing ontologies: A semantic-based service management model for the mobile publishing. In: 2nd IEEE International Conference on Information and Financial Engineering, pp. 580–584. IEEE Press, Chongqin (2010)
7. Uschold, M., Gruninger, M.: Ontologies: Principles, Methods and Applications. Knovledge Engineering Review 11(2) (June 1996)
8. Corby, O., Dieng-Kunta, R., et al.: Searching the Semantic Web: Approximate Query Processing Based on Ontologies. IEEE Intelligent System, 20–27 (January/February 2006)
9. Hejie, C., Dan, C.: Ontology driven semantic service system to support mobile publishing. In: 6th International Conference on Wireless Communications, Networking and Mobile Computing

Evaluating Quality of Chinese Product Reviews Based on Fuzzy Logic

Wei Wei[1,2], Yang Xiang[1], Qian Chen[1], and Xin Guo[1]

[1] College of Electronics and Information Engineering, Tongji University, Shanghai, China
[2] College of Electronics and Information Engineering, JingGangShan University, Ji'an, China
{weiwei,chenqian857,guoxinjsj}@163.com,
shxiangyang@tongji.edu.cn

Abstract. The prevalence of web2.0 makes e-Commerce an increasingly popular trend. Consumers can post their product reviews about product after buying products on the web. These product reviews can help online user make sensible decisions and enable business enterprises to improve their business strategies. To cope with the information overload problem, opinion mining is needed to extract useful expression and summarize important opinion for users. But the quality of product reviews at websites varies greatly. In this paper, we propose a method based fuzzy logic to evaluate the quality of Chinese product reviews. We define three fuzzy sets to represent the different types of product reviews. We determine the quality of product reviews under the proximity membership principle based the three fuzzy sets. Experiments based on an expert-composed product reviews corpus show that our method can achieve promising performance.

Keywords: Text Ming, Opinion Ming, Product review, Fuzzy logic.

1 Introduction

In the past few years, there has been the explosive growth of the user-generated content on the web. Web users use web logs or discussion forums to express personal opinions and share valuable information. Online shoppers often read the product reviews to gauge their shopping decision; business enterprises also examine product reviews to analyze custom opinions and predict market trends. Because users' opinions posted on the web have significant impact on consumer decisions [1], opinion mining has been attracting increasing attention from both academia and industry.

Opinion mining aims to extract and retrieve valuable information from product reviews. Opinion mining involves identifying and classification the opinionated document or sentences in product reviews. Opinion summarization and opinion retrieval are also important tasks of opinion mining. However, there has been a great progress in opining mining using natural language techniques; it is still challenging to acquire useful information from product reviews. Due to the fact that there is lack of editorial and quality control, the quality of product reviews varies enormously. Thus, it is urgent to have an effective quality evaluation mechanism for assessing the quality of product reviews. Some shopping websites already provide a vote mechanism of

G. Zhiguo et al. (Eds.): WISM 2011, CCIS 238, pp. 328–333, 2011.

ranking the product reviews. For example, Amazom encourages user to evaluate the helpfulness of product reviews and displays the product reviews based on the votes submitted by readers. Users can quickly find the helpful product reviews and make the consumer decision with less time. But the helpfulness voting mechanism is not perfect. The imbalanced vote bias, winner circle bias, and early bird bias affect the quality evaluation mechanism [2].

To address the above problem, we propose a method based fuzzy logic to evaluate the quality of Chinese product reviews in this paper. We first define the three types of product reviews, namely good product review, useful product review and bad product review, respectively. Then we construct membership function based fuzzy logic, using many aspects of product reviews. We apply the proposed approach to evaluate the quality of Chinese product reviews. The experiment results show that the proposed method can achieve a promising performance on the test set for evaluating the quality of Chinese product reviews.

This remainder of this paper is organized as follows. In Section 2, we review the related works .In Section 3 we introduce the types of product reviews and present our approach to evaluate the quality of Chinese product reviews. Section 4 reports our experimental results on test data. The conclusion and some possible direction for future research are given in section 5.

2 Related Work

Kim et al [3] measured the quality of review based the percentage of helpful votes. The quality of a review is high if many readers think the review is helpful. Zhang and Varadarajan [4] predicted utility of product reviews based on the helpfulness votes given by readers. SVM regression was employed to predict the utility of the product reviews. Liu [2] et al showed three types of bias of users 'votes and presented the specification on the quality of product reviews. They used SVM as the model of classification to categorize reviews as either high-quality or low-quality. Chen and Tseng [5] adopted an effective framework of information quality for assessing the quality of reviews. Experiment demonstrated that the method using information-oriented features improved the evaluation performance.

Different from extant studies that focus on traditional machine leaning techniques, in this paper we attempt to address the problem of evaluation the quality of product reviews based fuzzy logic [6]. We choose fuzzy logic to resolve the problem in that it provides a more straightforward way to describe the intrinsic fuzziness in the quality of product reviews.

3 Methods

In this section, we define three types of product reviews quality, according to different values of the product review to reader' purchase decision. Three types of product reviews quality, namely "good product review", "useful product review", "bad product review" are defined as follows.

Definition 1 (good product review). A good product review must contain complete and detailed information about a product. It can provide a large number of convincing comments on a product. In a word, a good product review can be taken as valuable clue to help the readers make purchase decisions.

Definition 2 (useful product review). A useful product review is relatively complete information about a product, but not as much enough comment on a product. Although readers may think such review is useful, it could not be used as recommendation for readers to make decisions.

Definition 3 (bad product review). A bad product review contains only a little information about a product, or incorrect specification about some aspects of a product. A bad product review can be considered a spam product review, as it may be a fake review or a duplicate review.

We define a feature domain, namely F, to identify the quality of the product review. The features in feature domain are listed as follows.

- The number of sentences in the product review f1.
- The number of product features in the product review f2.
- The number of sentences with product features f3.
- The number of positive sentences in the product review f4.
- The number of negative sentences in the product review f5.
- The number of opinion-bearing words in the product review f6.
- The rank of the reviewer f7.
- The rank of the product review f8.
- The similarity between the product review and the product description f9.
- The interval between the current product review and the first product review of the product f10.
- The number of helpfulness votes for the product review f11.
- The number of replies to the product review f12.

As the quality of product reviews is difficult to discriminate, there is not a clear boundary between the different types of product reviews. To solve the problem, we apply the fuzzy logic theory to discriminate the three types of product reviews. We define three fuzzy set to represent three types of product reviews.

Definition 4 (good product review fuzzy set). A good product review fuzzy set is defined as A_g, namely

$$A_g = \sum_{i=1}^{n} \mu_{A_g}(f_i)/f_i \,. \tag{1}$$

Where f_i denotes the feature of a product review, $\mu_{A_g}(f_i)$ represents the membership function of f_i in A_g that maps F to the membership space M.

Definition 5 (useful product review fuzzy set). A useful product review fuzzy set is defined as A_u, namely

$$A_u = \sum_{i=1}^{n} \mu_{A_u}(f_i)/f_i \,. \tag{2}$$

Where f_i denotes the feature of a product review, $\mu_{A_u}(f_i)$ represents the membership function of f_i in A_u that maps F to the membership space M.

Definition 6 (bad product review fuzzy set). A bad product review fuzzy set is defined as A_b, namely

$$A_b = \sum_{i=1}^{n} \mu_{A_b}(f_i)/f_i \,. \tag{3}$$

Where f_i denotes the feature of a product review, $\mu_{A_b}(f_i)$ represents the membership function of f_i in A_b that maps F to the membership space M.

Based on the above membership functions, we can calculate the grade of membership of a given product review in three fuzzy sets, and thus determine its quality under the principle of proximity membership. Given a product review B, ∃ i∈{g,u,b},if

$$N(A_i, B) = \max\{N(A_g, B), N(A_u, B), N(A_b, B)\} \tag{4}$$

Where N(A, B) denotes the fuzzy degree of nearness between fuzzy set A and fuzzy set B, namely

$$N(A,B) = \frac{\sum_{i=1}^{n}[\mu_A(f_i) \wedge \mu_B(f_i)]}{\sum_{i=1}^{n}[\mu_A(f_i) \vee \mu_B(f_i)]} \tag{5}$$

4 Experiments and Results

In this section, we evaluate the effectiveness of our method. However, to the best of our knowledge, there are no official benchmarks for the product reviews quality evaluation. We therefore construct our own product review corpus for performance evaluation. We selected 2000 Chinese product reviews about popular laptops and digital cameras from 360buy.com. The evaluation are based the Chinese product review corpus where the quality of each review is annotated by two human experts, with the label of "good", "useful", and "bad". Table 1 shows the annotation of the 2000 Chinese product reviews. The value of the kappa statistics between the two human experts are 0.7836. This shows the two human experts achieved highly consistent results, although they annotated independently. The Chinese product review corpus is good enough to evaluate our method.

Table 1. The statistics of the Chinese product review corpus

	Good	Useful	Bad	Total
Laptops	78	875	47	1000
Digital cameras	82	852	66	1000
Total	160	1727	113	2000

We randomly selected some reviews for determining the three fuzzy set based on the above features. The values of features in three fuzzy sets are obtained by observing manually. We ran the three fuzzy sets to evaluate the remaining reviews. Table 2 reports the accuracies of our method to evaluate the Chinese product reviews. The accuracy is defined as the percentage of correctly evaluated reviews.

Table 2. The accuracy of the evaluation

	Good	Useful	Bad
Laptops	81.3%	83.2%	83.1%
Digital cameras	82.4%	83.5%	82.7%

5 Conclusion

In this paper, we proposed a method based fuzzy logic for evaluating the Chinese product reviews. To the best of our knowledge, this study may be the first attempt to apply the fuzzy logic in evaluating the Chinese product reviews. We defined three fuzzy sets to describe the different quality of product reviews. The encouraging results of performance show the effectiveness of our method.

We hope to explore our future work in several directions. First, it would be interesting to investigate implicit feature about the quality of product reviews deeply. Second, it is definitely important to optimize the proposed membership functions for representing the fuzzy quality of product reviews.

Acknowledgments. This work is supported by the Science and Technology research project of Jiangxi Municipal Education Commission under Grant No. GJJ11178.

References

1. Chevalier, J.A., Mayzlin, D.: The effect of word of mouth on sales: online book reviews. Journal of Marketing Research 43(3), 345–354 (2006)
2. Liu, J., Cao, Y., Lin, C.Y., Huang, Y., Zhou, M.: Low-Quality Product Review Detection in Opinion Summarization. In: Proceedings of the 2007 Joint Conference on Empirical Methods in Natural Language Processing and Computational Natural Language Learning, pp. 334–342 (2007)
3. Kim, S.M., Pantel, P., Chklovski, T., Pennacchiotti, M.: Automatically Assessing Review Helpfulness. In: Proceedings of the 2006 Conference on Empirical Methods in Natural Language Processing, pp. 423–430 (2006)

4. Zhang, Z., Varadarajan, B.: Utility Scoring of Product Reviews. In: Proceedings of the15th ACM International Conference on Information and Knowledge Management, pp. 51–57 (2006)
5. Chen, C.C., Tseng, Y.D.: Quality evaluation of product reviews using an information quality framework. Decision Support Systems 50(4), 755–768 (2011)
6. Zimmermann, H.-J.: Fuzzy set theory and its applications. Kluwer Academic Publishers, Norwell (2001)

An Efficient Algorithm of Association Information Mining on Web Pages with Dynamic Scripts

Tao Tan and Leting Tan

School of Computer, China West Normal University, Nanchong 637002, China
tantao99132@163.com,
tanleting@yahoo.com.cn

Abstract. The hyperlink analysis algorithm is widely used by public search engines. But with the development of the websites with dynamic script, this algorithm is not fit to realize the efficient searching for these related pages, because there is not enough hyperlink information for these pages. The research on the association information mining on web pages with dynamic scripts is progressing gradually. This paper proposes an improved search framework which can be more efficient for the pages with dynamic scripts. Then, by building up state information tables which is in accord with page changes of the same URL for these pages and state transition chains for pages loading, the paper presents an analysis algorithm based on state-interrelated matching of these pages. Finally, the paper detailedly describes entire implementing process of the algorithm, and demonstrates the efficiency of the algorithm by experimental results.

Keywords: Dynamic Scripts, State-interrelated Matching, State Transition, Association Information Mining, AJAX.

1 Introduction

With the development of the internet, more and more people make use of the internet to publish and gain the information. Web has become to the main platform for producing information, publishing information and processing information. At present the service provided by major search engines is that searching the web pages associated with some topic by the way of keywords enquiry. In these search engines, hyperlink analysis algorithms represented by PageRank, HITS were used[2, 11], and the rankings were computed based on hyperlink topological structure information between web pages. Its basic point is that giving a quantization value to the web page according to the statistical result of the hyperlink information in large scale web pages, then sorting according to the sequences from high value to low value, and the result will be return to the user.

But with the popularization of web2.0 application, more and more websites adopt the dynamic scripts method to interact with the user. That includes AJAX (Asynchronous JavaScript and XML).The web pages using this technology don't have enough hyperlink information. URL (Uniform Resource Locator) is not the unique identification of the pages. For the AJAX application, several pages can share a URL. The internal important contents are loaded from web server through executing the client scripts (e.g., JavaScript). When executing a load, the contents of pages will

G. Zhiguo et al. (Eds.): WISM 2011, CCIS 238, pp. 334–342, 2011.

change. Thus, the page transition of the web pages with dynamic scripts not only controlled by the hyperlink, but also by various events triggered on random page elements. The traditional hyperlink analysis algorithms based on the hyperlink topological structure no longer adapts to analysis and ranking of the pages with dynamic scripts. When searching the associated information of these web pages, the main public search engines can not discovery its internal information very well.

2 Related Works

The association rule mining for web pages with dynamic scripts is a new research area. Recent advances about the relevant research and application of hyperlink analysis are presented in literature [2].In particular, some results and achievements about hyperlink analysis and its applications on Web searching and information discovery are surveyed in it. In literature [1] an efficient valid page crawling approach for websites with dynamic scripts is proposed. The state transition diagrams are built up to descript these websites. The XPath patterns of the page elements are generated and the events need to be triggered are recorded, then the page information are captured by using these results. The state transition diagrams are mentioned in this literature. It points out the transform process, but there is not a very detailed description of the state information. In literature [3] the ranking of the weak-linked documents such as AJAX application is introduced. A search engine framework is proposed for correctly retrieving and ranking weak-linked documents based on clustering. The clustering technology is applied to the ranking algorithm of these documents. For the AJAX application, it needs a high performance crawler. It is should be studied further that how to dig the rules from the crawling log by using clustering technology for accelerating the crawling.

The research of this paper mainly consists of those jobs: Because the URL is not the unique identification of the web pages with dynamic scripts, the state information table will be built up when loading a new page. The different state is considered as a basic unit of analysis. Then, a search approach based on association information between the pages with dynamic scripts is presented and an analysis algorithm StaM based on state-interrelated matching of these pages is proposed.

3 Basic Principles of Improved Search Framework

Because there is not enough hyperlink information in these web pages with dynamic scripts, many page transitions are triggered by the function in the script. It is hard to realize a high-efficiency dig by using the hyperlink analysis algorithm for these pages.

For the AJAX application, when the dynamic client scripts are executed, user will get the new page content at the same URL. These applications show a new state to users. So identifying the hyperlink targets using URL can not adapt to these documents. A better way is that the pages with dynamic scripts should be divided into different states, and a one-to-one correspondence between different states and different page contents should be built up. For the sake of discussion, we define the state as the content provided to users after their operation. In order to realize the efficient mining, the page states need to be considered as a minimum retrieval unit in the new search approach. At the same time, according to analysis on the users search habits, the search results are agree with the subject distribution of the pages with dynamic scripts[12].So the following methods are used for solving problems: Firstly, submit a given query, make use of the public search engines to retrieval information directly. And then compute according to the association degree of page states in the web pages with dynamic scripts, finally, ranking is presented in accordance with these results. The following is the description of the new search approach. Its framework is shown in figure 1.It is similar to the traditional search engines, and it is divided into two parts: part of online and part of offline.

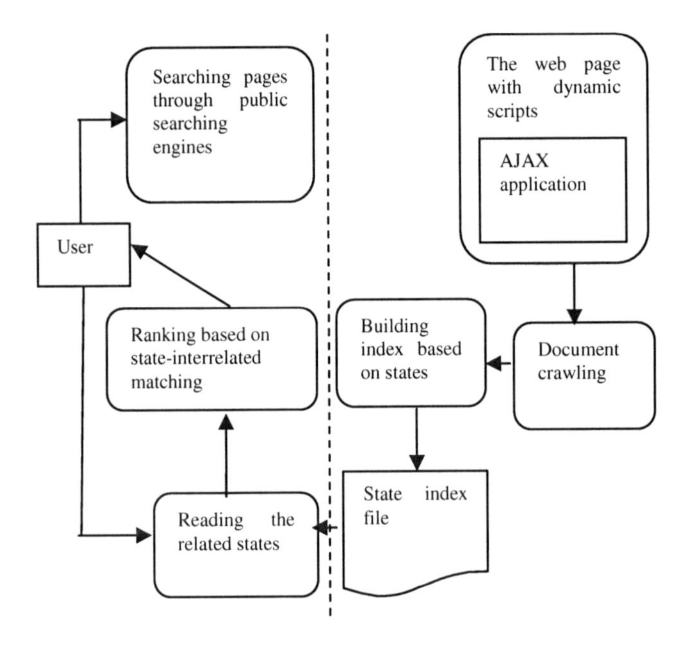

Fig. 1. The framework of new search approach

3.1 Public Search Engines Searching

At present the search engines represented by Google can search the static web pages very well. When a given query is submitted, if a page is more important and popular, its value is higher, the higher ranking will be. The subjects of the higher ranking illustrate the subject of this query on the World Wide Web. So, the high- quality page search results can be used to improve the ranking of the association pages. One direct method is that select the original documents by using the public search engine, and then the following work will be developed on that basis.

3.2 Document Crawling

In the AJAX application, the internal link and client script must be used in order to request content. When the original URL is given, the web crawler will simulate user behaviors and trigger the events which may be triggered by each user so as to get the whole contents of dynamic applications. The crawler will receive a new state when an event is triggered. In this process, the most difficult task of crawler is how to detect the repeating states. For the detection of the static URL, the crawler just needs to maintain a hash table of URL. But in the process of AJAX application crawling, the request from client to server is generated and executed dynamically. Thus, it is need to understand the script logic to identify these repeating states. Special AJAX crawler is used here. The crawler will get the call relationship diagram of AJAX scripts by way of analyzing the scripts. Then the parameters of HOT function nodes linked by network will be monitored and saved. The detection of repeating states is accomplished successfully. The crawling efficiency of crawler is greatly raised [4, 5].

3.3 Index Building

Because the document state is basic unit of searching, the traditional inverted file list should be expanded, and the states should be treated as basic units when index files are built. The three major data structures are shown in Table 1: The document ID, the document URL and the number of states included in the same URL are all stored in the document table[6].The state table records the basic information of states, including state ID, length, inner-url and inner-score. The term, state ID, score and position are included in the inverted file list. For example, the first state ID of the document which DocID=2 is 8, its length is 23, its inner-url is #3, and its inner-score is 0.7.The inverted file list of the index term "key" is that: state ID 5, score 0.5, position 12. At the same time, change of page contents means transition of states. The directed graph is used to describe this dynamic process. Let G=<V, E> be a binary ordered pair, V means states. E means the page changes caused by events triggering of these page elements. V is a set of states, V= $\{v_1, v_2, ...v_n\}$.E is a set of state changes. $E_{i\text{->}j}$ means a change from v_i to v_j.

Table 1. Three major data structures of state index

DocID	URL	SNum
1	http://bbs.travel.163.com/bbs/tuyou/178424684.html	18
2	http://softbbs.pconline.com.cn/12666863.html	22

DocID	StateID	Length	Inner-URL	Inner-Score
1	1	16	#1	0.9
1	2	25	#3	0.68
...	...	...	...	...
2	8	23	#3	0.7

Term	StateID	Score	Position
news	1	0.33	9
key	5	0.5	12
sport	8	0.28	22
...	...	...	...

3.4 Query Output

When the given query is submitted by users, firstly, the query is submitted to the public search engines. At the same time, crawl for the AJAX applications and index to the states. Then read the subject-specific states from state index file. Finally, compare state information with preceding search results according to the association degree matching algorithm and compute the final ranking.

4 Key Algorithm

In consideration of the preceding ideas, an analysis algorithm StaM based on state-interrelated matching of pages with dynamic scripts is proposed. In order to describe this algorithm distinctly, the following definitions are introduced first.

Definition 1. In the directed graph G of state transition, the page state changes are listed below[7]:

1) In (v_i): an in-degree set of state v_i, which points v_i $\{v_j | E_{j->i} \in E\}$.
2) Out (v_i): an out-degree set of state v_i, which is pointed by v_i $\{v_j | E_{i->j} \in E\}$.
3) Co-In (v_i): a combined-citation set of state v_i, $\{v_j | v_i \neq v_j \wedge E_{k->i} \in E \wedge E_{k->j} \in E\}$.
4) Co-Out (v_i): a combined-citation set of state v_i, $\{v_j | v_i \neq v_j \wedge E_{i->k} \in E \wedge E_{j->k} \in E\}$.

Definition 2. The two states v_i and v_j are both cited by the state V if and only if they are both pointed by state V or both point state V(the item 3, 4 in Definition 1).$<v_i, v_j>_v$ means a combined-citation pair of V. sim($<v_i,v_j>$) means the association degree of $<v_i,v_j>_v$. If v_i and v_j are not a combined-citation pair of V, let sim ($<v_i, v_j>$) be 0.

From the preceding definitions, it is easy to see that sim($<v_i,v_j>$) records the number of states including both states pointed by other states and other states point the both states. For each combined-citation pair, the association degree will be computed according to the state-interrelated information, and the other useful information (e.g. the related attributes of states) will be stored.

The directed graph G of state transition contacts one state with other states. It forms structured chain of states. A chain of states is a path consisting of two or more state nodes joined by the state transition edges[8,9]. Let it be $P=\{v_0e_0v_1e_1\ldots v_{n-1}e_{n-1}v_n\}$.from the view of state chain, the association degree between states v_i and v_j is closely related to the relative position between both states. The several states of being adjacent of the state v_i, they often have highest subject similarity. Based on this theory, the position association degree is computed by the following formula:

$$POS_V(<v_i,v_j>) = \begin{cases} 1 & |i-j| <= \lambda \\ e^{\frac{-(|i-j|-\lambda)}{2}} & |i-j| > \lambda \end{cases} \quad (1)$$

In the formula, i and j mean position numbers of v_i and v_j in the path P. Neighbour threshold λ is a dividing point of position relation.

Definition 3. For the two states v_i and v_j, the association degree within the path from v_i to v_j is defined as $PSim_v(v_i\text{->}v_j)$.Similarly, the association degree within the path from v_j to v_i is defined as $PSim_v(v_j\text{->}v_i)$.

These association degrees of combined-citation pairs in $POS_V(<v_i,v_j>)$ add up to the value of $PSim_v(v_i\text{->}v_j)$.But in the process of accumulation these restrictions are as follows: In order to avoid some vicious pages providing higher association degrees for the two states. Then high association degrees of the two unrelated states are given. So the maximum values of $PSim_v(v_i\text{->}v_j)$ need to be limited, too. The maximum should not exceed N, meaning that max threshold of association degrees[10].

$$PSim_v(v_i\text{->}v_j)=\min(N, \sum_{i=1}^{n}\sum_{j=1}^{n} POS_V(<v_i,v_j>)) \tag{2}$$

Procedures of this algorithm:

Input: query q, a public search engine E

Output: a sorted-sequence of pages with dynamic scripts J (q)

Step 1. The query q is submitted to the search engine E, the first n pages returned are D (q) = $\{d_1, d_2,\ldots, d_n\}$.

Step 2. In the results of Step 1, the sets of states including query q received after document crawling and index building mentioned before are S (q) = $\{s_1, s_2, \ldots, s_n\}$.

Step 3. The association implicit state s_i received from (2), according to the StaM algorithm, for the $s_i \in$ S (q), its association degree will be computed.

Step 4. According to the association degrees, rank in descending order, then the results J (q) will be returned.

5 Implementation of Algorithm and Analysis of the Results

The whole algorithm includes many inspired experience formulas. In order to give a reasonable value to each parameter threshold of these formulas, analysis of specific statistical data and results of empirical estimation are added together. For the neighbor threshold λ, through analyzing simulation on events of some websites with dynamic scripts, it is found that the number of states transition is often less than 5. So in this experiment, let λ be 4. For the threshold of path association degree N, by means of sample within these paths, it is found that only 6% of path association degrees are more than 8. So let N be 8. For the total association degree threshold T, in order to prevent the states of too low total association degree from ranking and outputting, let T be 2.0, the states which total association degrees are less than 2.0 will be discarded.

In order to verify the searching ability of the StaM algorithm for pages information with dynamic scripts, the simulation experiments are carried out. The algorithm is realized by using JAVA. Google is used as public search engine, because it can enhance the search results of high quality and through the API provided by Google the search results can be accessed easily. 10 queries are selected. This results of simulation experiments are compared with the results of experiments which done on the public search engines. The 10 queries belong to 10 subjects, including literature, technology, and enterprise portal and so on. For each query, based on the results of these experiments, the first 30 rankings generated by the two implementations judged by five testing staffs. This can be judged from two aspects: First, comparison of the number of related states in ranking. That means the results of one website with dynamic scripts will be added up for the first 30 rankings respectively. On the other hand, comparison of related precision. That means judging from the related precision. The associations between key words and search results can be divided into three cases: completely related, partly related and quite separate.

In the experiments, in view of the fact that the people often concern about the search results of higher ranking, the assessment criteria P@N is used as measuring the ranking performance which is widely adopted in the information retrieval (IR) field. P@N shows the accurate rates of search results in the position of N. It is defined as being: $P@N=|T \cap R|/|R|$. R is a set of the first N results, and T is a set of manual identifier and related query document. P@10, P@20, P@30 are used in the experiments.

The comparisons between two kinds of experiments are shown in figure 2 and figure 3. It shows that the first 3 results is nearly on the related precision. But when more results are introduced the testing shows a gradually expanding of differences on precision. Comparison on the number of related states shows that the number of related pages when StaM algorithm is introduced is more than the result of traditional search engines. That means the proposed algorithm is more effective and precise on searching of pages with dynamic scripts than public search engines.

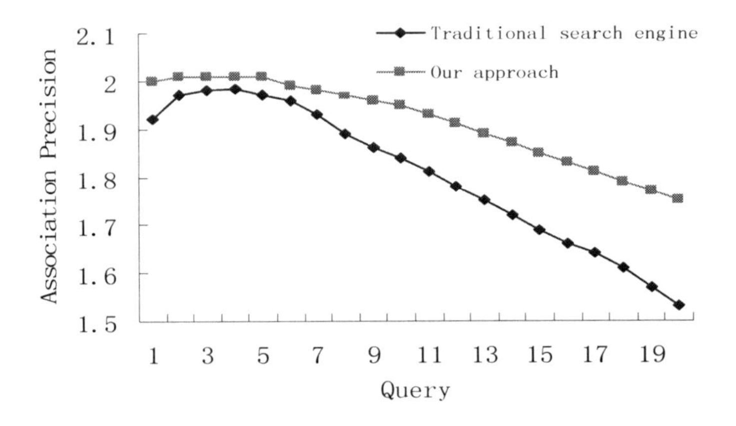

Fig. 2. Comparison of association precision between the contrast experiments

In order to test the operational performance of the algorithm, set of experiments is divided into eight blocks. Each block includes five query terms. Each time five query terms are added to the algorithm input. The size of data entered changes from 5 to 40. The operational performance of the algorithm is shown in figure 4. Here, the runtime is similar to linear growth largely because the time cost of IO has exceeded the running cost of the algorithm in the course of implementation.

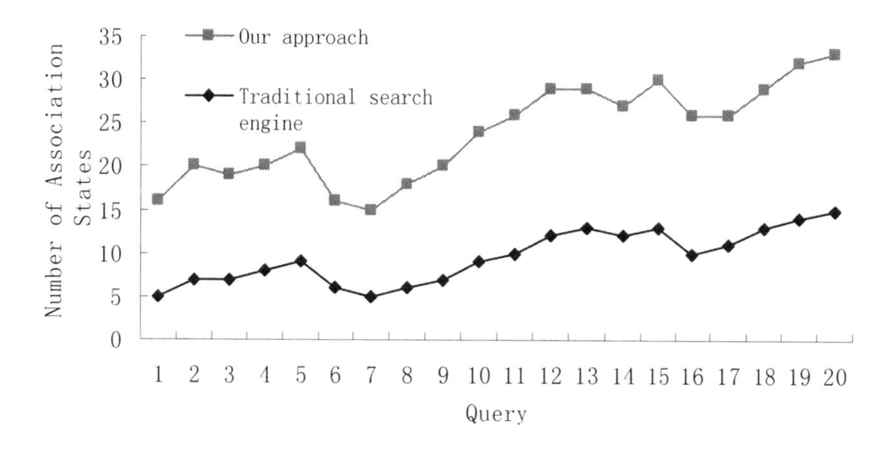

Fig. 3. Comparison of the number of association states between the contrast experiments

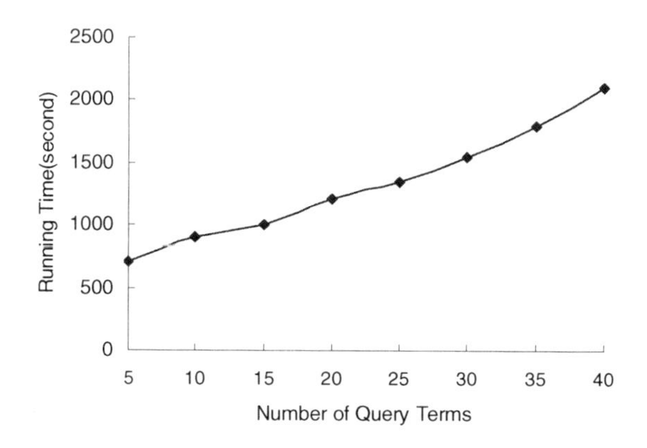

Fig. 4. Runtime of algorithm when number of query terms changes from 5 to 40

6 Conclusions

For mining association information on web pages with dynamic scripts, the creation in this paper is that states information table which is in accord with page changes of the same URL for these pages is presented, and state chain is built up for processes of state transition. Each different state is considered as a basic unit for analysis of information. And accordingly a search approach based on association information between the pages with dynamic scripts is presented and an analysis algorithm StaM based on state-interrelated matching of these pages is proposed.

Experimental results show that compared with the traditional public search engines for mining this information, the solution the paper presented has far more superiorities in search quantity, quality and accuracy for association information. But there is still much work to do so as to realize the discovery of application. Firstly, with the update of website contents with dynamic scripts and the state change of association information, how to track these changes and record these information are key issues for further study. In addition, different events of state pages may triggered and point to the same pages. It is need to consider that how to avoid impact of the repeat on efficiency of information mining effectively.

References

1. Xia, B., Gao, J., Wang, T., Yang, D.: An Efficient Valid Page Crawling Approach for Websites with Dynamic Scripts. Journal of Software 20, 176–183 (2009)
2. Wang, X., Zhou, A.: Linkage Analysis for the World Wide Web and Its Application: A Survey. Journal of Software 14, 1768–1780 (2003)
3. Zhou, C.: Document Clustering in Search Engine [Dr. Dissertation]. HuaZhong University of Science and Technology, Wuhan (2009)
4. Duda, C., Frey, G., Kossmann, D., Matter, R., Zhou, C.: AJAX Crawl: Making AJAX Applications Searchable. In: Proceedings of the 25th International Conference on Data Engineering, pp. 78–89 (2009)
5. Duda, C., Frey, G., Kossmann, D.,et al.: AJAX Search: crawling, indexing and searching web 2.0 applications. In: Proc. VLDB Endow, pp. 1440–1443 (2008)
6. Mesbah, A., Bozdag, E., van Deursen, A.: Crawling ajax by inferring user interface state changes. In: Proceedings of the 8th International Conference on Web Engineering, pp. 122–134 (2008)
7. Tombros, A., Ali, Z.: Factors affecting Web page Similarity. In: Proceedings of the 27th European Conference on IR Research, pp. 487–501 (2005)
8. Chirita, P.A., Olmedilla, D., Nejdl, W.: Finding related pages using the link structure of the WWW. In: Proceedings of IEEE/WIC/ACM International Conference on Web Intelligence, pp. 632–635 (2004)
9. Lin, Z., King, I., Lyu, M.R.: PageSim: A novel link-based measure of web pages similarity. In: Proceedings of the 15th International Conference on World Wide Web, pp. 1019–1020 (2006)
10. Sadi, M.S., Rahman, M.M.H., Horiguchi, S.: A new algorithm to measure relevance among web pages. In: Proceedings of the 7th International Conference on Data Mining and Information Engineering, pp. 243–251 (2006)
11. Fang, Q., Yang, G., Wu, Y., Zheng, W.: P2P Web Search Technology. Journal of Software 19, 2706–2719 (2008)
12. Kleinberg, J.M.: Authoritative sources in a hyperlinked environment. Journal of the ACM 46, 604–632 (1999)

Construction of the International S&T Resources Monitoring System

Yun Liu[1], Xiao-Li Wang[2], Wen-Ping Wang[1], Xuan-Ting Ye[1], and Wei Fan[1]

[1] School of Management & Economics, Beijing Institue of Technology, Beijing, China
[2] School of Economics & Management, Zhongyuan University of Technology, Zhengzhou, China
liuyun@bit.edu.cn, sxmdwxl@sina.com

Abstract. Based on the systematic research in contents, methods and technology of the international S&T resources, the article established the visual monitoring & service system of international S&T resources, proposed the "three first-classes" concept of international S&T resources. For different types of international science and technology information databases, by use of scientometrics, data mining, visual technology and other methods, we have designed and built a set of effective resource monitoring framework for international S&T resources, which provided the information support for the grasp of international distribution of technology resources, searching for a high level of international cooperative partners and more effective use of international S&T resources.

Keywords: International S&T resources, monitoring system, scientometrics, data mining.

1 Introduction

In China's strategic decision and project management of international S&T cooperation, policy makers, managers, and assessors frequently encounter the problem of failing to judge the quality and level of partner accurately. There is arbitrariness when Chinese part finding its international partners; there lacks of a high level of data center for international science and technology cooperation resources and a dynamic monitoring decision support system which can provide a guide for managers and executives. All these above will lead to that we seldom grasp the initiative and direction in international S&T cooperation and we cannot make effective use of global S&T resources to enhance the capability of China's indigenous innovation. Within the wide application of giant S&T Literature and Patent database system, the S&T data information has been increasing sharply worldwide. Useful knowledge has to been extracted quickly and effectively from huge database of information technology. Therefore, people propose the data mining methods to process large scale of S&T information and discover the hidden information behind them, which means to service S&T activities.

G. Zhiguo et al. (Eds.): WISM 2011, CCIS 238, pp. 343–355, 2011.

2 The Concept and Needs of the International S&T Resource Monitoring

2.1 The Concept of the International S&T Resource Monitoring

Early, the science and technology resource refers to social resources including technological talents, capital, infrastructure, operating environment and also known as generalized. Narrow technological resources are the ones including the three factors: knowledge, technology and talents. Currently, the S&T resource in scientific research refers to the narrow one.

The S&T resource monitoring means the method of dynamic monitoring, analyzing and evaluating scientific and technological activities with scientific and technological information and data analysis as the basic, with the advanced information science technology such as data mining, information extracting, knowledge discovery, data visualization technology, etc[1].

The S&T resource monitoring is a new interdisciplinary direction developed by the interaction of Science of Science, Scientometrics, R&D management, Information Science and so on, it is a method of combining the qualitative with quantitative analysis. It makes full and comprehensive use of modern advanced information technique and strategic intelligence of experts in all fields, does systematic research on the development of science, technology, economy and society from past through present to future. It aims at providing dynamic and accurate development state of science and technology for technological management and decision making in order to realize the scientific and technological hotspot, grasp the technological opportunity, lower investment risk and promote efficiency[2].

The core of Science and Technology Monitoring is analyzing and mining the carrier of scientific and technological activities with the use of advanced technology of information science and the combination of technology forecasting, technology assessment, expert expertise and offering effective support to technological managers and researchers of governments, scientific research departments and enterprises[3]. The Science and Technology Monitoring emphasizes on discovering those knowledge that hasn't been discovered from collections of data and realizes the process of acquiring information automatically from collections of data of scientific and technological activities[4].

2.2 The Needs of the International S&T Resource Monitoring

The development of science and technology can propel the development of economic and the improvement of general national power and promote the whole world to compete to develop applied science and technology, which is of paramount importance for effectively formulating scientific and technological policies, building healthy environment of S&T development and doing scientific evaluation. Because of the restriction of specific historical conditions and realities of countries and the difficulties in the data and method of quantitative analysis, research management activities are mainly based on expert evaluation method such as the peer review at present, which is evaluating the innovation and scientific of projects with the use of the collective wisdom of evaluators(experts) by means of grasping the present situation and forecasting the future situation of a certain technology domain[2]. But there are some flaws in the method of S&T assessment which uses expert assessment as the chief means, which are as follows:

(1) It is restricted by the extent, depth and timeliness of understanding the new result of relative technology domain of certain expert. The assessment and forecast which are based on intuition and limited knowledge domain are always lack of powerful proven cases and support of data and knowledge.
(2) It involves the work of evaluation which is relevant to resource allocation or interest, which is partly influenced by artificial and unjust factors. So there are stochastic and unbalanced problems.
(3) It mainly depends on expert knowledge from external system and does evaluation and forecast on the development of science, technology, economy and society of a long period of time in the future, which is overly dependent upon external source (experts) and hardly provides objective basis to scientific decision making and management.

So it results in the situation that many problems arise in the aspects of research evaluation, research policy making, research resource management, research project management, research planning, etc. With the wide application of large-scale S&T literatures and patent database system in scientific research, the amount of S&T information increases rapidly. Confronted with the huge science and technology information database, we need retrieve useful and effective knowledge rapidly from it. Consequently, we need take the Science and Technology Monitoring Method and combine it with expert intelligence to solve the problem.

3 The Goals and Objectives of the International S&T Resources Monitoring

The objectives of international S&T resources monitoring included in this paper is classified into five categories: the worldwide science frontier monitoring, worldwide technology frontier monitoring, worldwide-class R&D enterprise monitoring, world-class renowned scientist monitoring, international science and technology cooperation (for example of Sino-USA cooperation in this paper).

3.1 Worldwide Science Research Frontier Monitoring

The research frontier of science and technology represents ideological current of one research field and it is a relative concept. In scientific activities, whether it is frontier is based on its position in the current objectives areas. If it is in front of one subject area, we can call it frontier.

Monitoring the frontier will help researchers and research managers to quickly understand the research frontier and focus in a certain field, then get into the mainstream of scientific research in time and seize the high ground of technology. Meanwhile, the monitoring research frontier can timely provide a high level of international partners for scientists; can help governmental agencies to provide timely help for major research projects; or can provide basis for decision making of internal department of one company about the allocation of research resources, thus promote the development of science and technology and socio-economy.

The source of data used by world's technology frontier monitoring is mainly from ESI (Essential Science Indicators) database. ESI was econometric analysis database launched by Thomson Corporation based on SCI (Science Citation Index) and SSCI (Social Sciences Citation Index) documents to measure established performance of scientific research, tracking development trends of science.

3.2 Worldwide Technology Frontier Monitoring

We use the classification of patented technology as a basis for classifying the different fields, then carry out dynamic monitoring for "first-class technology", "first- class talents", "first-class institutions" of six great fields: Chemicals exc. Drugs, Computers & Comm., Drugs & Medical, Electrical & Electronics, Mechanical, Others.

Data mining and systematical analysis are applied to monitor the worldwide techniques. Based on that, we determine the content and evaluation model of three "first-class" and monitor them dynamically, we achieve the analysis functions of statistical information, cooperative patent licensing, technology hot, monitoring reports etc. Users can carry out dynamic tracking and analysis of a particular area according to themselves demand to grasp the development change, technology frontier, research focus of the field[5].

The source of data used by world's technology frontier monitoring is mainly from the patent database authorized by U.S. Patent, Trademark Office and patent information retrieval network and ESI database of some nation.

3.3 First-Class R&D Enterprises Monitoring

First-class R&D enterprises make decisions based on the First-class R&D enterprise announced by the UK Department of Trade and Industry as objectives, focusing on monitoring the following main elements:

(1) First-class R&D enterprise's intensity and trends of R&D, and the distribution of R&D intensity in all areas (regions) of countries.
(2) Get "which technology is world-class" "Who is the best talent "and "which companies or organizations to cooperate" for the world-class R&D enterprise by monitoring quantity and quality of patent.
(3) Using the world's leading IP management and analysis platform Aureka, release monitoring information of first-class R&D enterprise according to different themes, such as the visualization maps of patent, the citation trees of highly cited patent[5].
(4) Regularly publish monitoring reports. The source of data monitored by first-class R&D enterprise is mainly from the patent database authorized by U.S. Patent and Trademark Office patent information retrieval network of some nation.

3.4 World-Renowned Scientists Monitoring

World-renowned scientists are the ones of authoritative scholars in related field. They will be defined as winners of international authority award, academicians and scholars of international authority Academy of Sciences, and scientists in international scientific organizations.

Monitor unstructured data on the network, extract the relevant information of world-renowned scientists (including research results) to establish dynamic monitoring database of well-known scientists, provide users with information about world-renowned scientists to release, finally give reference for ending users to selecting partners in scientific and technological cooperation [5].

3.5 S&T Cooperation Monitoring between Sino-US

This part of paper mainly monitor the situation of SCI for the Sino-US cooperation, by data mining of SCI retrieving results, analyze the situation and development trends of participating countries, institutions, journals, subject areas and high-quality papers during the past five years of China-US scientific cooperation[5]. Main content s are as follows:

(1) Evolution of the number of paper each year. Monitor the annual number of cooperation and their evolution in the last five years.
(2) Evolution of the number of papers each objective. Monitoring the number of papers of each 17 subject in last five years.
(3) Evolution of the institution of each objective. Monitor participation in cooperation paper of each inter-objectives and China-US cooperative network in last five years.
(4) Evolution of cooperation scientists in various disciplines. Monitor the participation of scientists in each discipline and the number of papers in last five years and China-US cooperative network.
(5) Cooperation Funding Scheme of Various disciplines or about institutions. Monitor major funding program and its changes of each subject in last five years [5].
(6) Participation of third-party countries in all subjects. Monitor distribution of third-party countries participated in China-US cooperation in various disciplines, and its evolution.
(7) High-citation papers of all subjects. Monitor high- citation papers within each subject each year.

The source of data monitored by China-US cooperation is mainly from retrieve data of SCI(Science Citation Index) papers of China-US scientific and technological cooperation.

4 The Process and Methods of International S&T Resources Monitoring

4.1 The Process of S&T Resources Monitoring

According to the basic theory and methods of scientific and technological resources to monitor, requests needed to monitor the objectives, the general processes and steps to monitor the scientific and technological resources are as follows:

(1) To determine objectives of monitoring scientific and technological resources, delineation of the scope of research topics. For example: To monitor the latest hot

spots and key researchers of the electronic information industry. We must choose a database of electronic information industry, electronic information industry journal articles, conference papers, theses and other patent documents for the study.
(2) Data collection. To collect the possible full document information within the scope of the research, so that research results can be more authoritative and convincing, then put all the collected information on a specific database.
(3) Data integration, data pre-processing to remove noise words, to clean and convert. Remove from the target data set the obviously wrong data and redundant data, and remove noise or irrelevant data, and then do data cleansing. And through various conversion methods to convert the data into an effective form, prepare for data mining in the future work [6].
(4) Data mining and analysis. According to the visual data mining goals identified in the first steps, select a specific data mining algorithms (such as technical group(s) automatic identification, technological innovation indicators, natural language processing and competitive analysis, fuzzy clustering, etc.), extract the data model from the database, and use some of the ways to express the model into something easy to understand (knowledge)[7]. According to some interestingness measurement, to interpret the model (knowledge) that has been found, and to evaluate and assess the value, if necessary, return to the previous process to get repeated extraction [8].
(5) Knowledge representation and formation of a series of reports and charts. To make conclusion through data collection, collation, mining and other steps, and then with visualization techniques to write report and display in a way for users easy to understand.

4.2 The Methods of International S&T Resources Monitoring

4.2.1 Bibliometrics

Bibliometrics is based on the mathematical and statistical methods, an interdisciplinary science to analyze quantitatively. It's a comprehensive quantitative knowledge system of mathematics, statistics, and philology as a whole. The main measuring objects are the volume of documentation (publications, particularly mostly journal articles and citations), the number of authors (individual group or groups), the number of words (the literature identified, of which thesaurus the majority) the most essential features of Bibliometrics must be in the "quantity"[9].

Bibliometrics indicators include: statistics of keywords frequency, the source of agency, the source of international and so on. Bibliometrics visualization Indicators itself does not point the trends of the study, but through the various statistical indicators, graphics to show the changes in the timing, in order to reflect the fact of the statistical data, and the final decision is due to users or experts, as one of the basics for determining [10].

Visualization evaluation techniques and methods based on bibliometrics analysis, in accordance with different regions and areas, form a number of different evaluation techniques and methods. Such as citation analysis, a total of citation analysis, multivariate statistical analysis, word frequency analysis, social network analysis, etc. Applications software platform are: Bibexcel, SPSS, WordSmith Tools, Pajek, Ucinet, CiteSpace II and so on.

4.2.2 Text Mining

Technology Resource monitoring is a kind of data mining, aiming at the information mining, and using the information technology to make in-depth data mining, so as to predict and evaluate Technology hot and direction of the development.

With the increasing amount of data, data complexity and information diversity, information through data mining algorithms may be difficult to understand or may not be correct, and thus experts propose the use of visual data mining techniques, that is to use graphics, charts and other visual expressions which are easy to understand for people, to represent complex data information, or to require the users to participate in the data mining process, through setting the parameters to control excavation progress and quality, so that users can enhance the understanding of complex data and information to ensure the correctness of data results.

Visual data mining is a particular step in knowledge discovery (KDD: Knowledge Discovery in Databases) process, and provides a communication interface between the user and the computer to help users find useful information, theory and technology, which is unknown and potential, from the database or data warehouse. Data mining in visual Data Mining KDD process relates with model assessment. First, dig out the information through the mining algorithm or data warehouse, and then display through man-machine interface in a form easy to understand, so that the users can understand the mining results well, it also can be that users interact with the data mining process fully through man-machine interface, make real-time observation to dig out information in order to promptly correct the erroneous data model [11].

Text mining is a branch of the field of data mining in recent years, and in the international arena, it is a very active area of research. Technically, it is actually two discipline intersect of data mining and information retrieval.

Traditional data mining techniques are mainly for the structured data, such as relational databases. With the rapid development of information technology, a large number of different forms of complex data (such as structured and semi-structured data, hypertext data, and multimedia data) continue to emerge. Most text data is unstructured, after conversion the feature of the data will reach tens of thousands or even hundreds of thousands. How to extract simple, refined, understandable knowledge needed quickly from heterogeneous data sources from a large-scale text information resources relates to text mining [12].

At present, there are mainly two types of text mining tools, one is a web-based text mining tools, such as: Chemical Abstracts Service, Results Analysis feature subsystems of Web of Science; the other is desktop-based text mining tools, such as : Vantage Point, Aureka, TDA (Thomson Data Analyzer) and so on.

4.2.3 Geographic Information System

Geographic Information System which means the GIS, is a discipline that developed with geographic science, computer technology, remote sensing technique and information science. It's also a new discipline which combines the computer science, information technology and geography together. Then it treats the geographic space database as the research target. GIS is supported by the computer software and hardware, utilizes system construction and information science theory, science management and synthesized analysis that contain space connotation of geographic data to provide necessary space information system of planning, managing and decision.

GIS is able to combine various information with geographic position and related graphics together, then treat the geography, geometry, computer science and various application, Internet, multimedia technology and virtual realistic technology as a whole, utilizes computer graphic and database to gather, store, edit, display, transfer, analyze and output geographic images and their property data. At the same time, it is capable of sending the words with image to users according to their needs. It also provides convenience to constructing managers in analyzing, designing, checking and so on as foundation.

Normal visualization of S&T Monitoring is to use the monitored S&T information to display the information status and mutual relationship of monitoring objects by statistics, images associated chart. As the S&T resource in the world most is related with production, representation and including places, so GIS has fully certificated this point. On the base of utilizing the normal S&T resource monitoring visualization, integrates the advantages that GIS system is able to represent geographic place of the monitored objects, GIS has been one of the most important S&T monitoring research direction.

The usually GIS system domestically can be divided into the GIS based on single core, such as MapGis in Wuhan University, SuperMap Objects and so on. Based on Web application, such as Arc Gis, mapinfo, ArcGIS, SuperMap. net and so on.

Besides the three usually S&T resource monitoring technology describing above, there are many other technology, such as pattern calculation and so on. However, the technology tools they use are almost the same as document calculation, only have difference in different functions of those tools. This paper will not over-talk about that.

5 Design for the Framework of the International S&T Resources Monitoring

5.1 The Present Situation of International S&T Information Service

For a long time, large sum of the S&T documents, patent data and S&T information is stored in various professional database, however, there is no direct channel between the final users and these database during the necessary moments; Maybe some part of primary data has been gained, because there is no scientific and convenient information processing tools, the data information which can be used for helping decision is not able to be formed directly. It can be list as following.

(1) The Purchasing of S&T Information Database
Currently, most of the units who buy various S&T information database domestically mainly are some science institutes, large-scale research centers and companies. The cost of access to these databases is a large sum of money while the normal companies which are in urgent need are not able to afford.

(2) The Use of S&T Information Database
There appear to be more problems after the researchers obtain the necessary S&T data information by the authorized S&T databases, it is necessary for them to study the new software tools in the information processing field. For example, although some social network analyzing software are open sourced, but there needs some procedures during the acquiring and learning the target software, for some of the users, this is impossible.

(3) The Usage of the S&T Information Processing Tools

Currently, for the S&T information processing tools which are used widely, we would better acquire some level of basic theory knowledge and methods of utilizing so as to apply them in more flexible way, this maybe a big problem for the none-professional work. On the other side, though the web database can be processing directly, most of the tool software is processing the information in the form of single machine. The information after processing by the tool software may need the test of web users' or users' network processing requirements, in this situation, some transformation is necessary. The S&T information transformation has different levels of hardness for normal users which may not be accomplished because of the limit of one's working field.

There may be some problem factors not mentioned above. For example, various S&T information visualization requirements are not able to be satisfied, this leads to the situation that current S&T information can be put into wide and quick use, at the same time, S&T researchers keep working in this field continuously. Among the advancement, the TDA in Thomason company has special improvement in this field, but the cost is very high, the none-structure handling, database field and format limit, data information result have not been individuation, such as the S&T information resource real-time geological distribution as so on. The most important factor is that it is necessary to use the uniform database format to deal with the information or under one kind of application software platform to realize the transformation of different data format.

5.2 Intelligent S&T Information Service Solution

Aiming at the S&T information service existing problems mentioned above, this article proposes some basic consideration for the solution. The core target of solution in this article is to construct a low cost and easy formed solution pattern. The detailed train of thought is as following:

(1) Construct User Service Platform

The service platform can be constructed by S&T department or research institute, users may utilize the method of registering membership to load in the system, and then choose different S&T service information according to their needs. In this way, most of the company users can save a lot of money for strategy consulting. The user service platform structure can be as the figure one. For the operating method of figure 1 in applying server, please see figure 1.

(2) **Construct the Application Serving Channel**

After the institutes acquire the right of using S&T information database, they are able to search discrete database information at any time. Most of the S&T departments and research center are able to have these conditions. The system can read the necessary information from the S&T database then import its own hardware automatically, but the whole service is still in construction. For the technology nowadays, this problem can be solved very easily.

The key point in solving the problem above is constructing self-adapting and application servers, founding certification service between departments' application servers and professional S&T information providers, allowing the direct connection of department servers to access S&T information database, then transferring to the information database that can be structuring dealt with. This is also a change in the current professional S&T of providing service. Comparison between two service patterns, as the figure 2:

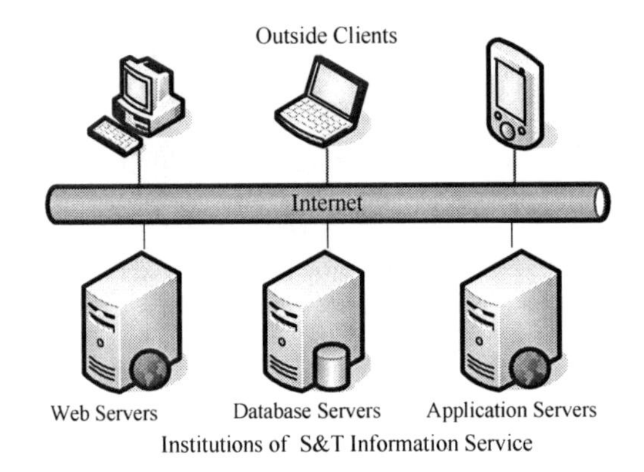

Fig. 1. Structure of S&T Service Platform

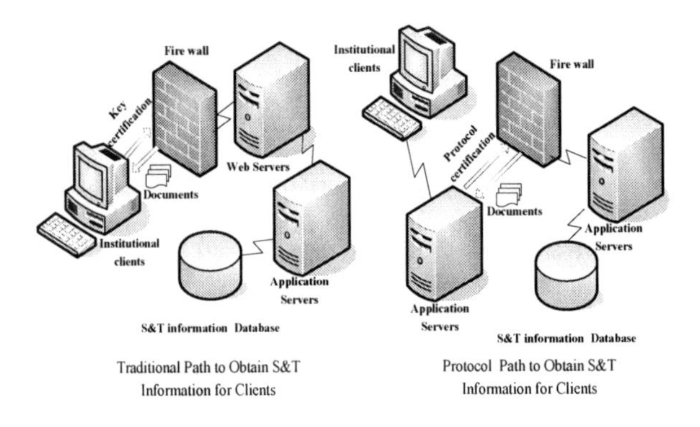

Fig. 2. Patterns Comparison of Clients Obtaining S&T Information

(3) Develop and Design the Self-Adaptive Processing Software

Current software, such as the social network analyzing software UCINET, Pajek, CiteSpace and so on are all single machine dealing pattern, they are not able to satisfy the network users' requirements convenience. But most of them are opened structure, for the software field, it is comparably easy to develop the processing software based on network service.

There is some networking processing software like Results Analysis which is the function subsystem of Web of Science, Vantage Point, and TDA. Their fatal shortcomings are pointed to the limited source of database, because they serve their own professional database. But nowadays, the scientific research departments who purchase the database resource are always tens in number, or even more database providers.

Since most of the software is open sourced, the referenced pattern is based on the past. Then according to that base, self-developing will take place while referring to some operation of professional software and related service pattern which has realized the network processing, this method can be put into use absolutely. In the article, the self-adaptive software and department application server function can be as following of figure 3:

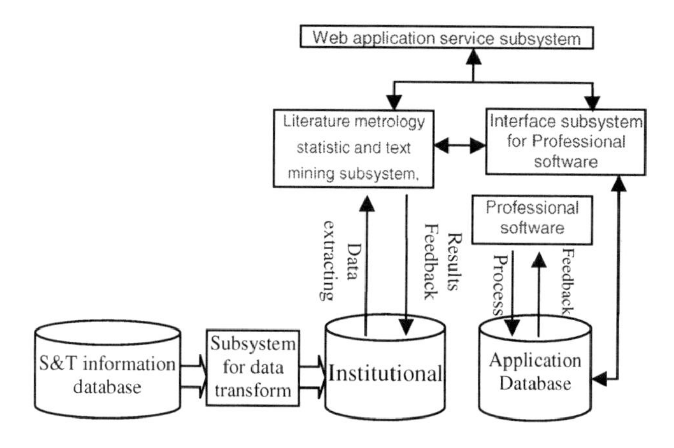

Fig. 3. Function Structure of Institution Application Servers

5.3 Example Analysis

Set the project subsidized by this article as the prototype, a portal is constructed first to show a basic platform for serving the users as figure 1. Then relevant S&T information database inquire and utilizing right should be bought and gained, set up the framework protocol as figure 2. Point at the special needs in this project, the application servers' function structure as the figure 4.

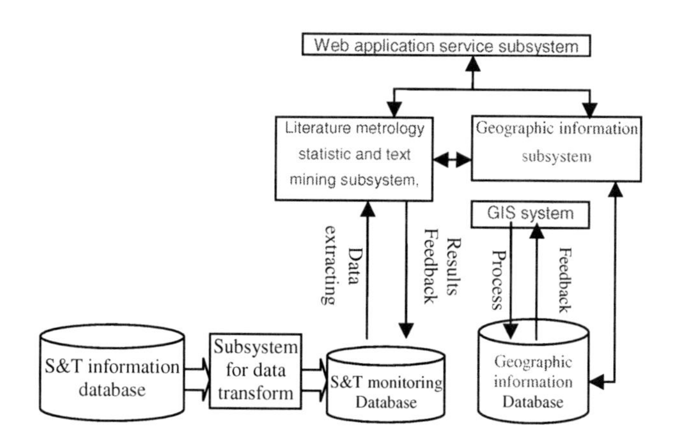

Fig. 4. Function Structure of International S&T Resources Monitoring Application Server

The main design idea involved in the figure 4 is: after user sending technical information service request by web server, the server collects a mass of data about technical literature, patent database and web text, and then processes them to target data, saves them to technical monitor database. On one hand, the sever uses social net analysis software such as Ucinet software, and text mining software, TDA software to do technical information statistics and analysis, On the other hand, by the Supermap geographic processing system, etc. The S&T information and its owners' geographic data in the monitoring database can be connected, then make S&T users grasp S&T information gathering and distributing state & geographic distributing information at all times to support companies' innovational strategic decision.

6 Conclusions

The article aimed at the research in the international S&T resources monitoring, and carried out systematic analysis of the objectives, means and methods of international S&T resources monitoring. We defined a set of effective intelligent solutions and analyzed science project, according to the current international S&T resources monitoring and service.

In the fields of international S&T resources monitoring, there are still much more techniques and methods to be developed and explored further; and in practice, we should adjust and improve accordingly. Therefore, there leaves much to be deeply researched by scholars in international S&T resources.

Acknowledgements. Thanks for supporting by international cooperation project of the MOST, 'research on the model and method of international S&T cooperation resources monitoring, based on data mining', and thank all the research members for their great cooperation work as well.

References

1. Kontostathis, A., et al.: A survey of Emerging trend Detection in textual Data Mining. A Comprehensive Survey of Text Mining, ch.9. Spring-Verlag, Heidelberg (2003)
2. Kostoff, R.N., et al.: Literature-related discovery (LRD): methodology. Technological Forecasting and Social Change (75), 186–202 (2008)
3. Sun, t.: International Science and Technology Cooperation-Oriented S&T Resource Monitoring. Beijing Institue of Technology, Beijing (2010)
4. Taotao, S., Yun, L., Wenping, W.: Competitive Technical Intelligence Analysis Based on Patents Coupling. In: 2010 International Colloquim on Computing, Communication, Control and Management Proceedings, Yangzhou, China, pp. 102–105 (August 2010)
5. Taotao, S., Yun, L.: Development of Virtual Reality Technology Rearch via Patents Data Mining. In: 2010 Third International Symposium on Intelligent Uniquitous Computing and Education Proceedings, Beijing, China, September 18-19, pp. 428–431 (2010)
6. Meng, X., Gao, Y.: Electric systems analysis, pp. 3–21. Higher Education Press, Beijing (2004)
7. Small, H.: Co-citation in the scientific literature: a new measure of the relation between two documents. Journal of American Society for Information Science 24(4), 265–269 (1973)

8. Coates, V., Fardoque, M., Klvans, R.: On the future of technological forecasting. Technological Forecasting and Social Change 67(1), 1–17 (2001)
9. Chen, C.: CiteSpace II Detecting and Visualizing Emerging Trends and Transient Patterns in Scientific Literature. Journal of the American Society for Information Science and Technolog 57(3), 359–377 (2006)
10. Moody, J., Mcfarland, D., Demoll, S.B.: Dynamic Network Visualization 110(4),1206 – 1241 (2005)
11. Navarro, J.G.C., Lario, N.A.: Building co-operative knowledge through an unlearning context. Management Research Review 34(5), 1–24 (2011)
12. Franco, M., Magrinho, A., Silva, J.R.: Competitive intelligence: A research model tested on Portuguese firms. Business Process Management 17(2), 1–27 (2011)

A Component Clustering Index Tree Based on Semantic

Chunhong Wang and Yaopeng Ren

Department of Computer Science and technology, Yuncheng University,
Yuncheng, Shanxi, China
ycuchwang@126.com, ryaopeng@163.com

Abstract. The reasonable classification of components is the basis and key of component efficient retrieval. In order to overcome the shortcomings of faceted classification method widely used, we adopt a method combing faceted classification with full-text retrieval to describe components. Based on that description, a component cluster index tree is proposed which uses cluster analysis technique and semantic analysis technique. And the experiments prove that the index tree is feasible, which can effectively overcome the shortcomings of faceted classification method. Meanwhile to some extent, it can achieve the component semantic retrieval and has higher component recall ratio and precision ratio. Moreover, the description of retrieval conditions is no longer limited by restrictive terms so as to be convenient for general users.

Keywords: faceted classification, cluster analysis, semantic analysis, index tree.

1 Introduction

In recent years, along with the rapid development of the object-oriented technology and the software component technology, software reuse has been the research hot and is considered as an effective way of improving the software's Productivity and quality[1]. Meanwhile, as a core technology to support the software reuse, software component technology has obtained high importance. In the component library system, there are two main problems: component classification and component retrieval[2]. Effective component retrieval can reduce the cost of component finding and understanding so as to promote the component reuse. Moreover, the reasonable component classification is one effective way and method to achieve rapid retrieval.

Currently, there are many component classification representation methods. From the view of the component representation, the methods are divided into three kinds by W.Frakes: artificial intelligence method, hypertext method and information science method[3]. Among them, facet classification method belonged to the kind of information science method has been applied more successfully. But, the facet classification method has its own shortcomings which possibly lead user can't retrieve required components, for example this method needs to build and maintain the term space artificially and has strong subjective factors[4]. Therefore, in this paper, the method combining the facet classification and the full-text retrieval is used to represent components. On the foundation of that, a many binary tree component index mechanism based on semantic is proposed, which adopts the clustering analysis technology and semantic.

G. Zhiguo et al. (Eds.): WISM 2011, CCIS 238, pp. 356–362, 2011.

2 Component Representation

In this paper, the representation method combining the facet classification and the full-text retrieval is used to describe the components. Firstly, before the component inventory, the component description text is divided into corresponding facet value according to some facet classification scheme. Secondly, extract feature words for each facet value and build the component vector space. Then, achieve the component classification by the clustering analysis according to the full-text retrieval method. The component classification chart is shown blow as Fig.1.

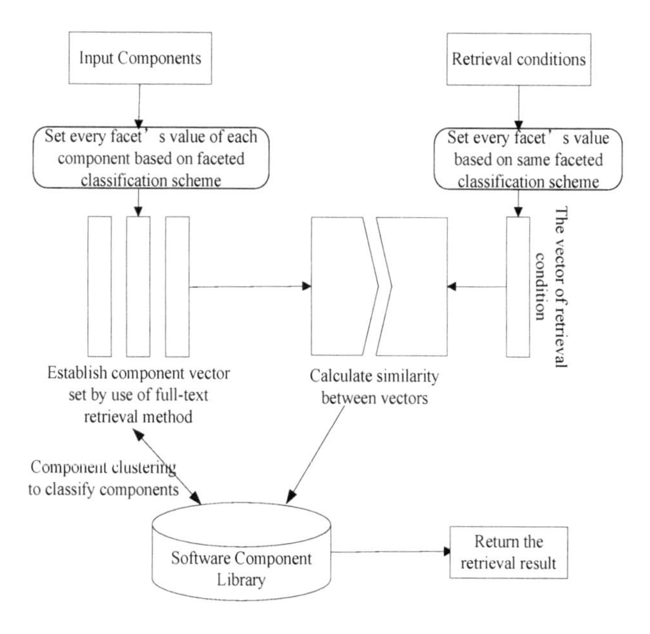

Fig. 1. Component Classification System

Meanwhile, the following five facets are defined in the facet classification method from the view of the integrity and independence.

- Function: component functional description, application fields;
- Operation objects: the input and output objects of the component;
- Use environment: the software and hardware environment;
- Representation method: the component form (such as class, architecture, frame, model etc.) and representation (such as graphics, pseudo code, language etc.);
- Performance: the component reliability evaluation.

The representation method combining the facet classification and the full-text retrieval has two advantages: For one thing, it convert the facet value from the traditional controlled words to the text, which reduces the subjectivity of artificial building and maintaining the term space; For another thing, after the facet classification, the content of the text is focused (such as function text, performance text) in favor of the similarity calculation of the component texts so as to promote the component clustering effect and achieve more reasonable component classification.

3 The Component Clustering Based on Semantic

On the foundation of the above component description, a component clustering algorithm based on semantic similarity and optimization[5] is adopted to achieve the component classification. The basic thought of the algorithm is that: describe the components by the method combining the facet classification and the full-text retrieval; calculate the similarity between the component texts from the view of the semantic; then cluster the components by the method which combines the nearest neighbor clustering and genetic algorithm.

The process of this clustering algorithm can be described as: firstly, use the ith (i is a control variable, and i=1) component text to initialize the first clustering center, calculate the similarity between the other component text and the class component text through the similarity calculation method based on semantic, cluster the component texts by the nearest neighbor clustering algorithm and meanwhile update the clustering center; and then use GA to optimize and merge the clustering results which are produce by the nearest neighbor clustering algorithm. The whole algorithm is a big circle whose termination condition is that the value of the control variable i is bigger that the number of the component texts or the change of the objective function value is smaller that a some setting threshold t.

From the view of the semantic, this algorithm can obtain the better clustering results, which combines the nearest neighbor algorithm and the genetic algorithm. As a result, it can achieve the more reasonable component classification and then promote the component retrieval. Meanwhile, this algorithm can obtain the feature words of each component and the class feature words of each class (the clustering center), which can be helpful to build the component clustering index tree based on semantic.

4 The Component Clustering Index Tree Based on Semantic

The clustering tree is a kind of tree structure which represents the state of the data clustering by layer[6]. A clustering tree can organize the data as a layered structure according to the clustering information on the different layers, and can provide a effective index structure. The parent node on the clustering tree includes the pointer pointing its child nodes and the information of its child nodes. And the root represent the raw data set.

The forming process of the clustering tree can be simply described as: firstly, through the clustering algorithm, classify the raw data set into some classes, and each class is as a child node of the root; and then for the child nodes repeat the above process until the child nodes satisfy some condition, and thus a clustering tree is formed. The child nodes on the clustering tree include the pointer pointing the data in order to easily find the data. The structure of the clustering tree is shown blow as Fig. 2.

On the foundation of the introduction of the clustering tree, the following are the definitions related the component clustering index tree.

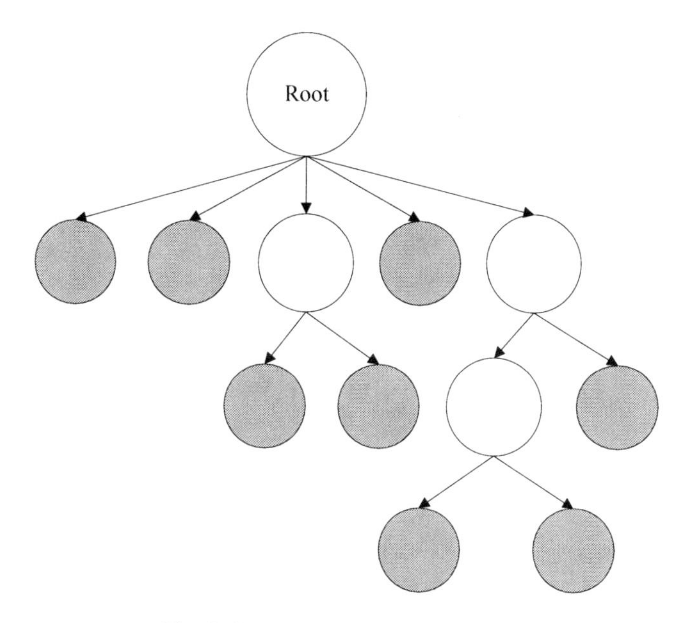

Fig. 2. Structure of the Clustering Tree

4.1 The Related Definitions

Definition 1. The component feature words set (CS) is defined as the feature words set which can effectively represent the component attributes. It can be represented as CS= {Ck1,Ck2,Cki,…Ckn}, and generally the value of n is 3~5.

Definition 2. The cluster feature words set(CTS) is defined as the feature words which can well reflect the characteristics of all components in the cluster and can effectively distinguish from the other sets. It can be represented as CTS={CTk1,CTk2,CTki,… CTkm }, and generally the value of m is 3~5.

Definition 3. The component cluster index tree (CCIT) is a non-empty tree which satisfies:

(1) only one root that represents all components in the component library;
(2) the parent node including the pointer pointing its chide nodes and the information of its child nodes;
(3) the child nodes including the pointer pointing some specific component;
(4) the tree's depth is four. The first layer is the root; the second layer is the facets; the third layer is the clusters; the fourth layer is the components. And the cluster layer includes the pointers pointing all components in this cluster, and meanwhile includes the information of the cluster feature. And the component layer that is the leaves layer includes the information of the component feature except the content in the (3). The structure of the component clustering index tree is shown blow as Fig.3.

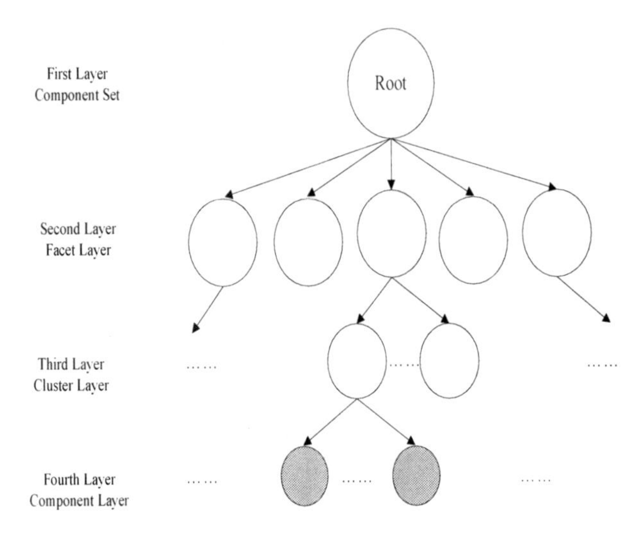

Fig. 3. Component Clustering Index Tree

For the construction of the general clustering index tree, from the beginning the data set are clustered until satisfy some condition such as the number of the leaves is not more than some threshold. While, for the construction of the component clustering tree, the components are clustered from the second layer and the number of the leaves is equal to the number of the components in the component library.

4.2 The Construction of the Component Clustering Index Tree

The construction algorithm of the component clustering index tree based on semantic is described as following.

Input: All components in the component library which is represented by the ComponentSet. Here is the component text set.

Output: The tree of the component clustering index tree based on semantic which is returned by the Root.

```
BuildCCIT ( ComponentSet )
{ The root represent the ComponentSet;
According to some facet classification scheme, classify the root firstly which obtains Root1, Root2, …, Rootn;
    for(i=1; i<=n; i++)
    { Cluster (Rooti);}
}
Cluster (Rooti)
 {Data= Rooti
  Classify the Data by the use of the component clustering algorithm based on semantic analysis and optimization;}
Each sub-cluster is processed as
```

```
{ Extract the CTS;
 Exstract the CS;
     Output all the components in this cluster, that is the leaf nodes;
}
```

5 The Results of the Experiments and the Analysis

In this paper, the KNN retrieval algorithm[7] is used to prove the effectiveness of the component clustering index tree based on semantic. The experiment data is some components about the user interface from the Shanghai Component Library[8] and some classes from the Java Class Library as components. And the components are described by the use of the method combining the facet classification and the full-text retrieval. These components can be divided into six subjects: Button, Menu, Tree controller, Word Process, Java's Color Class and Java's Input/Output Class., Here, we adopt two indicators to evaluate the results of the experiments, which are the precision and the recall rate. The two indicators are defined as following:

Pecision=the number of the similar components and the accurate components in the result set/the total number of the components in the result set.

Recall Rate= the number of the similar components and the accurate components in the result set/the total number of the components which are similar to the retrieval condition in the component set.

Here, three retrieval test are processed: K=1, K=3, K=7. And the numbers of the feature words in the CTS and in the CS are all five. The precision and the recall ratio of the three experiments' results are given blow as Fig.4.

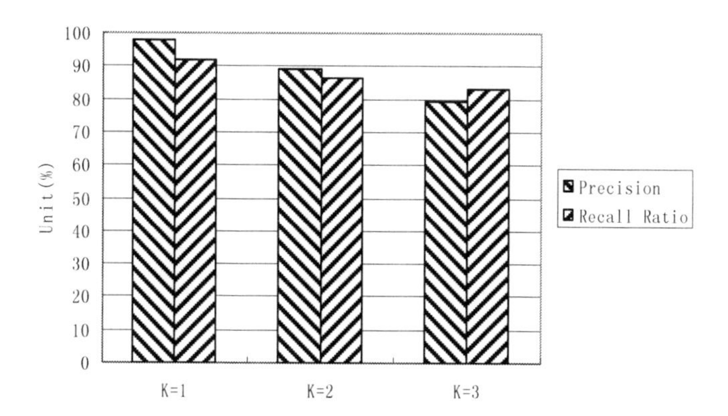

Fig. 4. Experiment Result

From the fig.4, we can find that the component clustering index tree based on semantic is effective and feasible. Based on this component index tree, the precision and the recall ratio are all above 80%. Moreover, when K=1, the precision and the recall ratio are even above 90%. Meanwhile, semantic analysis technology is introduced to the index tree so as to overcome the subjectivity of the facet classification to some extent.

6 Conclusion

Aiming at the shortcomings of the facet classification method which is widely used currently, a component representation method combining the facet classification and the full-text retrieval is adopted to describe the components. On the foundation of that, a component clustering index tree is propose by the use of the clustering analysis and the semantic analysis technology. And through the experiments, the feasibility and effectiveness are proved. This tree can not only avoid the subjectivity of the facet classification to get the more reasonable component classification, but also to some extent can achieve the component semantic retrieval for users to retrieve components easily and meanwhile has higher component precision and recall ratio. But, this clustering tree depends on the development of the semantic analysis technology, and the component retrieval method based on this clustering tree needs further research.

References

1. Yang, F., Mei, H.: The Component Software Design and Achievement. Tsinghua University Press, Beijing (2008)
2. Zhong, C., Guo, G., Zhang, Y.: Research on Matching Algorithm for Faceted-based Software Component Query Through Different Component Repositories. Computer Engineering 32(21), 82–84 (2006)
3. Pan, Y., Zhao, J.F., Xie, B.: The Research and Development of the Technologies in Component Library. Computer Science 5, 90–93 (2003)
4. Liu, D., Zhao, L., Wang, Z.: A Component Classification Retrieval Method based on the Facet Classification and Clustering Analysis. Component Application 24, 89–90 (2004)
5. Zhang, Y., Ren, Y., Chen, L., Xie, B.: Component Clustering Algorithm Based on semantic similarity and optimization. Computer Engineering and Design 31(11), 2531–2534 (2010)
6. Liu, Y., Kuang, Y.: Approximate K-NN Search in High-Dimensional Spaces Based on the Clustering Index Tree. Modern Computer 255, 18–21 (2007)
7. Liang, J., Feng, Y.: CSA-Tree: An Improved High-Dimensional Index Tree for Main Memory Access. Computer Science 30(3), 415–423 (2007)
8. Shanghai Component Library, http://www.sstc.org.cn/

Application of Shuffled Frog-Leaping Algorithm in Web's Text Cluster Technology

Yun Fang[1] and Jianxing Yu[2]

[1] Computer Science and Technology College
Taiyuan University of Technology
Taiyuan, Shanxi, China
fyun63@126.com

[2] Computer Science and Technology college
Taiyuan University of Technology
Taiyuan, Shanxi, China
yuchi0529@qq.com

Abstract. With the rapid development of Internet, more and more massive information, search engine technology developed rapidly, but the search engine's search results don't not meet the search requirements, The k-means clustering algorithm are introduced to gather web documents class, in order to improve the clustering performance, the introduction of leapfrog algorithm selection of k value aiming to improve the accuracy of search results and to increase the search engine returns results associated with the query topic.

Keywords: Clustering algorithm; Text clustering algorithm; Leapfrog algorithm.

1 Introduction

The rapid development of Internet makes the number of sites, more and more unstructured increasing web document information, Internet users to be able to collect valuable information, become the major research focus in the search engine technology, the proposed clustering technique is Find the exact search engine provides a very important role, allowing users to quickly find the needed information effectively has become an urgent problem. In this paper, we are clustering technology to the text on the handling of search engine results.

Clustering in data mining web document also plays an important role, document clustering can reveal the internal structure of the document collection, the discovery of new information, document clustering document collection is divided into several clusters, the cluster content of the document requested Similarity as large as possible, while the cluster similarity between the documents as small as possible.

In this paper, a document clustering algorithm[1] is widely used K-means based segmentation algorithm, K-means clustering algorithm is the clustering of a dynamic clustering method, K-means algorithm based on the predetermined value K, to be together Class samples are divided into K classes, so that all the samples cluster in the domain of the square of the distance to the cluster center and the smallest. However, the clustering algorithm selected by the number of cluster centers of K, especially for

G. Zhiguo et al. (Eds.): WISM 2011, CCIS 238, pp. 363–368, 2011.

the class number of the sample set, K value of the options to be specified and a random person, in order to obtain better clustering results, Usually test the different K values. K are chosen in view of the uncertainty, this K value selected by Leapfrog algorithm[2] to avoid the use of random numbers as the initial cluster centers resulting from the same K value of the clustering effect of instability.

2 Cluster

2.1 Text Clustering Ideas and Steps

The main idea of clustering[3] is through a specific algorithm, according to the similarity of text data and difference, the text is divided into several categories, in the same category as similar as possible differences in different categories. The main process: First, the clustering of the text to pre-processing, segmentation of words and the exclusion of irrelevant words, and the frequency of statistical terms, that each of the text processing, and generates the text of the feature vector space, Then for each particular feature vector space decomposition of its extraction, as far as possible to represent the text extracted feature vector, and calculate the similarity between the text and then choose the clustering model to test and evaluate the clustering quality, and finally The results displayed.

2.2 K-means Clustering Algorithm

Clustering algorithm for text clustering algorithm K-means algorithm is a clustering algorithm based on segmentation, segmentation-based clustering algorithm can be simply described as; to construct a set of objects form a partition of K clusters, makes the evaluation function is optimal. Because the number of K clusters will not change. The algorithm to K clusters start to the end of K clusters, each iteration, each instance of the original, if not retained in the cluster, the other is to be assigned to a cluster, this process is repeated until it meets a Stop criteria.

X_j (j = 1,2,3, ... N) is divided into K classes Gk (k = 1,2,3 .... k), all samples of each class form a group, find the center of each cluster, and Allocation vector A [1], A [2], ..., A [n], makes the non-similarity (or distance) index of the value function (or objective function) to a minimum.

The value is the set of cluster centers of all the mean vectors:

Here, this calculation is repeated basis, until the evaluation function below a threshold or 2 adjacent close iteration. This calculation is repeated basis, until the evaluation function below a threshold, or 2 adjacent. The difference is lower than a threshold, given the following K-means clustering algorithm pseudo code. The difference is lower than a threshold, given the following K-means clustering algorithm pseudo code.

K-means clustering algorithm K-means clustering algorithm

1: procedure KMeansCluster (X1, ..., XN, K)
2: A [1], A [2], ..., A [n] initial cluster⇓initial cluster assignment
3: repeat
4: Change false

5: For i = 1 to N do 5: For i = 1 to N do
6: function (xi, ck)
7: if A [i] is not equal k then
8: until change is equal to false return A [1], A [2], ..., A [n]
9: end procedure

K-means clustering algorithm is an important aspect of the choice of initial cluster centers, in addition to generating k-cluster, but also generates the center of each cluster, cluster performance and choose the initial cluster centers, and that most choose to be the first K sample set cluster samples as initial cluster centers. K-means clustering algorithm is an important aspect of the choice of initial cluster centers, in addition to generating k-cluster, but also generates the center of each cluster, cluster performance and choose the initial cluster centers, and that most choose to be the first K sample set cluster samples as initial cluster centers. but this selection is not scientific and increase a great deal of time and space overhead, the introduction of leapfrog algorithm for this iteration K, aims to improve the clustering performance. But this choice is not scientific and increase a great deal of time and space overhead, the introduction of leapfrog algorithm for this iteration K, aims to improve the clustering performance.

3 Experimental Result Analysis

Purpose of this object to web document clustering is a form of unstructured data, in order to apply clustering algorithm, the document must be expressed to form a structured data form. We use the vector space model (VSM) to represent the document in the form of VSM We use the vector space model (VSM) to represent the document in the form of VSM.

3.1 Data Preprocessing and Feature Extraction

Purpose of this object to web document clustering is a form of unstructured data, in order to apply clustering algorithm, the document must be expressed to form a structured data form. We use the vector space model (VSM) to represent the document in the form of VSM We use the vector space model (VSM) to represent the document in the form of VSM.

V (di) = (Ti1, Wi1; Ti2, Wi2; ..., Tin, Win) Where T for the entry, Wij to Tij in the document in the weights di, TF is term frequency, TF greater, indicating the term the greater weight in the article. Where Tij for the entry entry, TF is term frequency, TF greater, indicating the term the greater weight in the article. IDF is the document frequency, the classification of the document to be centralized, The document contains a number of S words. Where W (t, di) for the word t in the text di in weight, t (t, di) for the word t in the text di of word frequency, N is the text of the total number, ni is the training of the text focus appears t the text of the number of denominator is a normalization factor. Where W (t, di) for the word t in the text di in the weight, t (t, di) for the word t in the text di in the word frequency, N is the text of the total number, ni is the training of the text focus appears the number of text t, the denominator is a normalization factor. log (N / nt +0.01) actually represents the IDF. log (N / n t +0.01) actually represents the IDF.

XML documents can be expressed by the title, keyword, abstract and so on. XML documents can be expressed by the title, keyword, abstract and so on. So we modified the weight of the important position of the weight of larger vocabulary. To this end we correct weight, an important position to make the words bigger weight. We are introducing a random number that represents an important position in terms of the degree of value Where w (t, di) is the weight, w (t, di) to calculate the original TF-IDF weight, It for the term t in the document title, keyword, abstract, URL, and other important positions on the number of times. Extracted from the document that best represents the content of the document vocabulary, to improve the running efficiency and reduce the impact of noise words. Extracted from the document that best represents the content of the document vocabulary, to improve the running efficiency and reduce the impact of noise words.

Text similarity calculation is an important part of preprocessing on the text similarity measure, we use the Euclidean distance calculation method used to calculate the similarity between vectors, Where xi, and xj, respectively, the text refers to the characteristics of the two vectors, xi, and xj are respectively characteristic xik and xjk k-dimensional elements in the first, |xi| is the dimension of feature vector xi. Where xi, and xj denote the two texts are the eigenvectors, xi and xj are, respectively, characteristics of xik and xjk k-dimensional elements in the first, xi is the dimension of feature vector number.

3.2 Leapfrog Algorithm K-Means Clustering Algorithm

Conditional (or observed) variable belong to the category of evidence cliques, while the cliques which contain only target (or label) variables belong to the category of compatibility cliques. Obviously, in the example, the cliques in the flat conditional Markov networks indicated by the thin lines belong to the category of evidence cliques, while the cliques specified by the relational clique templates indicated by the bold lines belong to the category of compatibility cliques.

The initial position of cluster[4] centers will affect the performance of K-means algorithm, the initial location of cluster centers so the choice is based on iterative leapfrog algorithm selected K values, so that the location of the initial cluster centers to be scattered clustering of several groups of artificial selection of the document. The first is the formation of the initial population, followed by frog design groups: According to F frogs (solutions), randomly generated frog populations; each frog for a specific calculate the target function value; in descending order based on the objective function value F S frogs were divided into subgroups. for each sub-group of frogs are found in one of the best individual and the worst individual to identify the best individual groups; for each subgroup, in descending order according to the objective function value of the individual, re-distribution and mixing operations; termination conditions are met, the end of iteration, the optimal objective function value of the output information, or turn to the original sequence. The first is the formation of the initial population, followed by frog design groups: According to F frogs (solutions), randomly generated frog populations; each frog for a specific calculate the objective function value; descending order according to the objective function F frogs into S sub-groups according to; for each sub-group of frogs are found in one of the best individual and the worst individual to identify the best individual groups; for each subgroup, in descending order according to the objective function value of the individual, re-distribution and mixing operations; termination conditions are met, the end of iteration, the optimal objective function value of the output information, or turn to the original sequence.

Leapfrog algorithm parameters: maximum step size Dmax = 20, the total number of frogs F = 2000, every subgroup of evolution algebra N = 1200, the number of m = 8 subgroup iteration Iter = 1000, the number of sub-group of the frog n = 250, for the parameter settings are currently no guiding principle, most of them are obtained by experimental test, K values between 1 and maxnum, leapfrog algorithm iteration the value of K must satisfy this condition. Leapfrog algorithm parameters: maximum step size Dmax = 20, the total number of frogs F = 2000, every subgroup of evolution algebra N = 1200, the number of subgroups m = 8 iterations Iter = 1000, sub-group n = number of frogs 250, set the parameters there is no guiding principles, most of them are obtained by experimental test, K values between 1 and maxnum, leapfrog algorithm iteration the value of K must satisfy this condition.

3.3 Leapfrog Algorithm K-Means Clustering Algorithm

The design of fitness function evaluation of clustering results is very important. A good clustering should be a large class of distance; class within the distance is small; the number of samples between classes evenly. A good clustering[5][6] should be a large class of distance; class within the distance is small; the number of samples between classes evenly. Between the samples which NumDifference that the number of different types of statistics, K-means algorithm performance depends on the value of the initial cluster centers selected good or bad, when the initial cluster centers were randomly selected, for the same value as the initial center of K different fitness function value will lead to the same parameters in different results. Between the samples which NumDifference that the number of different types of statistics, K-means algorithm performance depends on the value of the initial cluster centers selected good or bad, when the initial cluster centers were randomly selected, for the same value as the initial center of K different fitness function value will lead to the same parameters in different results. leapfrog algorithm is applied to the selection of K value is more reasonable to reduce the time and improve efficiency. Leapfrog algorithm is applied to the selection of K value is more reasonable to reduce the time and improve efficiency.

4 Conclusion

Experimental results show that adding leapfrog algorithm with the original K-means algorithm K-means algorithm is basically the same running speed, the average accuracy than in the original algorithm has generally improved, especially in the right when the specified number of clusters K, the average accuracy improved by nearly 29%, indicating that the K means clustering algorithm using leapfrog algorithm, can improve the clustering performance[7].

References

1. Jiang, M.F., Tseng, S.S.: Two phase Clustering process for outliersdetection. Pattern Recognition Letters 22, 691–700 (2001)
2. Rahimi-Vahed, A., Dangchi, M., Rafiei, H., Salimi, E.: A novel hybrid multi-objective shuffled frog-leaping algorithm for a bi-criteria permutation flow shop scheduling problem, May 5 (2008)
3. Hearst, M.A.: Clustering versus faceted categories for information exploration. Communications of the ACM 49(4), 59–61 (2006)
4. Jiang, Y.-Q., Zhang, Y., Zhou, Y.: K-means algorithm for optimizing the number of clusters based on particle swarm optimization. Control Theory and Applications 1175-1179 (October 2009)
5. Li, T.: Document clustering via Adaptive Subspace Iteration. In: Proceedings of the 12thth ACM Intrnational Conference on Multimedia (2006)
6. Makoto, I., Takenobu, T.: Hierarchical Bayesian clustering for automatic text classification
7. Bruce Croft, W., Metzler, D., Strohman, T.: Search Engines Information Retrieval in Practice. Pearson Education, London (2010)

Speed Up Graph Drawing for Social Network Visualization

Wei Liu and Xing Wang

Institute of Scientific and Technical Information of China, Beijing, 100038
Information Resource Centre
Beijing, China
{liuw,wangxing}@istic.ac.cn

Abstract. Though classical forced-directed algorithms have many advantages on graph drawing, such as good-quality results, flexibility, simplicity, etc., the high running time makes them impractical to real applications. To this end, we propose a community-based approach to speed up graph drawing for social networks under the assumption that the position of a vertex in the graph is mainly determined by the vertexes which are in the same community with it. Experiments indicate that, comparing with the classical force-direct algorithm, our approach can significantly reduce the running time without the loss of graph readability.

Keywords: Graph drawing, Force-direct algorithm, Social network.

1 Introduction

Information visualization is any technique for creating images, diagrams, or animations to communicate a message on computer, which can help people to find useful and potential knowledge from the visualized data. Visualization techniques have been widely used in various application areas, such as science, education, engineering, interactive multimedia, medicine, etc. The explosive growth of information greatly pushes forward visualization techniques over the last few years. One of the most important issues in this field is graph drawing. Graph drawing, as a branch of graph theory, applies topology and geometry to derive two-dimensional representations of graphs. Graphs are usually represented pictorially using dots to represent vertices, and arcs to represent the edges between connected vertices. Arrows can be used to show the orientation of directed edges.

As is well-known, very different layouts can correspond to the same graph. However, the arrangement of these vertices and edges impacts understandability and aesthetics. Therefore, most researchers mainly paid their attentions to the graph readability. To this end, a number of algorithms, such as the radial algorithms[1] and the force-directed algorithm[2], have been developed. Because the force-directed algorithm can express the close degree of the relation between two vertexes in a reasonable way[3], recent empirical researches[4, 5, 6] focused on improving the force-directed algorithm. In addition, several aesthetic criteria[7] have also been proposed and generally accepted to measure the graph readability, including the following:

G. Zhiguo et al. (Eds.): WISM 2011, CCIS 238, pp. 369–376, 2011.

Fig. 1. A typical social network drawn by forced-direct algorithm

- Small number of edge crossings
- Even distribution of vertices
- Uniform edge length
- Small drawing area
- Maximum angular resolution

These criteria were originally proposed based on human intuition. In Section 2, we will briefly review the important literatures in field of graph drawing.

However, in many cases, the graph is very large and complex(maybe hundreds of thousands of vertices and edges), such as social networks. The efficiency of graph drawing is turning out to be more important in practice. Especially, some applications require the graph has to be drawn online due to the randomness of user query, which makes the graph drawing time must be tolerable for users.

In fact, most of the real world data has social property. That is, the graph actually shows a social network by one or more specific types of interdependency, such as friendship, common interest, or co-authorship and citations in scientific papers. Fig. 1 shows a typical graph on co-authorship network using force-directed algorithm, and there are a number of large or small communities in this graph. As can be seen from Fig. 1, there are often hundreds of thousands of vertexes in a graph. One of the disadvantages of force-directed algorithms is the high time: the typical force-directed algorithms are generally considered to have a running time equivalent to O(V3), where V is the number of vertexes of the input graph. To speed up the graph drawing process, we study the efficiency improvement problem for graph drawing in this paper. A community-based approach is proposed to reduce the number of computation on forces in each round. The intuition behind our approach is that, suppose the network has been divided into a group of communities which are organized into a flat or hierarchical structure, the position of a vertex in the graph is mainly influenced by those of the vertexes which are in the same community with it. The position of a community can also be determined in a similarly way. As a result, we can draw the graph with a two stage strategy. As the low level, the layout of the vertexes in one community is determined. Note that a vertex only has the relative coordinate in its community. As the high level, the layout of communities is determined. Different to vertex, any community occupies a nonzero area in theory. As a result, a new algorithm is needed to treat such nonzero-area "vertexes". At last, several measures are used to embellish the current graph, such as size adjusting and angle rotation for subgraph.

Overall, the contributions in this paper are summarized on four aspects. First, we argue that the finial position of a vertex in the graph is not necessarily determined by a small number of vertex rather than all other vertex. Second, aiming at the large scale social data, we propose a new community-based approach to speed up the graph drawing process, which can be applied to both flat-community social network and hierarchical-community social network. Third, two optimizations are introduced to improve graph readability, which includes scaling and rotation.

The rest of this paper is organized as follows. In section 2, we briefly review the classical force-directed algorithm. Section 3 introduces our proposed community-based approach for graph drawing. The experimental results are reported in section 4. Finally we conclude in section 5.

2 The Classical Force-Directed Algorithm

Force-directed algorithms have been studied extensively[2, 4]. They view the graph as a virtual physical system, where the vertices of the graph are bodies of the system. These bodies have forces acting on or between them. Often the forces are physics-based, and therefore have a natural analogy, such as magnetic repulsion or gravitational attraction. The attraction force pulls the connected vertices toward to each other, while the repulsive force pushes the unconnected vertices apart. A force-directed graph can involve forces other than mechanical springs and electrical repulsion; examples include logarithmic springs, gravitational forces, magnetic fields and electrically charged springs. In the case of spring-and-charged-particle graphs, the edges tend to have uniform length due to the spring forces), and vertices that are not connected by an edge tend to be drawn further apart due to the electrical repulsion.

Starting with random layout of the initial graph, the algorithm calculates the combined force for each vertex and moves it according to the force' magnitude and direction. This process will be repeated for a more or less times until the termination condition is met. To the best of our knowledge, a number of more sophisticated techniques[9, 10, 11] have been developed for various purposes.

3 Our Community-Based Approach

As a result, to make the drawing process more efficiently, we propose an improved force-directed algorithm for flat-structure communities. The algorithm for communities is formally depicted first(see Fig. 2), and then we give the discussion for this algorithm.

In this algorithm, the force-directed algorithm is actually performed twice. For the first time, each community is regarded as a small network, and we employ the classical force-directed algorithm to draw the graph for each community. For second time, each community is regarded as a vertex, and we employ the classical force-directed algorithm to draw the graph for the whole network. However, two parameters, spring and mass, in the second time are different with those in the first time because a community has to occupy a certain area while a vertex has not definite size. The basic technological process of the classical force-directed algorithm has been introduced in section 2. Below we make the descriptions for spring and mass in the second time.

```
Algorithm: Force-directed Layout for Communities(FLC)
   set up initial vertex velocities to (0,0)
   set up initial vertex positions randomly // make sure no 2 vertices are in exactly the
same position
   for each community
  do
     total_energy := 0
     for each vertex in this community
        net-force := (0, 0)
        for each other vertex in this community
           net-force := net-force + repulsion-force( this_ vertex, other_ vertex)
      next other vertex
        for each spring connected to this vertex
           net-force := net-force + attraction-force( this_ vertex, spring )
        next spring
        this_ vertex.velocity := (this_ vertex.velocity
                                + timestep * net-force) * damping
        this_ vertex.position := this_ vertex.position
                                + timestep * this_ vertex.velocity
        total_energy := total_energy + this_ vertex.mass
                          * (this_ vertex.velocity)^2
     next vertex
  while total_energy is less than some small number
   next community
   do
    total_energy := 0
    for each community in this network
      net-force := (0, 0)
      for each other community in this network
         net-force := net-force
                      + repulsion-force( this_ community, other_ community)
     next other community
      for each spring connected to this community
         net-force := net-force + attraction-force(this_community, spring')
      next spring
      this_ community.velocity := (this_ community.velocity
                                 + timestep * net-force) * damping
      this_community.position := this_community.position
                                 + timestep * this_ community.velocity
      total_energy := total_energy + this_ community.mass'
                      * (this_ community.velocity)^2
     next community
    while total_energy is less than some small number
```

Fig. 2. The algorithm for flat-structure communities

In recent years, there have been numerous and various complex social networks with development human society and science&technology. Many types of social networks have been widely studied, such as friendship, common interest, or co-authorship and citations in scientific papers. In a social network, any node in it represents an entity in the real world, and a group of nodes which are tightly connected are considered to form a community, such as a research team in the co-authorship network.

For the researchers on community study, they mainly pay their attentions to two aspects: the relationship among the entities in one communities and the relationship between communities. In other words, there is less study value for the entities which belong to different communities. Motivated by this, we propose a novel approach to address the low-efficiency of force-directed algorithms. The basic idea of the proposed approach is that, the position a vertex is determined by the vertices which are in the same community with it.

3.1 Algorithm

As a result, to make the drawing process more efficiently, we propose an improved force-directed algorithm for flat-structure communities. The algorithm for communities is formally depicted first(see Fig. 2), and then we give the discussion for this algorithm.

In this algorithm, the force-directed algorithm is actually performed twice. For the first time, each community is regarded as a small network, and we employ the classical force-directed algorithm to draw the graph for each community. For second time, each community is regarded as a vertex, and we employ the classical force-directed algorithm to draw the graph for the whole network. However, two parameters, *spring* and *mass*, in the second time are different with those in the first time because a community has to occupy a certain area while a vertex has not definite size. The basic technological process of the classical force-directed algorithm has been introduced in section 2. Below we make the descriptions for *spring* and *mass* in the second time.

3.2 Discussions for Parameters Spring and Mass

The parameter *spring* is to make the connected vertices move toward to each other. However, different to vertices, a community occupies a certain area in the graph. In this paper, we use the minimum cycle which cover the whole community to represent the area, which is denoted as C(r, (x, y)), where r is the radius and (x, y) is the coordinate. From the example we can find that there are two distances between two communities: center distance and border distance. We argue that the later is more reasonable than the former since two large communities may be very close even if the center distance is large. As a result, two communities may be overlapped if the center distance is used to represent the parameter *string*. Based on such consideration, we use border distance as the parameter *string*, which can be computed with the following formula:

$$string=\frac{\sqrt{(x_1-x_2)^2+(y_1-y_2)^2}-r_1-r_2}{\sqrt{(x_1-x_2)^2+(y_1-y_2)^2}} \quad (1)$$

where (x_1, y_1) is the coordinate of the center of community 1, r_1 is the radius of community 1, (x_2, y_2) is the coordinate of the center of community 2, and r_2 is the radius of community 2. From the formula, we can know string is always between 0 and 1.

The parameter mass is the important factor to calculate the repulsion force and the total energy. Intuitively, the mass of a community is proportional to the repulsion force and the total energy. In this paper, we use the density-like way to represent the community's mass, which can be computed with the following formula:

$$mass = \frac{1}{Z} \frac{|V|*|E|}{\pi r^2} \quad (2)$$

where $|V|$ is the total number vertices in the community, $|E|$ is the total number edges in the community, r is the radius of the community, and Z is the normalization factor which makes *mass* is always between 0 and 1. The formula indicates that, the higher the volume of unit area of the greater population density, larger *mass* is.

4 Evaluations and Examples

In this section, we evaluate the proposed approach based on several popular aesthetic criteria. Further a group of examples are presented to make people have visual feelings. Experiments were performed on a 3GHz computer with 2GB RAM. The running time and the number of iterations were recorded.

4.1 Test Data and Criteria

We conduct the experiments on two special types of dense ($|E|>|V|$) connected graphs listed below.

Coauthor undirected graph (CA for short): CA graphs are a kind of popular social network which can reveal the knowledge sharing among researchers. Totally, there are 101 vertices and 1157 edges in this graph.
Author Citation directed graph (AC for short): AC graphs are a kind of popular social network which can reveal the knowledge flow among researchers. Totally, there are 396 vertices and 1988 edges in this graph.

For each type of graph, we manually assign community labels to the vertices. In practice, the labeling process can be done automatically by computer.
The criteria we used for graph measurement include:

1. Number of crossings
2. Average size of crossing angles
3. Standard deviation of crossing angles
4. Iterations
5. Running time (sec.)

The first three ones are to measure the graph readability, and the left ones are to measure the efficiency of graph. We do not use some criteria (such as average edge length and the standard deviation of edge length) in many previous works because we argue that edge length is an effective way to reflect the relationship strangeness between the connected vertices.

4.2 Results

The results are shown in Table 1 and 2. Table 1 shows the evaluation results on readability criteria, and Table 2 shows the evaluation results on efficiency criteria.
As can be seen from Table 1, the algorithm FLC proposed in our approach is very close to the classical force-directed algorithm(FL) on graph readability. Further, it is

surprising that slight improvement can also be found. These experimental results indicate that at least our approach does not make the graph quality decline.

As can be seen from Table 2, the algorithm FLC proposed in our approach is superior to the classical force-directed algorithm(FL) on efficiency criteria. The efficiency has been improved about 70% on CA graph and more than 100% on AC graph. These experimental results indicate that our approach can reduce the running time greatly. Through community discovery is also a challenging issue, this process can be done offline while social network visualization is usually an online process.

Table 1. Experimental results on readability criteria

Graph type	Number of crossings		Average size		Standard deviation	
	FLC	*FL*	*FLC*	*FL*	*FLC*	*FL*
CA	3704	3752	68.33	68.50	9.06	9.41
AC	10229	11538	70.46	71.82	6.25	7.04

Table 2. Experimental results on efficiency criteria

Graph type	Iterations		Running time	
	FLC	*FL*	*FLC*	*FL*
CA	2955	3867	1.05	1.72
AC	6098	14436	2.81	5.94

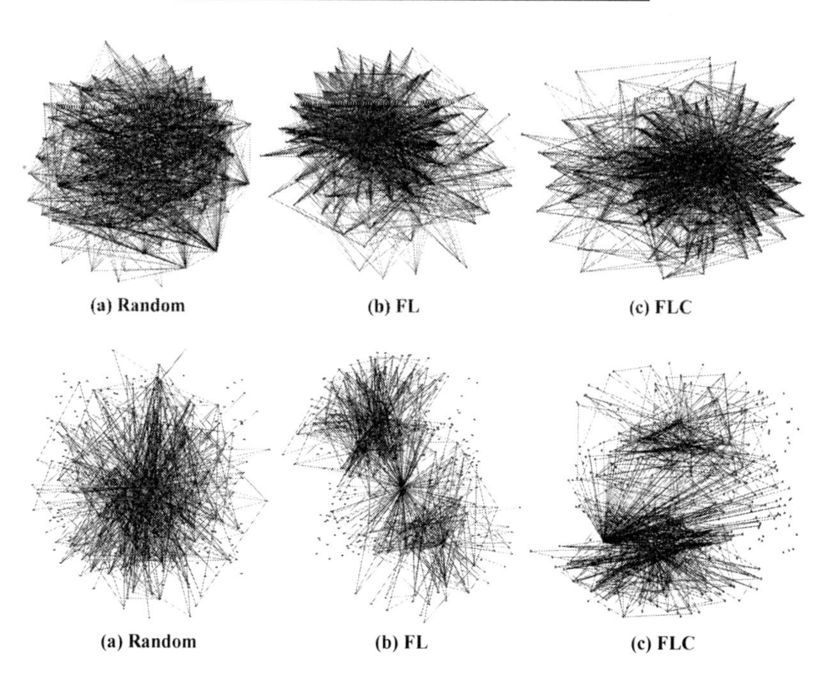

Fig. 3. Drawing AC and CA graph with different algorithms

4.3 Examples

To make people more intuitive understanding of our approach, examples are presented in Fig. 3, which are taken from our experiments.

As can be seen from Fig. 3, both FL and FLC can present the community property of graphs. Visually, there are no obvious advantages and disadvantages between them.

5 Conclusion and Future Work

This paper introduced a community-based algorithm to reduce the running time. The experimental results demonstrate that this algorithm can improve the efficiency of graph drawing without the loss of graph readability. We only focus on the flat-structure communities in this paper. However, we notice that communities are organized in hierarchical structure in many scenarios. As the future work, we will extend the algorithm to be suitable for hierarchical-structure communities.

References

1. Battista, G., Eades, P., Tamassia, R., Tollis, I.: Graph Drawing: Algorithms for the Visualization of Graphs. Prentice-Hall, Englewood Cliffs (1999)
2. Eades, P.: A Heuristic for Graph Drawing. Congressus Numerantium 42, 149–160 (1984)
3. Brandenburg, F.-J., Himsolt, M., Rohrer, C.: An Experimental Comparison of Force-Directed and Randomized Graph Drawing Algorithms. Graph Drawing, 76–87 (1995)
4. Thomas, M., Edward, M.R.: Graph Drawing by Force-directed Placement. Softw., Pract. Exper. 21(11), 1129–1164 (1991)
5. Holten, D., Jarke, J.: Force-Directed Edge Bundling for Graph Visualization. Comput. Graph. Forum 28(3), 983–990 (2009)
6. Eades, P., Huang, W., Hong, S.-H.: A Force-Directed Method for Large Crossing Angle Graph Drawing CoRR abs/1012.4559 (2010)
7. Helen, C.P.: Metrics for Graph Drawing Aesthetics. J. Vis. Lang. Comput. 13(5), 501–516 (2002)
8. http://en.wikipedia.org/wiki/Force-based_algorithms_(graph_drawing)
9. Hong, S.-H., Merrick, D., Hugo, A.: Automatic visualisation of metro maps. J. Vis. Lang. Comput. 17(3), 203–224 (2006)
10. Gansner, E.R., Hu, Y., Kobourov, S.G.: GMap: Drawing Graphs as Maps. Graph Drawing 405–407 (2009)
11. Itoh, T., Muelder, C., Ma, K.-L., Sese, J.: A hybrid space-filling and force-directed layout method for visualizing multiple-category graphs. PacificVis, 121–128 (2009)

ORMF: An Ontology-Based Requirements Management Framework for Networked Software

Jianqiang Hu[1,2,3], Gang Wu[2], and Qian Zhong[2]

[1] School of Information Science and Technology, Hunan Agricultural University
410128 Changsha, China
[2] Department of Computer Science & Technology, Tsinghua University
100084 Beijing, China
[3] State Key Laboratory of Software Engineering, Wuhan University
430072 Wuhan, China
jqhucn@gmail.com

Abstract. Networked software, a special kind of applications in service-oriented computing and ultra-large-scale systems, is a complex software system deploying on network environment. Currently, traditional requirements management systems are far from meeting demands of networked software, due to low degree of automation, coarse-grained management and limited support for requirements modeling activities. Facing the challenges, based on meta-modeling frame called RGPS (Role-Goal-Process-Service) international standard, an ontology-based requirements management framework called ORMF is presented in this paper. In addition, an open requirements management architecture is designed to fulfill ORMF framework. Finally, a requirement registration and query tool called RGPS Browser is implemented to provide supporting services for open requirements process of networked software. Combined with an on-demand Q/A case in transport domain, it displays fine-grained retrieval in visualization.

Keywords: Networked software; Meta-modeling frame RGPS; RGPS Browser; Ontology-based requirements management framework.

1 Introduction

Networked software (NS) can be viewed as a special kind of applications both in service-oriented computing (SOC) and ultra-large-scale (ULS) systems. It is a complex software system deployed on network environment, whose topology structure and behavior can evolve dynamically. Networked software has the characteristics of loose coupling and dynamic evolution, due to users' uncertainty and continuous growing requirements. As for Networked software, a domain-oriented and user-centered development manner is advocated, and a unified requirements meta-modeling frame RGPS (Role/Goal/Process/Service) for networked software named RGPS is proposed by merging functional and nonfunctional (context and trustworthy) requirements [1,2]. RGPS provides a common standard for domain modeling,

G. Zhiguo et al. (Eds.): WISM 2011, CCIS 238, pp. 377–384, 2011.

requirements elicitation, requirements analysis, and requirements V&V, and other phases in the requirements lifecycle of networked software.

Traditional requirements management methods have the following characteristics: software requirements are insufficient in semantic information and has no unified ontology-based meta-model; requirements management mainly includes change control, version management and requirements tracking management, but it can not provide perfect support for requirements elicitation, requirements analysis and requirement V&V, and other activities; the majority of requirements management tools (such as QSS's DOORS[3], IBM Rational's RequisitePro[4] and TBI's Caliber-RM[5]) store requirements specifications and related information in multi-user database; requirements management relies on users to build the tracking relationship among requirements since it is of low degree automation; requirements are low reuse and aren't suitable for applied to complex software systems including networked software.

Compared to traditional requirements management, requirements management of networked software shows new characteristics: requirements management meets the users including network users and service providers to the greatest extent; requirements of networked software are personalized and uncontrolled; requirements of networked software are managed throughout the entire lifecycle of requirements, including their induction period–growth period–maturity period–decline period; requirements changes are more complex and difficult to be controlled; requirements management needs to be more refined to achieve the finer granularity reuse.

Facing the challenges, based on meta-modeling frame RGPS, an ontology-based requirements management framework called ORMF, including ontology-based requirements management mechanism layer, ontology-based requirements management service layer, and open requirements management prototype layer, is first presented. Then, an open requirements management architecture is designed to fulfill ORMF framework. Finally, we initially build prototype tool called RGPS Browser.

The rest of the paper is organized as follows. Section 2 introduces the research background of requirement management of networked software; Section 3 gives the ORMF framework; Section 4 describes the architecture for fulfill ORMF framework in detail; Section 5 builds RGPS Browser; Section 6 draws some conclusions.

2 Background

In this section, we introduce the Meta-model frame RGPS, and discuss the fundamental role that requirements management plays in requirements engineering of networked software.

(1) Meta-model Frame RGPS

The ultimate goal of networked software is to realize mass customization at a low cost in a short time. There are many challenges in requirements engineering of networked software, such as no unified domain knowledge for mass requirements customization, no unified requirements description method for personalized preference and so on. Furthermore, requirements of networked software is uncertainty, and single requirements modeling method is difficult to deal with personalized, mass, diversified requirements of networked software.

Meta-model frame RGPS is a hierarchical and cooperative frame, which sums up multi-level and multi-granularity requirements as Role layer-Goal layer-Process layer-Service layer and the relationship between these four layers.

In detail, the Role layer characterizes the organization, role, and actors and describes the interaction and cooperation in problem domain; the Goal layer depicts the decomposition of goals and determines the constraint relationship among goals; the Process layer distinguishes atomic processes and composite processes, and defines the Inputs/Outputs/Preconditions/Effects of processes; the Service layer guides the construction of service chains and aggregation of service resources. RGPS is committed to describe personalized requirements from multi-role, multi-objective, multi-process and multi-services and is used to guide dynamic requirements modeling in an orderly manner [2].

(2) Requirements Management of Networked Software

The purpose of requirements management of Networked software is as much as possible to meet personalized requirements of users. Requirements management initiative in response to the requirements changes subordinated to the users' preferences. Requirements management of networked software is a more complex process to construct a system of networked software. In addition to considering the above features of traditional requirements management, it needs to provide supporting services for other requirements activities including requirements elicitation, requirements analysis, and requirements validation, and so on, following Meta-model frame RGPS. The majority of traditional requirements managements are text-based document management, because software has not a unified meta-model, and computer can not understand the content of a requirement; query on text-based documents is using the query method based on keywords which is difficult to reuse content; the relationship between requirements are constructed by manual way or information retrieval technologies. Correspondingly, requirements management of has Meta-model Frame RGPS and needs semantic content management with RGPS requirements; keyword-based matching is not reusable and needs to improve the granularity and accuracy of query; Information retrieval technologies to be used to build the relationship between RGPS requirements which may not be applicable to requirements tracking of networked software and needs more automation.

3 An Ontology-Based Requirements Management Framework

As for characteristics of requirements management of networked software, a hierarchical and ontology-based requirements management framework called ORMF is presented (See Fig.1), including ontology-based requirements management layer, and ontology-based requirements management services layer.

(1) Ontology-based requirements management layer

By ontology-based content management technologies, this layer focuses on solving the storage mechanisms of RGPS requirements. It gets rid of the document-centric management mode of traditional requirements management, and meets the manageability of massive RGPS requirements under the complex network environment. It has the following features: treating triple as data model to solve the scalability of RGPS requirements and achieve effective cluster depository; resolving the rapid location and enhancing the access performance of RGPS requirements by a hypergraph-based comprehensive indexes [6]; using tags to enhance the reasoning ability and improve the accuracy of requirements match [7].

By ontology-based requirements interoperability mechanisms, this layer enhances heterogeneous RGPS requirements to share and reuse. Because network environment is uncertain and domain ontology is distributed and complex, RGPS requirements easily exists heterogeneity. Ontology mapping is an effective method to improve the semantic share of domain ontology. Therefore, this layer relies on flexible and effective ontology mapping strategy to improve the mapping accuracy and mapping efficiency, and realizes the maximize reuse of heterogeneous requirements.

By ontology-based relation mining methods, this layer can find out the potential relationship among RGPS requirements, which provides support for requirement traceability of networked software. Requirements traceability is concerned with documenting the life of a requirement. Traditionally, it is difficult to construct the relationship between RGPS requirements by manual way because it is not applied to fine-grained changes of RGPS requirements. Through ontology-based relation mining between RGPS requirements, including machine automatic analysis, cluster analysis and statistical analysis, semantic relationship and measurable relationship are found out to support requirements change analysis and requirements version management.

(2) Ontology-based requirements management services layer

Requirements management services are defined as a set of common services for requirements modeling, requirements evolving, requirements V&V, and requirements reconstruction.

Requirements registration and query service is one of most common services. However, traditional keyword-based query is not enough for RGPS requirements, which easily leads to low reuse rate. Query service of ORMF achieves query with the any arbitrary level (Role/Goal/Process/Service) and any granularity on basis of ontology-based requirement management mechanisms. Furthermore, query service can be customized through well-defined interfaces to support other requirements activities.

Requirements tracking service needs to achieve tracking capabilities in traditional horizontal and vertical directions. Through ontology-based requirements management, this service can quickly locate requirements clips, and rely on machine to automatically create requirements relationship and generate some corresponding metadata (tracking ontology). Therefore, complex tracking problem is converted into a process of tracking ontology generation and maintenance, what's more, which can ensure the tracking granularity. It can trace back to the origin of each requirement and every change made to the requirement.

Version management service is one of the important services to achieve the stability of RGPS requirements. Version management of RGPS requirements is more complex than simple document management. Besides similar CVS mode of requirements document, version management service provides the fine-grained version management and ensures that requirements can be rolled-back edit.

Change management service is critical to requirements management. On basis of the information of requirements tracking, this service adopts the unified change controlling method to manage requirements changes. Instead of limiting the users to change requirements, this service induces them to improve the quality of RGPS requirements.

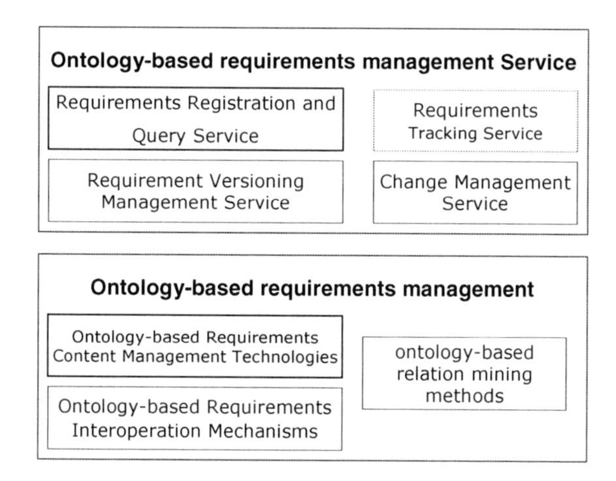

Fig. 1. ORMF: An Ontology-based Requirements Management Framework

4 An Ontology-Based Requirements Management Architecture

Following RGPS and MFI [1] (Meta-model Framework for Interoperation) specification, under the guidance of ORMF, we design an architecture for ontology-based requirements management as shown in Fig.2.

(1) Ontology-based requirements manager
As for project metadata of networked software, RGPS requirements and attributes information (priority, stability, and version etc.), RGPS requirements manager adopts ontology-based content management technologies to achieve semantic management. Some modules are listed as follows:

In RDF Indexes-based accesser, RGPS requirements (OWL file) are imported and saved with triple, which is a RDF hypergraph in nature. Comprehensive index (node-value index and triple-order index) is used to enhance the performance of location and access in RGPS requirements.

OWL Lite-based reasoner, combined with the characteristics of hypergrah, rules of OWL Lite are applied to reason and match of RGPS requirements.

In MFI mapper, multi-adaptive ontology mapping algorithm called RiMOM (Risk Minimization Based Ontology Mapping) is used to automatically combine the matching strategies to achieve semantic interoperability cross-cutting heterogeneous requirements.

In Project-related metadata manger and requirements attributes manager, ontologies represent metadata of project of networked software and attributes information of requirements documents. Thus, they are easy to be formalized as OWL files. Generally, these OWL files are too simple and these managers read/write them by OWL editor.

(2) Supporting service layer
This layer provides four services, including requirement registration and query service, requirements tracking service, requirement versioning management service and changes management service, which will support for requirements modeling, requirements evolving, requirements V&V, and requirements reconstruction.

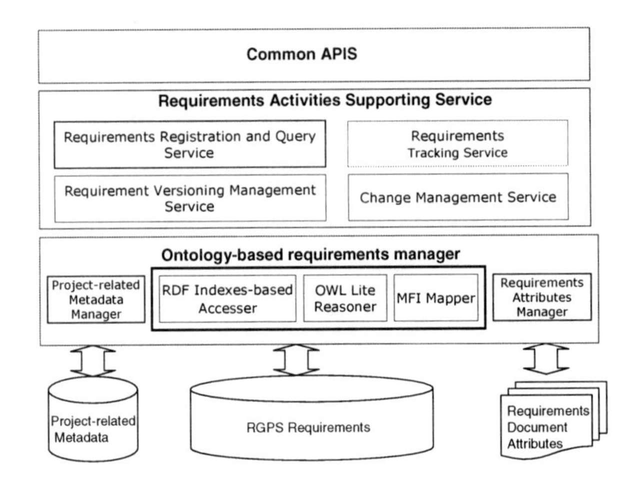

Fig. 2. An architecture for ontology-based requirements management

Take Requirement query service as an example, by four types of query (i.e., simple query, advanced query, hierarchical query and relation query), it provides any granularity query on RGPS requirements on basis of Ontology-based requirements manager.

(3) Common interface layer

This layer provides well-defined and rich APIs for integration and invocation services of requirements management when it is needed by requirements activities. For example, requirements query service gives four main interface, i.e., SimpleQuery(), AdvancedQuery(), HierarchicalQuery(), and RelationQuery(). These interfaces shield complex implementation of services and guarantee transparency when requirements activities access RGPS requirements management system.

5 RGPS Browser

Requirements management prototype system embodies open characteristics as follows: following the RGPS and MFI (Meta-model Framework for Interoperation); providing well-defined interfaces for integration and invocation of other requirements activities.

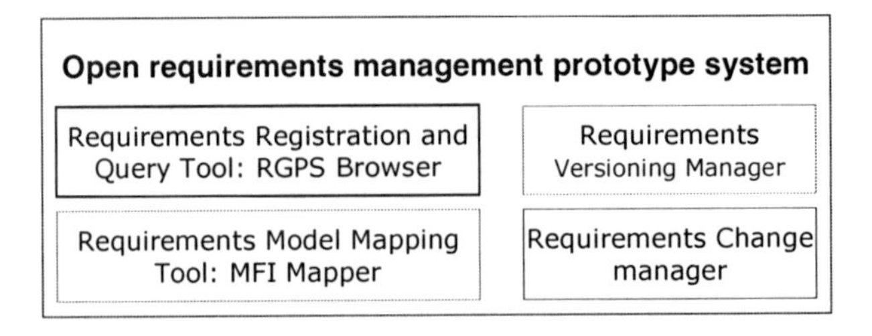

Fig. 3. The Composition of open requirements management prototype system

In order to facilitate development, this prototype is divided into four tools (See Fig.3), including RPGS requirements registration and query tool, requirements model interoperability tool, RGPS requirements version management tool, and RGPS requirements change management tool. These tools can guarantee rational and concentrative development of some function module. Ultimately, these tools are integrated as a whole to provide a flexible and scalable prototype system.

We illustrate an on-demand Q/A system case in travel domain. We construct the domain ontology of travel domain and the instance of RGPS requirement modeling.

Take Guangzhou Asian Games 2010 traveling planning as an example, Mr. Li choose a suitable planning in accordance with the Asian Games schedule, weather conditions, traffic information, and the distance between his hotel and corresponding gymnasium. The On-demand Q/A system are constructed to give the answer of traveling planning via Mobile message. RGPS is committed to describe personalized requirements from multi-role, multi-objective, multi-process and multi-services and is used to guide dynamic requirements modeling.

Under the guidance of RGPS and ORMF, we implement RGPS Browser which is a requirements registration and query tool. Fig.4(a) gives a hypergrah in view of the role of generation information Operator (i.e. On-demand Q/A system) from RGPS Brower. For broad acceptance, it defines unified registration and query APIs, which allows requirements activities to access RGPS requirements. Furthermore, it shields the implementing complex of SPARQL language. We are easy to find each level (Role/Goal/Process/Service) and suitable granularity via unified query APIs. Fig.4(b) gives the view of AdvancedQuery(), whose parameter is SPARQL statements. Furthermore, RGPS Browser enhances heterogeneous RGPS requirements to share and reuse based RiMOM.

Although we only implement this tool at present, there are other three tools under implementing. Based on RGPS and ORMF, we hope to complete them as soon as possible and integrate all the tools into a whole, i.e., networked software requirements management prototype system.

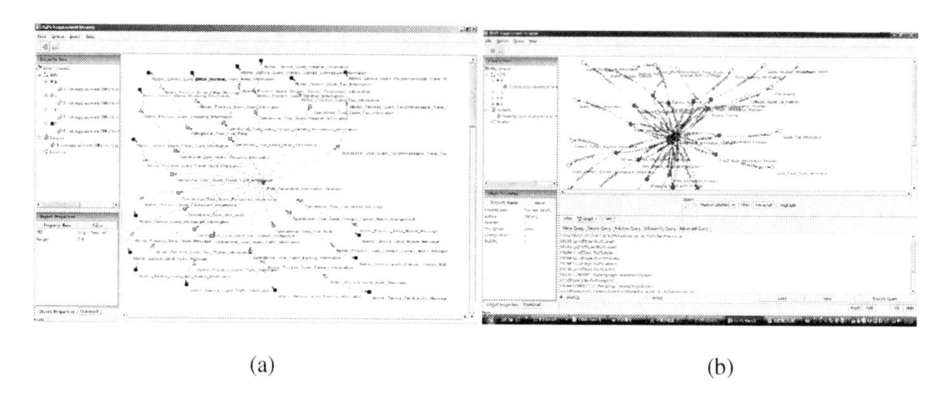

(a) (b)

Fig. 4. (a) Hypergraph of a RGPS Requirement in RGPS Browser; (b) AdvancedQuery of a RGPS Browser

6 Conclusion

Networked software brings forward many new challenges to requirements management by taking a domain-oriented and user-centered development manner. In this paper, following RGPS of networked software, an ontology-based requirements management framework called ORMF is presented. In addition, an open requirements management architecture is designed to fulfill ORMF framework. At last, we initially build a requirements registration and query tool called RGPS Browser.

In future, based on the RGPS and ORMF, some ontology-based requirements relation mining methods needs to be further investigated and researched, and corresponding tools are also under construction.

References

1. Metamodel Framework for Interoperability (MFI), http://metadata-stds.org/19763/index.html
2. Wang, J., He, K., Gong, P., Wang, C., et al.: RGPS: A Unified Requirements Meta-Modeling Frame for Networked Software. In: Proceedings of IWAAPF 2008 at ICSE 2008 Workshop, Leipzig, Germany (2008)
3. DOORS (dynamic object oriented requirements system), http://www.telelogic.com/index.cfm
4. Rational RequisitePro, http://www-306.ibm.com/software/awdtools/reqpro/
5. Borland CaliberRM, http://www.borland.com/us/products/caliber/index.html
6. Wu, G., Li, J., Wang, K.: System II: A Hypergraph Based Native RDF Repository. In: Proceedings of the 17th International Conference on World Wide Web (WWW 2008), Poster Track, Beijing, China (2008)
7. Wu, G., Li, J., Feng, L., et al.: Identifying Potentially Important Concepts and Relations in an Ontology. In: Bizer, C., Joshi, A. (eds.) Proceedings of 7th International Semantic Web Conference, Karlsruhe, Germany. CEUR, vol. 401 (2008)

A Ranking Model Based on Credit for Social News Website

Guanglin Xu[1], Xiaolin Xu[2], and Jiali Feng[3]

[1] School of Mathematics and Information, Shanghai Lixin University of Commerce, Shanghai 201620
[2] College of Computer and Information, Shanghai Second Polytechnic University, Shanghai 201101
[3] College of Information Engineering, Shanghai Maritime University, Shanghai 200135

Abstract. Social news websites were popular for a time, but recently the traffics of some social news websites have been slowing down, even some of them went out of business. It's caused partly by the emergence of massive spam news and the news publication algorithm heavily relying on the ranking of senior users while ignoring the ranking of ordinary users. In order to solve these problems, this paper proposes a ranking model based on credit. On the technology, utilizes the qualitative mapping as the credit threshold activation model, uses the K-means method for classifying the users and attribute ranging model to calculate the value of the user credit. The simulation experiment shows this model can effectively reduce the spam news and pay more attention on regular users.

Keywords: Attribute Theory, K-means, Ranking Model.

1 Introduction

Since the birth of Web 2.0 websites, they are very popular. Among them, those types of social news websites such as Laikee, 21dig, Digg and Reddit etc. are very fashionable for a while because users of these websites can be dominant in news publication and the published news always are of high quality. However from the latest statistics [1], the traffics of the social news websites are not increasing as that of Facebook and Twitter, but had been decreased to some extent, some of them, like Laikee, 21dig and Techtag are even out of business. This situation is caused by various reasons, one of which is related to ranking model.

There have three stages of development of the ranking model in the social news websites. The algorithm of first stage considered the contribution weight of the votes by new and old users is all the same. This algorithm was too simple to avoid the spam news. For example, after a user submits a piece of news, he can create many new users by proxy servers and then login onto the website to vote the news submitted by him. Or he can contact with his friends to do so, or attract others from some BBS websites. When the votes reach a certain amount, this news will be published and listed in the home page so that the website having this news will absorb a lot of clicks for

G. Zhiguo et al. (Eds.): WISM 2011, CCIS 238, pp. 385–390, 2011.

advertisement benefits. To avoid this, the ranking model of these social news websites has been improved. That is, to set the weight hierarchy for users with much higher weight for senior users than regular users. Although this prevented the spam news efficiently, it brought another problem. The regular users would leave for their mini weight in the algorithm so that they do not their power of determining which news should be published or not, only a few senior users are stay.

In order to prevent the spam news to be published effectively at the same time that maintaining the weight of experienced users and the value of votes by regular users, this paper establishes a ranking model based on credit combining previous study results[2].

2 The Establishment and Implementation of the Ranking Model Based on Credit

2.1 The Establishment of the Ranking Model Based on Credit

Definition 1. The principle of the ranking model based on credit is to adopt credit as the conception, let ***U*** $\{u_i \mid i=0,\cdots n-1\}$ be a set of users having n distinct elements, let ***C*** $\{c_i \mid i=0,\cdots n-1\}$ be a set of the credit value of each user, let ***P*** $\{p_i \mid i=0,\cdots m-1\}$ be a set of the credit possessed by m pieces of news. Supposing the credit value of user u_h is c_h, and the votes by u_h are of authority and are totally reliable, so that the news voted by u_h are totally reliable and can be published. Upon this suppose, the selection of being published or not can be the presented by the qualitative mapping [3] τ (1) with the qualitative criterion basis of $[\alpha,+\infty]$:

$$\tau(p[\alpha,\beta]) = \tau(p \underset{?}{\in} [\alpha,\beta]) = \begin{cases} publish & x \in [\alpha,\beta] \\ 0 & x \notin [\alpha,\beta] \end{cases} \quad (1)$$

2.2 The Implementation of the Ranking Model Based on Credit

The key to the ranking model based on credit is how to choose the proper user u_h. If the choosing criteria for u_h is too loose, the spam news will not be avoided. If the choosing criteria for u_h is too strict, regular users will lose their power of discourse. This paper divides users into four types, i.e. senior users, vice- senior users, regular users and newbie. The credit value of u_h can apply the average value of all senior users $\overline{c}$, as the formula (2). Only if the credit value exceeds $\overline{c}$, the news can be published.

$$\overline{c} = \frac{\sum_{i=0}^{n_1} c}{n_1} \tag{2}$$

Where n_1 is the group amount of the senior users.

2.2.1 The Clustering of the Users

Generally using clustering analysis methods divides users into four groups. The common methods of cluster analysis are partitioning method, hierarchical method, density-based methods and grid-based methods. K-means classification method is a kind of partitioning method that applies Euclidean distance to calculate the similarity and *n*umber of clusters needs to be pre specified. Because K-means classification method are of simple and fast [4], this paper applies it to cluster the users. Next is the detail for its procedure.

To make user groups after clustering more reasonable reflects the user's credit, seven attributes of a user are adopted, which are user's online time (ot), number of submitted news by user (sn), number of the submitted news that being published (snp), number of comments (cn), the number of votes (vn), number of news published after being voted (esp) and the times of being read of the submitted news (snr). Where the ot is a continuous variable. To facilitate the calculation, discretize and map it on a scale from 1 to 5.

Each user has these 7 basic attributes, i.e. they are a 7-dimentioned set. Let A = {***ot***, ***sn***, ***snp***, ***cn***, ***vn***, **esp**,***snr***} be the attribute domain set of U. And S = ***ot***×***sn***×...×***snr*** represents a 7-dimentioned space. By using Euclidean distance formula (3) can calculate out the similarity among all users.

$$U(X,Y) = \left\{ \sum_{i=0}^{6} \left| x_i - y_i \right|^2 \right\}^{\frac{1}{2}} \tag{3}$$

Where x and y are the clustering objects, that is, the different users.

The squared error criterion function applied by K-means method is as the formula (4)

$$E = \sum_{j=0}^{3} \sum_{i=0}^{n_j - 1} \left\| x_i - m_j \right\|^2 \tag{4}$$

Where E means the sum of the squared error of all clustering objects, m_j is the average value of the clustering objects for Cluster c_j, then constantly revised by recursive computation as formula (5):

$$m_j^{(t+1)} = \frac{1}{\left| c_j^{(t)} \right|} \sum_{x_i \in c_j^{(t)}} x_i \qquad j = 0,1,2,3 \tag{5}$$

For K-means method, to the way of selection of the "seeds" of k points is very sensitive. Different initial values may lead to different clustering results. To make the clustering results more accurate and avoid partial optimization, choose the initial point from the set (6), i.e. pick a newbie as the first seed and set the characteristic ot value to 1 and vn v to zero.

$$\{m_1 \in U \mid (ot_1 = 1) \cap (vn_1 = 0)\} \tag{6}$$

Pick the user whose ot value is 5 and esp value of the largest as the second seed, i.e. to fulfill formula (7).

$$\left(\arg_U \max(esp_i) \right) \cap (ot_i = 5) \tag{7}$$

Pick the user whose ot value is 2 and esp value of the 1/3 of that of largest as the third seed. Pick the user whose ot value is 3 and esp value of the 1/3 of that of the second seed as the fourth seed. After selection of the seeds, use K-means method to do clustering for the users. The user group after clustering that including the second seed is the senior user group.

2.2.2 The Establishment of the Credit Ranking Model

After the senior user group is clustered, the next step is how to calculate the average credit value of the senior users. This paper applies the attribute coordinate evaluation to solve this problem. Attribute coordinate evaluation is based on qualitative mapping theory and similar to the human's thinking so that it is able to reflect expert preferences and the preference curve. By applying attribute coordinate evaluation model, according to the 7 attributes of users the credit value of each user can be calculated out. The formula (8) is the mathematical expression to calculate out the credit value by using coordinates evaluation function.

$$\mathrm{C(x,z)} = \left(\frac{\sum_{i=1}^{m} x_{ij}}{\sum_{j=1}^{m} X_j} \right)^S * \exp\left(\frac{\sum_{j=1}^{m} w_j \left| x_j - b(x^h(z_j) \right|}{\sum_{j=1}^{m} w_j (b(x^h(z_j) - \delta_j)} \right) \tag{8}$$

Where w_j is the weight and $\left(\frac{\sum_{i=1}^{m} x_{ij}}{\sum_{j=1}^{m} X_j} \right)^S$ is the conversion factor from partial satisfaction to overall satisfaction.

After get credit value of each user by formula (8), through formula (2), credit value $\overline{c}_h$ of the senior user u_h can be calculated out which is the credit threshold of the news that needs to be published.

2.2.3 The Establishment of News Publication Model

After the calculation of the credit threshold $\overline{c}_h$, the question of whether to publish the news after being voted has been converted into the question that whether the credit value after vote is larger than the credit threshold $\overline{c}_h$. If the former value is larger than the latter, the news will be published. Otherwise, not be published. However, because the original data is discretized and normalized, the credit value calculated by formula (8) is distorted to some extent. By inverse shape mapping (9), the credit value can be revised.

$$\eta_i(x) = \frac{1}{1 + \exp\left(-\frac{|x - \xi_i|}{\delta_i}\right)} \tag{9}$$

At last, by using formula (1), the qualitative mapping model of the news being published is:

$$\begin{aligned} \tau(p[\alpha,\beta]) &= p \underset{?}{\in} [\alpha,\beta] \\ &= \begin{cases} publish & \mathrm{x} \in [\alpha,\beta] \\ 0 & \mathrm{x} \notin [\alpha,\beta] \end{cases} \\ &= \begin{cases} publish & \sum_{i=0}^{m-1} w_i c_i \in [c_j, +\infty] \\ 0 & \sum_{i=0}^{m-1} w_i c_i \notin [c_j, +\infty] \end{cases} \end{aligned} \tag{10}$$

3 Simulated Experiments

In the simulation experiments by using the social news website ranking model given in this paper, a dataset containing 2,500 users and 6,000 pieces of news including 500 piece of spam news is tested, the result of the simulation experiment shows the publication rate of spam news dropped from 18.5% to 9.2% so that the quality of the published news is secured. At the same time, some news that can not be published without votes by senior users before can be published now, which pays more attention on regular users.

4 Conclusion

This paper provides a ranking model based on credit for social news website. This model is able to prevent the spam news publication to some extent to ensure the quality of the news publication and provide a new model for the social news websites and other websites with similar ranking functions. The simulation experiment in this paper shows the reasonability of this model. At the same time, the ranking system for the news publication is dynamic. How to automatically adjust the credit threshold according to the user feedback, user preference and various characteristics of all types of news is the next direction for further study.

References

[1] O'Dell, J.: Digg's Decline by the Numbers: Plummeting Traffic, Waning Power (September 24, 2010), http://mashable.com/2010/09/24/digg-traffic-stats/

[2] Xu, G., Liu, N., Feng, J.: A voting algorithm based on attribute theory for social news sites. Transactions on Intelligent Systems 4(2), 118–121 (2009)

[3] Feng, J.: Qualitative Mapping Orthogonal System Induced by Subdivision Transformation of Qualitative Criterion and Biomimetic Pattern Recognition. Chinese Journal of Electronics, Special Issue on Biomimetic Pattern Recognition 85(6A), 850–856 (2006)

[4] Kanungo, T., Mount, D.M., Netanyahu, N., Piatko, C., Silverman, R., Wu, A.Y.: An efficient k-means clustering algorithm: Analysis and implementation. IEEE Trans. Pattern Analysis and Machine Intelligence 24, 881–892 (2002)

[5] Duan, X., Liu, Y., Xu, G.: Evaluation on 3PL's core competence based on method of attribute theory. Journal of Shanghai Maritime University 27(1), 41–43 (2006)

Study on ODMRP Protocols and Its Application

Jian Chun Ye

Shanxi Vocational and Technical College of Coal
P.R. China, 030031
yejch@163.com

Abstract. Ad Hoc network is a multi-hop temporary autonomous system, which consists of a group of mobile terminal with a wireless transceiver device and whose topology structure changes constantly due to the high dynamic of mobile autonomous network. ODMRP, as the name implies, is an on-demand multicast routing protocol which could better fit in Ad Hoc network. On the basis of this protocol, the NCR-ODMRP protocol the paper proposes, based on its stability of path, applies MRP (Multipoint Relay) mechanism to limit flooding in ODMRP and selects three relatively stable routing by means of the change rate of a neighbor node and optimizes multicast forwarding grid, which reduce the protocol overhead, and enhance the robustness of the protocol.

Keywords: Ad Hoc, ODMRP, multicast routing protocol.

1 Introduction

Researches on Ad hoc network, formerly known as PRNET (Packet Radio Network), spring from the needs of military communications and have continued for nearly 20 years. In 1972, the U.S. DARPA (Defense Advanced Research Project Agency) started a project on PRNET to study its application in the field of Data Communication at war. In 1994, DARPA launched the Global Mobile Information System (GMIS) project, based on the research achievements of PRNET, with being content with the needs of military applications, fast start and high invulnerability, had been studied in-depth and was continuing today [1].

The particularity of Ad Hoc Network determines the particularity of its multicast routing protocol. Research shows that ODMRP (On-Demand Multicast Routing Protocol) has a good overall performance. It provides invulnerability of link failure by redundant links to reduce unnecessary overheads of path re-search brought by individual link failure, and it is no need to trigger the maintenance of the path as the busy nodes increase. Thus it will not increase the network control overhead, which is more suitable for Ad Hoc network in such frequently movable nodes. However, since ODMRP protocol is by periodical flooding JOIN-QUERY message to establish of multicast forwarding grid, when the load is very large, the control overhead of searching and updating route increases rapidly, and competition and conflict of shared channel intensify, which make transmitting efficiency decrease. Up till now, focuses on the improvement of ODMRP are mainly how to limit the flooding information of route, without considering the stability of the redundant paths. In existing improved

G. Zhiguo et al. (Eds.): WISM 2011, CCIS 238, pp. 391–395, 2011.

schemes, generally using the method of motion prediction, the source node can re-select the path when the expected path failed. However, the node of redundant links can also move and be instable. Therefore, this paper, based on the flooding in the limit, introduces the path stability factor NCR and proposes path-based Multicast Routing Protocol of Mobile Ad Hoc network, NCR-ODMRP [2].

2 Characteristics of Ad Hoc Network

Ad hoc network is a special kind of wireless mobile networks, all nodes in which are in an equal status, without setting any central control node. Node in Ad hoc network not only has the functionality required by ordinary mobile terminals, but also has the ability of datagram forwarding. Compared with the ordinary mobile networks and the fixed networks, it has the following characteristics.

2.1 Acentric Distribution

Ad hoc network has no strict control center. All nodes are in an equal status, that is, an P2P (peer-to-peer) network. Nodes can join and leave the network at any time. With strong survivability, failure of any node will not affect the operation of the whole network.

2.2 Self-organization

The establishment or spread of the network depends on no any default network facilities, because nodes coordinate their behavior by layered protocols and distributed algorithms. After the startup the node can quickly and automatically form an independent network.

2.3 Multi-hop Routing

When the node communicates to the node outside its spectral range, multi-hop transmission of the middle of nodes is needed. Being different from fixed network multi-hop, multi-hop routing in Ad hoc the network complete by common network node, rather than by a dedicated routing devices (such as routers).

2.4 Dynamic Topology

Ad hoc network is a dynamic network. Network node can move anywhere, and boot and shutdown anytime, which could change the network topology structure at any moment. These characteristics make the Ad hoc network have significant differences from ordinary cellular mobile communication network and fixed communication network in architecture, network organization, protocol design and other aspects.

3 NCR-ODMRP Protocol

NCR-ODMRP aims to maintain the information transmission with the minimum network overhead, and obtain a relatively stable path under the situation of constantly moving node to increase the transmission success rate, and recover path transmission in a short period of time as the path fractures.

3.1 Limit Flooding and Optimize the Grid

First, apply MPR (Multipoint Relay) to flooding limit for ODMRP protocol, that makes the node selectively forward packets with using less data forwarding overhead and obtains the same data transmission effect with the whole network flooding. Define that node N that must be through 2 hops and at least 2 hops to reach is the two hop neighbors of N. Each node N periodically forward Hello message in which contains a list of its own neighbors, through whether itself is contained in its neighbor list in which Hello message is received from a neighbor node determines whether the link is a two-way link, to obtain its two-way neighbor list and the two-hop neighbor set. MPR information in node N is also sent in the Hello message. Thus when the N's neighbors received, they can identify themselves as multipoint relay neighbors (MPR-Neighbor) or non-multipoint relay neighbors (NMPR- Neighbor)selected by N. And only the MPR-Neighbor received flooding data of N node and started to relay through neighbors N selected, all two-hop neighbors of N can receive the data [3] [4].

In addition, divide the nodes in Ad Hoc the network into two categories: upstream node and downstream node. Defined as follows: when node i is forwarding the JOIN-QUERY packet to node j, we call node i the upstream node, and node j the downstream node. Thus only the upstream node can forward the JOIN-QUERY packet while the downstream node can not forward the JOIN-QUERY packet to its upstream node, and the same way is in the JOIN-REPLY packet, which not only limits the node packet forwarding the route requests, but also inhabits the JOIN-REPLY flooding.

3.2 Predict Stability and Select Stable Routing

Each node receives the Hello message from neighbor node through flooding limit program. By comparing differences among the node set and calculating computing neighbor change rate of node, local topology changes of node were perceived. As the route establishes, node with little change of local topology should be chosen as much as possible to participate in the new routing. In accordance with the cumulative value of neighbor change rate of each node in different paths, the node forwards data by selecting few hops of forwarding and the path with the minimum changes rate of local topology as the primary path, and selecting other two paths with the relative stability and irrelative to the node of the main path as a backup path, to enhance data transmission reliability and movement adaptability.

The route irrelevant to node, also known as completely irrelevant route, between which there is no shared node except the source node and the destination node.

The route irrelevant to node has a high Fault-tolerant ability and for the independence of the link would not produce a chain reaction. Since ODMRP protocol has multiple routes, in order to select stable routes and reduce the cost of the routing table of each node, on the choice of multiple paths in this article, we only choose three paths irrelevant to node, even though more than three paths are found in the route discovery process. We compromise by selecting the three paths for efficiency and cost because it is relatively the best [5].

NCR-ODMRP protocol through limit flooding, adopts the selection method of stable route based on neighbor change rate to extend ODMRP. The main differences between ODMRP and NCR-ODMRP are followed:

(1) NCR-ODMRP protocol, by reducing overhead of the JOIN-QUERY message in periodic flooding, optimizes the multicast forwarding grid and inhibits the JOIN-REPLY. The optimization of the latest and the greatest connectivity neighbor increases of the HELLO message overhead, but just a small part, which is determined mainly by the network-intensive, without relation to the size of the multicast group.

(2) In JOIN-QUERY packet the routing metric for the increase of the neighbor change rate represented as NCRpath field, and this field is updated in accordance with the formula (2) when intermediate nodes forward the JOIN-QUERY packets. Furthermore, PathRecord field is increased and the nodes the route passed are recorded, to select the route irrelevant to node and determine the upstream nodes and the downstream nodes.

(3) The requirement for that the node updates the routing table entry and forwards the message is having bigger destination sequence number or having the same destination sequence number but larger value of NCRpath. The discovery and the maintenance process of route in NCR-ODMRP is as follows:

When the multicast source node has data to forward but no multicast routing in Ad Hoc network, the multicast source node periodically broadcasts a JOIN-QUERY message, which contains a sequence number, forwarding hops, the neighbor change rate and routing metric, such as NCRpath, PathRecord and other fields, and are used separately to make the duplicate packet inspection, record the message forwarded hops, label the stability of paths and record the nodes the route passed. This message can constantly update node information and routing information. Forwarding node records its own upstream node ID, and records this node and the current value of NCRpath in the group, to establish the reverse path. The destination node of the reverse path is the source node generating JOIN-QUERY packet, and the node of the next hop is the neighbor node that forwards JOIN-QUERY message to this source node. For the JOIN-QUERY message in the follow-up receive, only whose sequence number does not repeat, and the TTL (Time-To-Live) value is greater than zero, or sequence numbers are the same but greater value of NCRpath, the forwarding node updates the routing table and forwards the packet. In addition, if the NCRpath value of the first received JOIN-QUERY message is 0, and the NCRpath value of the follow-up JOIN-QUERY message is also 0, the destination node will maintain the current path unchanged. Because in this case, End-to-End Delay of the corresponding path to the first received JOIN-QUERY message is usually smaller than the back received JOIN-QUERY message.

Destination node may receive more JOIN-QUERY messages from the same source node, and when received, the destination node does not immediately respond, but wait for a while for the cache of all route requests from the same source node and the constitution of routing collection from the source node to the destination node. In accordance with the route request packet in the current routing collection, the NCRpath field value is requested, in which the maximum value is selected, that is the most stable route as the first transmission path. The destination node informs all the intermediate nodes on this path from the PathRecord field in the route request packet, and forwards JOIN-REPLY message to the source node. This message contains the node list of whole route. At the same time, make a comparison between the node in PathRecord field of the route of the maximum NCRpath value in the current remaining route request packets and the selected route. If the values of

NCRpath field are the same among many routes, the route of the minimum hop is selected, and if there is the same node, the route request packet is discarded, or the route reply packet is forwarded to the source node. Repeat this process until you find three paths irrelevant to node.

Local repair and route reconstruction are used to maintain the route. When a link break, local repair is used first, and if the local repair fails then route reconstruction is used. When an intermediate node detects the link until a break, it is first to do the cache of data flow from the source node and check whether the alternate route is in the route table. If yes, use directly the alternate route to forward data packets, and if no, forward route request to the neighbor node. In a limited time, if the destination node receives the request, the route reply is returned and route repair is successful, which is used continually to forward data packet, otherwise forward the route reconstruction packet to the source node in upstream node, and after receiving this information the source node, in accordance with the established way of routing, re-establishes routing. If the source node receives the route reconstruction packet and the route reply packet at the same time, route reply packet is discarded and route discovery needs to be restarted.

4 Conclusion

Based on the application of MRP flooding limit, NCR-ODMRP, starting from the stability of the link, make a preference selection on the stable path of which has a small change in local topology and less forward hop number, to forward the data, and recover in a short time in the phase of maintenance. The simulation results show that: the protocol in mobile scenarios can effectively reduce interrupt times and end-to-end delay of data packet, and improves network transmission efficiency and robustness, and reduces the overhead. It can be said that the aspects of transmission performance, scalability and robustness are improved considerably.

References

1. Camp, T., Boleng, J., Davies, V.: A survey of mobility models for ad hoc network research. Wireless Communication & Mobile Computing: Special issue on Mobile Ad Hoc Networking: Research, Trends and Applications 2(5), 483–502 (2002)
2. Xu, H., Mu, D., Dai, G., et al.: Prediction on Multicast Routing Algorithm in The Link State Based on ODMRP. Computer Engineering and Applications 43(5) (2007)
3. Zhao, Y., Xiang, Y., Xu, L., Shi, M., et al.: Improvement on MANET Multicast Route Protocols Based on Radio Relay. High Technology Letters, 2 (2004)
4. Zhao, Y., Xu, L., Shi, M.: On-demand multicast routing protocol with multipoint relay (ODMRP-MRP) in mobile ad-hoc network. In: Proceedings of the International Conference on Communication Technology, pp. 1295–1300. Institute of Electrical and Electionics Engineers Computer Sotiety, Beijing (2003)
5. Nasipuri, A., Castaneda, R., Das, S.R.: Performance of multipath routing for on-demand protocols in mobile ad hoc networks. ACM Mobile Networks and Applications, 339–349 (May 2001)
6. The Network Simulator - ns-2, http://www.isi.edu/nsnam/ns/

Design and Realization of the Cloud Data Backup System Based on HDFS

Dong Guo[1], Yong Du[1,2], Qiang Li[1], and Liang Hu[1]

[1] College of Computer Science and Technology, Jilin University, Qianjin Street 2699, Changchun, China
[2] Heilongjiang Technology and Business College, Harbin 150080, China
{guodong,duyong,li_qiang,hul}@jlu.edu.cn

Abstract. Based on cloud storage software HDFS, this paper has designed a cloud data backup system. Clients are divided into several groups and served by different servers so as to build backup/restore load balance. When backup server upload data to HDFS cluster or download data from HDFS cluster, it takes restore priority, conflict detection upload's strategies to reduce the network transmission pressure. To meet the feature that HDFS is suitable for large file's storage, backup server has combined small files to upload by setting a threshold, thus enhancing system performance. The HDFS-based cloud backup system designed by this paper has certain advantages on the aspects of safety, extendibility, economic efficiency and reliability.

Keywords: cloud computing, cloud storage, backup system, HDFS.

1 Introduction

As a data security strategy, backup is the last and fundamental way to avoid data missing [1] and many various types of data backup systems have been designed and realized to reach this goal. Currently, common network data backup systems can be divided into the following categories: the LAN-Free structure and the Server-Free structure based on the NAS (network attached storage) architecture, the LAN (local area network) architecture and the SAN architecture [2].

Cloud storage [3] technology has provided a new solution of data backup. Through the use of cloud storage, storage can become a service on demand, and the storage service can be customized in accordance with the demand of the user; it needs neither upfront investment nor later cost of construction, thus drastically reducing the expense that the user needs to spend on the hardware and the cost of daily maintenance.

The cloud storage has the following characteristics suitable for data backup: complete data storage service for users to get intellectual backup software and well-managed storage capacity; only data backup, without worrying about the control on former data; advantageous price to backup the same scale of data, which is far cheaper than building up data center by purchasing storage device [4].

G. Zhiguo et al. (Eds.): WISM 2011, CCIS 238, pp. 396–403, 2011.

Apache Hadoop [5] is a cloud computing [6] software framework that supports data-intensive distributed applications under a free license. The infrastructural component of Hadoop—HDFS (Hadoop Distributed File System) [7] is a distributed file system specifically developed for cheap hardware design inspired by Google File System (GFS) paper [8], and it has built in data fault-tolerant capacity on the software layer, which can be applied in the creation and development of the cloud storage system.

Based on cloud storage software HDFS, this paper has designed a cloud data backup system which applies existed cheap computers to set up a data backup cluster.

2 Architecture

The precondition of this design is to deploy and establish a data storage and restore system without increasing any investment in hardware or causing any change to current network connection. Based on the considerations mentioned above, this paper has designed a HB-CDBS (HDFS Based Cloud Data Backup System), which physically consists of the three parts of clients, backup servers and the HDFS cluster, and the architecture is shown in Fig. 1.

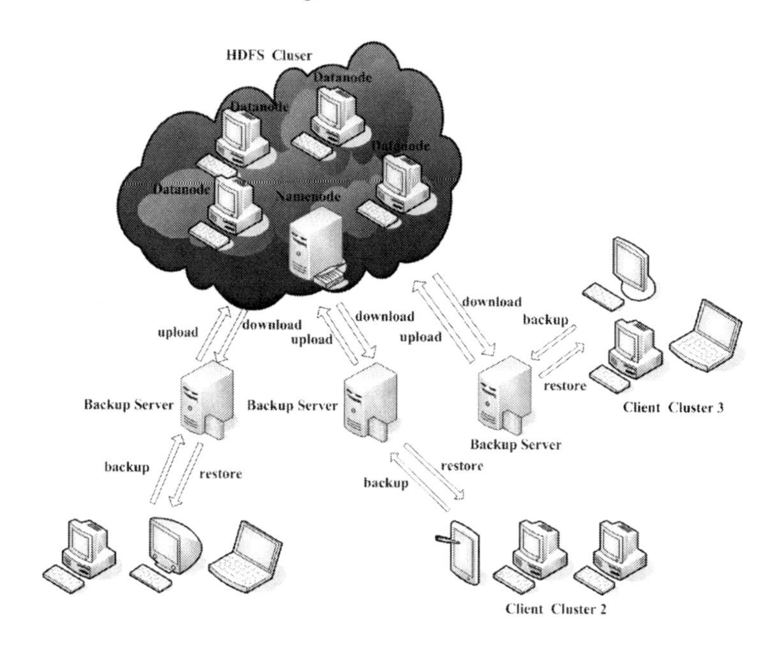

Fig. 1. The architecture of HB-CDBS

Clients are computer nodes that need data backup/restore service, which can be divided into various clusters in accordance with different types of regions; when data backup or restore is needed, they make a request to the backup server responsible of this cluster, and then it conducts the backup and recovery operation after getting the approval.

As the bridge of data backup/restore between the client and the HDFS cluster, backup servers consists of multiple high-performance servers with huge storage capacity, of which each server is responsible of one client cluster. They receive the request of backup and restore from the client, buffers the backed up data from the client, combines them and compresses them in accordance with different conditions of the backed up data, and uploads them to the HDFS cluster to conduct backup; in the meantime, the mapping table of the backup file from the client should be saved, and when the client makes a request of restore, the backup file should be read from the HDFS cluster, and it will be sent to the client in accordance with the file mapping table.

The HDFS cluster consists of the computers installed with the HDFS, and under the architecture of the HDFS, uploading and downloading services can be provided to multiple backup servers through configuration to realize the core function of the system.

Application of backup servers as the bridge between clients and the backup cluster is due to the following considerations: backup servers can shield the client's direct access to the backup cluster to increase safety of the backup cluster, and in the meantime data safety is realized between backup servers and clients through various technologies such as the firewall and secure channel, which can further ensure safety of the whole system; backup servers can temporarily store the data and decides to upload the data in a proper time in accordance with the load condition of the backup cluster and the network situation, which can ensure the load balance of the backup cluster; although under special circumstances, due to a large amount of backup/restore requests from clients, backup cluster might become a system bottleneck, this kind of situation can be avoided to the greatest extent by adopting the high-performance server as the backup server and through reasonable dispatch on the client side; uploading and downloading to the HDFS cluster requires to install specific modules of hadoop, which is unrealistic for the large amount of customers of different levels, and by collecting the data that the user needs to back up on the backup server and install the modules of hadoop on it to realize the backup and restore functions, the functions of HDFS can be easily realized and given full play to.

3 System Design and Realization

3.1 Design of the Client Module

The client module is designed for users to backup and restore data. Before backing up the data, all files should be packed into one backup file by using tools such as tar and winrar, and the package file should be named by following the rule of "Client Id-file name-bak"; in the meantime, compression should also be made to save storage space and reduce time of restore. After restore data, unpacking and uncompressing will be done to obtain the the original files.

When backup or restore is required, corresponding request will be sent to the backup server which provides service to this node.

3.2 Backup Server

The backup server includes the following modules:

1. Backup Management Module: core function module of the system, which is responsible for backup;
2. Restore Management Module: it is responsible for restore;
3. Safety Management Module: functions of this module include controlling the transmission safety and storage safety of the files as well as authentication and authorization of the client;
4. Catalogue Management Module: this module is responsible of the client management and catalogue management of the backup files. The file backup information table is responsible of the catalogue management of the backup files, and the client information table is responsible of the management of all clients that the backup server is responsible of.
5. User Interface Module: it provides friendly user interface which is used to display and configure the backup or restore operation information so that user can choose the backup or restore method in accordance with his/her need.

3.2.1 File Backup

After receiving the backup request from the client, the backup server will first conduct identification and authentication of the client. After the backup file has been uploaded, the backup server will temporarily store the backup file after adding time stamp, then it will record the information of the backup file into the file backup information table (its field contents are as shown in Table 1), and then the file name will be used as a parameter to upload data to the backup cluster by invoking the CDUA (Cloud Data Upload Algorithm) algorithm.

Table 1. The fields of data's backup table

file name	client id	file size	upload flag	upload file name
fname	client-id	fszie	upload-flag	upfname

Design of CDUA is based on the following consideration:

HDFS is a storage system specifically designed for large files, which is applicable to the storage of files with the size hundreds of MB, of GB or TB and to the circumstances under which you can write once and read for multiple times. Therefore, in order to ensure backup efficiency, CUDA has adopted the uploading strategy triggered by the threshold: first of all, set up an uploading threshold—th-size, and the uploading can only be made when the backup file uploaded from the client is bigger than th-size; otherwise, local temporary storage will be conducted to the file to wait for the uploaded file(s) from other clients, and when the size of all the un-uploaded files is bigger than the threshold, it will uploaded to the HDFS cluster. In general, the backup file temporarily stored in the backup server will not be larger than th_size, which thus will not cause storage burden to the backup server.

The threshold should be decided by Formula (1), in which, hdfs_szie refers to the size of the storage space of the HDFS cluster, cc_n refers to the number of clusters on the client side, rp_n refers to the number of the data backup copies in the HDFS cluster, c_n refers to the number of client nodes in the default cluster, c_{size} refers to the quota of the backup space size on each client side, and λ is the adjustable parameter.

$$th\_size = \lambda \times \left(hdfs\_size / (cc_n \times rp_n \times c_n \times c_{size}) \right) \quad (1)$$

The CDUA algorithm will first check whether the uploaded file's size is smaller than th_size, and if it is, it will upload this file to the HDFS cluster; after it has been successfully uploaded, change the flag field in the file backup information table to be true, fill in with the name of the uploaded file (at this moment, the file name should be the same with the name of the file uploaded from the client), and delete the file from the backup server. If the file's size is smaller than th_size, then it will read the file backup information table to obtain the information of all un-uploaded files and calculate their total size, and if it is no smaller than th_size, it will pack all un-uploaded files into one file, and name this file as "file name 1-file name 2…-file name n"; after it has been successfully uploaded, change the corresponding flag field in the file backup information table to be true, fill in with the names of the uploaded files, and delete the files from the backup server; if the size of all uploaded files is still smaller than th_size, then it will exit.

3.2.2 File Restore

After receiving the restore request from the client, the backup server will first conduct identification and authentication of the client, and after the client has passed the authentication, it will read the file backup information table, and if the backup file is temporarily stored locally, the file will be sent to the client from the backup server; if the backup file is stored in the HDFS cluster, it will be downloaded from the HDFS backup cluster and then sent to the client. If the backup file consists of multiple packed files, the files should be unpacked first, and then the file will be sent to the client.

3.2.3 Uploading/Downloading Strategy of the Backup Server

The HDFS cluster is in a 24-hour working state, therefore, the backup server can conduct data backup/restore operation at anytime. In order to ensure the backup performance of the system, the strategy of restore priority and collision detection uploading was adopted to conduct backup and restore, i.e., when the backup server needs to download data, it will be executed immediately; when it needs to upload data, if there is no other backup server uploading data, it will be uploaded immediately; otherwise it is regarded as causing collision, and it will wait for a while to check again and decide whether to upload, and the duration of the waiting time is determined by the backoff algorithm. The procedure of the backoff algorithm is as the following:

1) If there is a collision during first detection, set up the parameter L=2;
2) Take a random number between 1 and L as the backoff interval;

3) If there is a collision during repeated detection, double the parameter L. The maximum value of L is 256, and when L has been increased to 256, it cannot be increased anymore.
4) If the testing times have been more than 8, then data will be uploaded unconditionally and immediately.

By using the backoff algorithm, When there are many collisions in the backup servers, the possibility of a long waiting time is bigger, and it will ensure as less detections to the system as possible when the system is under heavy load; in the meantime, it can guarantee fairness by immediately uploading files when there are more than 8 times of backup server backoffs.

3.3 HDFS Backup Cluster

The HDFS backup cluster has adopted the Master/Slave structure, which is composed of one Namenode and a certain number of Datanodes.

A high-performance server serves as the NameNode in the cloud to realize high-efficiency management of metadata and also to avoid performance bottleneck. The NameNode is the centerpiece of an HDFS file system. It keeps the directory tree of all files in the file system, and tracks where across the cluster the file data is kept. It does not store the data of these files itself.

DataNode is used to store data, which consists of many cheap computers inside the cloud, and dynamic expansion can also be conducted in accordance with the scale of the backup data. During the backup, the file is divided in to one or more data blocks, and these blocks are stored in a group of Datanode.

4 Performance Evaluation

4.1 Performance

HDFS Based Cloud Data Backup System designed in this paper has the strengths in security, expansibility, economy, and reliability:

Security: through user identification, authorization and restricting access to systems, backup server can guarantee the security between clients and himself. Through the security mechanism of Hadoop, the safety of communication and data transmission between HDFS cluster and backups server is assurance.

Expansibility: by use of powerful storage and calculation expansibility of Hadoop, the scale of HDFS cluster can be expanded at any time to enhance backup capacity of the system.

Economy: HDFS is the distributed file system designed for cheap hardware, with good compatibility so that any computer can enter into this backup clusters by installing HDFS. Therefore, large amount of unoccupied computer resources can be used to cut down the expenses of procuring devices.

Reliability: backup files in HDFS cluster are mainly preserved by copies. And also the number of copies can be increased according to importance of files to enhance the reliability.

4.2 Comparison with Existing Data Backup System Backup Cluster

Comparing with the network attached storage system (DAS-Base), the HDFS-based cloud data backup system has overcome the deficiency of less manageable storage devices, inconvenience to share the backup system and not applicable to large data backup of the DAS-Base, and it can also provide real-time backup service.

Comparing with the LAN-Based backup system, due to the adoption of multiple-backup server to provide services to clients in different regions, the backup system designed by this paper has balanced the backup load on the one hand, and on the other hand, due to the uploading monitoring strategy of the backup service, it can avoid the situation in which more than one server transmits a large amount of data to the backup cluster at the same time, which might cause decrease of the network performance. Therefore, the system designed by this paper not only has the advantages of efficient investment and central backup management similar to the LAN-Based backup system, it has also overcome the shortages of a large network transmission pressure and fast decrease of the LAN performance when the backup data is large or the backup frequency is too high.

Comparing with the two SAN-Based backup solutions, the backup system designed by this paper uses the existing cheap equipment of the user to conduct establishment, and it does not require purchasing of expensive special communication and storage devices while at the same time ensuring fast and high-efficiency data backup, which thus drastically reduces the cost of data backup; in the meantime, due to adoption of software to establish the backup cluster, it also reduces the complexity of technique implementation, and does not need to consider compatibility problem of the manufacturer.

5 Conclusion

By introducing the cloud storage technology into the field of data backup, this paper has designed and realized the HDFS-based cloud data backup system. Under the precondition of not increasing equipment investment, this system has adequately used the existing hardware infrastructure to establish the data backup cluster based on the strong capacity of the HDFS. In this system, the users are divided into multiple groups, and each group has different backup server to realize its data backup and restore service, which thus realizes load balance of the data backup services; by using the backup server as a bridge between the client and the HDFS backup cluster, the backup server is responsible of uploading and downloading data to the cloud storage system to avoid direct facing to the cloud storage system by the user, which is not only in the convenience for use, but also can further improve the system safety; in accordance with the characteristic that the HDFS apply to the operation of large files, the merging uploading strategy of small files are adopted in the backup server to give full play to the capacity of the cloud backup system. The HDFS-based cloud backup system has certain advantages on the aspects of safety, extendibility, economic efficiency and reliability.

Acknowledgments. This work is supported by National Natural Science Foundation of China under Grant No. 61073009, supported by New Century Excellent Talents in University (NCET-09-0428) and Basic Science Research Fund in Jilin University Grant Nos. 201103253,201003035. Corresponding author: Liang Hu, Phn:86-431-85168277, E-mail:hul@jlu.edu.cn.

References

1. Li, W.J., Li, L.X., Li, L.L.: etal: Design and Implementation of Backup System Based on P2P Network. Journal Of Information Engineering University, 351–355 (2010)
2. Wang, G.X., Shi, M.R.: Data storage backup and disaster recovery. Electronic Industry Press, Beijing (2009)
3. Storage Networking Industry Association and the Open Grid Forum, http://www.snia.org/cloud/CloudStorageForCloudComputing.pdf
4. China, T.T.: http://www.cloudcomputing-china.cn/Article/luilan
5. The Apache Software Foundation, http://hadoop.apache.org/index.pdf
6. Cheng, K., Zheng, W.M.: Cloud computing: System instances and current research. Software Journal, 1337–1348 (2009)
7. Shvachko, K., Kuang, H., Radia, S., Chansler, R.: The Hadoop Distributed File System. In: 26th IEEE Symposium on Massive Storage Systems and Technologies, pp. 1–10. IEEE Press, Chicago (2010)
8. Ghemawat, S., Gobioff, H., Leung, S.: The Google file system. In: 19th ACM Symposium on Operating Systems Principles, pp. 29–43. ACM Press, New York (2003)

The Process Modeling of Collaborative Manufacturing Program Business Based on UML

Si Sun, Shan Zhao, and CongGang Wei

School of Computer Science and technology, Henan Polytechnic University
Jiaozuo 454000, Henan, China
zhaoshan@hpu.edu.cn

Abstract. With the globalization of manufacture and the network-based manufacture, the traditional centralized organization mode of manufacture will be replaced by distributed network-collaborative organization mode which is project-driven and based on modern information technology, especially network technologies. In order to manage the collaborative manufacturing network project (collaborative projects, hereinafter) effectively, this paper aims to make modeling language based on UML for the actual network-collaborative manufacturing process of a key product component in the hope of providing a good method for the realization of interactive view, co-operation and unified data management cross enterprises.

Keywords: UML network collaborative projects workflow Meta-Process Modeling.

1 Introduction

In order to realize the effective management of collaborative project progress, we should model the service process of collaborative projects based on analysis. The traditional central process analysis and its modelling method can not meet the Hypothesized Enterprise's distributivity and strong autonomy, meanwhile, the members of Hypothesized Enterprise will change according to different projects or production processes, they have strong collaborative ability although they are loose on the surface, so we must adopt the dynamic maintenance, the collaborative cooperation, and the cross-organization boundary service process in the management or control. Workflow technology has been widely applied in the Hypothesized Enterprise process modelling because it can model and control the complex service process effectively so as to realizes the collaborative management of business.

2 The Workflow Management of Collaborative Project

The collaborative project is composed of many activity flows in different levels which are correlated with each other in the actual operation. The main flow includes the task decomposition, the partner choice, the manufacturing as well as the delivery and so on, each main flow may decompose into many sub-flows which are related each

G. Zhiguo et al. (Eds.): WISM 2011, CCIS 238, pp. 404–411, 2011.

other, this process goes on until the activity is reached. While, the aim of the workflow management system is to realize the effective business operation by means of the reasonable information and source distribution in order to collaborate different activities in business operation. Therefore, the workflow management system can manage activities of projects effectively.

The present workflow product succeeds to some degree in the automation of business operation and information management. With the rapid development of market, the customers' demand tends to be diversified and idiosyncratic, the process defined beforehand can not meet the need of enterprise in the mass production, therefore, workflow managemnet system must provide more flexible process model, must own more dynamic features in order to adjust business operation according to different circumstances. In addition, for the sake of the large-scale workflow management under the isomerism computation environment between different Hypothesized Enterprises and the interior of themselve, we must expand the ability of workflow management system from many aspects to enhance its reliability, expandable and fault-tolerant abilities, etc.

3 The Business Process Model of Collaborative Project

In order to realize the management of collaborative projects workflow, a project operation process model must be established based on the analysis of its business process and the essential characteristics of business process acquired from the analysis. Process model must be featured by highly dynamic abstraction, the reconstruction, reusable type, scalability, etc., only in this way, can it meet the higher requirements of the characteristics of cooperative projects. So far, researchers in workflow management system research and development only put forward various descriptive methods for different business processes, such as activity network diagram[1], Petri nets[2], object-oriented[3], language or behavior theory[4], events - conditions - movement rules (ECA) [5], state and activity diagrams[6], etc. These descriptive methods of business processes have advantages in descriptive ability and flexibility of the model per se, different enterprises and developers will choose different software products according to their own needs, thus, it is difficult to view the cooperation and unified management of data across enterprises, so a unified business process description is needed urgently now in order to provide good support for cross-enterprise interactive process. UML was adopted by Object Management Group (OMG) in object-oriented technology and was made as the standard of the modeling language, which will provide a good method for solving this problem.

Based on the definition of process meta-model defined by the Workflow Management Alliance and the previous process hierarchical analysis of cooperation program business[7], the authors will establish a perfect Meta-process model by means of integrating global business process, sharing business process, related properties and objects together to form a complete object. The core of meta process is Activity, other processes involved in this task are Processes, Parameters, Resources, Transition, TransitionRestriction, Participant, Accesslevel, etc. The formula below can show clearly the meta_process: <meta_process>::=<Process,Activity,Participant, Resource,Parameters,Transition,AccessLevel,TranstionRestriction>

The logical relation of the basic objects in Meta-Process model is shown in figure 1.

Meta-process is the basic division of business processes which can provide the basic operation of the business process. It includes Processes, Parameters, Resources, Transition, Transition Restriction, Participant, Accesslevel, etc. which usually needs one or more associated business roles, among them, the Parameters and Transition Restriction are responsible for planning related resources according to the performance requirements of tasks to accomplish some specific tasks.

Different objects describe Meta-Process model in different aspects as follows:

(1) Process specifies the name, participants, access and process control parameters, etc. of business process. A business process includes one or more activities. The subscript for the 0, and other common scientific constants, is zero with permeability of vacuum subscript formatting, not a lowercase letter "o".

(2) Transition Restriction is used to specify the triggered condition for related activities in any situations, when the suitable condition comes, then the correspondent activity will begin to execute. Restriction can be single or compound. A graph within a graph is an "inset", not an "insert". The word alternatively is preferred to the word "alternately" (unless you really mean something that alternates).

(3) Activity is a process of the business process which includes names, types, input conditions of restriction, etc.

(4) Participants are subjects implemented by activities, in other words, it is the specific gent in Agent system which can be implemented in the local or through the CORBA/SOAP remote access.

(5) Parameters are responsible for planning and controlling implementation sequence in the process, which include the state of the role of activities, the times for the beginning and end of activities. Parameters need to connect related activities.

(6) Accesslevel specifies the sharing business process or private business process.

Meta-process model is abstract description of collaborative projects production, the above six meta processes describe related information in production activities respectively, such as the premise and subsequent conditions of the activity (Transition), data structure, parameters, and operation mode, etc. The collaborative projects is a complex process composed of some meta processes.

Meta-process model established on the basis of UML is needed to transformed into certain expressive way which can be understood by computers and can be controlled by Agents flexibly. XPDL is a kind of workflow modeling Language established by WFMC on the basis of XML. So XPDL has some properties of XML, for example, XPDL also puts more emphasis on information than form. XPDL is an universal framework in that it can help any developers to achieve any purposes, no matter what modeling methods and what way to achieve the developers adopt, they can use the same description as long as the external interface meet the standard of XPDL.

So XPDL can describe business process information clearly and support collaborative projects process modeling. The Meta-Process process model shown by Figure 1 can be described by XML Schema format in Table 1.

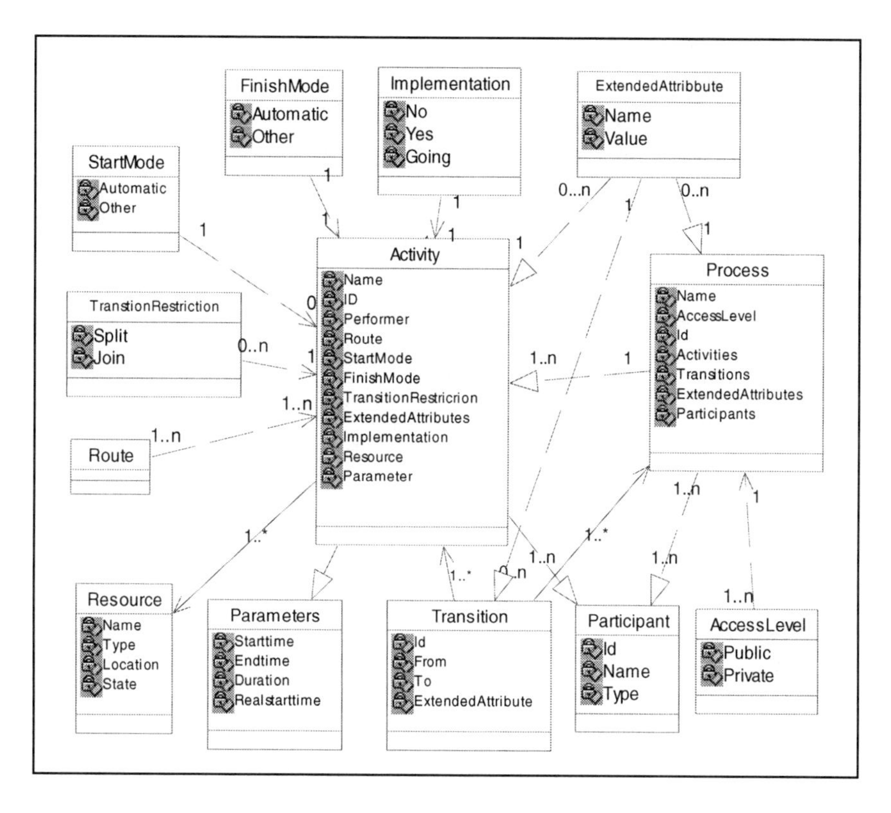

Fig. 1. UML Diagrams of Meta-Process model

Table 1. XML Schema of Meta-Process Model

```
<?xml version="1.0" encoding="UTF-8" standalone="yes"?>
<!--W3C Schema was established by XMLSpy v in 2006 U
(http://www.altova.com)-->
<xs:schema xmlns:xs="http://www.w3.org/2001/XMLSchema"
xmlns="http://www.nwpu.edu.cn-hwp"
targetNamespace=http://www.wfmc.org/2002/XPDL1.0
elementFormDefault="qualified">
<!-- Activities declarations-->
<xs:element name="Activities">
<xs:element name="Activity">
   <xs:complexType>
      <xs:sequence>
         <xs:element ref="Implementation" minOccurs="0"/>
         <xs:element ref="Performer" minOccurs="0"/>
         <xs:element ref="Route" minOccurs="0"/>
         <xs:element ref="StartMode"/>
```

```
            <xs:element ref="FinishMode"/>
            <xs:element ref="TransitionRestrictions" minOccurs="0"/>
            <xs:element ref="ExtendedAttributes"/>
            <xs:element ref="Resource" minOccurs="1"/>
            <xs:element ref="Parameters"/>
         </xs:sequence>
         <xs:attribute name="Id" use="required">
         <xs:attribute name="Name" use="required">
      </xs:complexType>
   </xs:element>
   <xs:element name="Automatic" type="xs:string"/>
   <xs:element name="ConformanceClass">
   <xs:element name="Created">
   <xs:element name="ExtendedAttribute">
      <xs:complexType>
         <xs:attribute name="Name" use="required">
         <xs:attribute name="Value">
      </xs:complexType>
   </xs:element>
   <xs:element name="ExtendedAttributes">
   <xs:element name="FinishMode">
   <xs:element name="Implementation">
   <xs:element name="Join">
      <xs:complexType>
         <xs:attribute name="Type" use="required">
      </xs:complexType>
   </xs:element>
   <!--Resouce declarations -->
   <xs:element name="Resource">
      <xs:complexType>
         <xs:attribute name="Name" use="required">
         <xs:attribute name="State" use="required">
         <xs:attribute name="Type" use="required">
         </xs:complexType>
      </xs:element>
   <!--Parameters declarations -->
   <xs:element name="Parameters">
      <xs:complexType>
         <xs:attribute name="Starttime" use="required">
         <xs:attribute name="Endtime" use="required">
         <xs:attribute name="Duration" use="required">
         <xs:attribute name="Realstarttime" use="required">
      </xs:complexType>
   </xs:element>
   <xs:element name="No" type="xs:string"/>
   <xs:element name="PackageHeader">
   <!--Participant declarations -->
   <xs:element name="Participant">
      <xs:complexType>
      <xs:attribute name="Id" use="required">
         <xs:attribute name="Name" use="required">
      </xs:complexType>
   </xs:element>
   <xs:element name="ParticipantType">
```

```
    <xs:complexType>
      <xs:attribute name="Type" use="required">
      </xs:attribute>
    </xs:complexType>
  </xs:element>
  <xs:element name="Participants">
  <xs:element name="Performer">
  <xs:element name="ProcessHeader">
    <xs:complexType>
      <xs:sequence>
        <xs:element ref="Created"/>
      </xs:sequence>
      <xs:attribute name="DurationUnit" use="required">
    </xs:complexType>
  </xs:element>
  <xs:element name="RedefinableHeader">
    <xs:complexType>
      <xs:attribute name="PublicationStatus" use="required">
    </xs:complexType>
  </xs:element>
  <xs:element name="Route" type="xs:string"/>
  <xs:element name="Split">
    <xs:complexType>
      <xs:attribute name="Type" use="required">
    </xs:complexType>
  </xs:element>
  <xs:element name="StartMode">
  <!--Transition declarations -->
  <xs:element namc="Transition">
    <xs:complexType>
      <xs:sequence>
        <xs:element ref="ExtendedAttributes"/>
      </xs:sequence>
      <xs:attribute name="From" use="required">
      <xs:attribute name="Id" use="required">
      <xs:attribute name="To" use="required">
    </xs:complexType>
  </xs:element>
  <xs:element name="TransitionRef">
  <xs:element name="TransitionRestriction">
  <xs:element name="Transitions">
  <xs:element name="Vendor">
  <!--WorkflowProces declarations-->
  <xs:element name="WorkflowProcess">
    <xs:complexType>
      <xs:sequence>
        <xs:element ref="ProcessHeader"/>
        <xs:element ref="RedefinableHeader"/>
        <xs:element ref="Participants"/>
        <xs:element ref="Activities"/>
        <xs:element ref="Transitions"/>
        <xs:element ref="ExtendedAttributes"/>
      </xs:sequence>
      <xs:attribute name="AccessLevel" use="required">
```

```
        <xs:attribute name="Id" use="required">
        <xs:attribute name="Name" use="required">
    </xs:complexType>
   </xs:element>
   <xs:element name="WorkflowProcesses">
   <xs:element name="XPDLVersion">
</xs:schema>
```

After being defined, Meta-process can be stored in sharing resource repository of collaborative Projects for all member enterprises, and their specific process information (the actual process of business) can be shown by XPDL document which is based on the correspondent Schema. Some source codes of description for a simplified key product component manufacturing process are shown in Table 2.

After the standardized description of business processes, workflow engine can analyze this descriptive process and specify the triggered conditions and executive order regulation in accordance with the definitions of specific business processes, then, it will start up corresponding activities to implement collaborate projects processes correctly at the right time. Workflow engine can be encapsulated in the Agent which can help the workflow engine to realize its functions.

Table 2. Some source codes of description for manufacturing

```
<?xml version="1.0" encoding="UTF-8"?>
<Package    Id="ManufacturingProcess"    xmlns="http://www.nwpu.edu.cn-hwp"
xmlns:xpdl=http://www.wfmc.org/2002/XPDL1.0
xmlns:xsi="http://www.w3.org/2001/XMLSchema-instance"
  <PackageHeader>
    <XPDLVersion>1.0</XPDLVersion>
    <Vendor>Together</Vendor>
    <Created>2006-07-12 16:28:54</Created>
  </PackageHeader>
  <WorkflowProcesses>
    <WorkflowProcess AccessLevel="PUBLIC" Id="ManufacturingProcess_
    Wor1" Name="process">
        <ProcessHeader DurationUnit="D">
          <Created>2006-07-12 16:29:26</Created>
        </ProcessHeader>
        <RedefinableHeader PublicationStatus="UNDER_TEST"/>
        <Participants>
          <Participant Id="ManufacturingProcess_Wor1_Par1"
          Name="Champion enterprise">
              <ParticipantType Type="ROLE"/>
          </Participant>
        <Participant Id="ManufacturingProcess_Wor1_Par2"
        Name="collaborative partner">
      <ParticipantType Type="ROLE"/>
        </Participant>
     </Participants>
  <Activities>
          <Activity  Id="ManufacturingProcess_Wor1_Act1"
          Name="task decomposition">
```

```
        <Implementation>
            <No/>
              </Implementation>
             <Performer>ManufacturingProcess_Wor1_Par1</Performer>
             <StartMode>
                <Automatic/>
             </StartMode>
          <FinishMode>
            <Automatic/>
          </FinishMode>
          <ExtendedAttributes>
           <ExtendedAttributeName="ParticipantID"
                          Value="ManufacturingProcess_Wor1_Par1"/>
               <ExtendedAttribute Name="time limit " Value="3"/>
           <ExtendedAttribute Name="time for start" Value="07-12"/>
           <ExtendedAttribute Name="time for end" Value="07-14"/>
           <ExtendedAttribute Name="resources"/>
        </ExtendedAttributes>
    </Activity>
    <Activity Id="ManufacturingProcess_Wor1_Act3" Name="partner selection">
        <Implementation>
                        <No/>
                   ......
```

On the basis of studies mentioned above, the authors analyze the business process modeling for collaborative projects in detail by means of the workflow technologies, then describe Meta-process model for business processes of collaborative program based on UML technologies and standards, it is hoped that this paper can provide theoretical and technical supports for the schedule, progress tracking, schedule control, and scheduling implementation of collaborative projects in the future.

References

1. IBM Corp.IBM MqSeries Workflow:Concepts and Architecture.Version 3.3 Reference Book (March2001), http://www.ibm.com/software/ts'mqseries/workflow
2. van der Aalst, W.M.P., ter Hofstede, A.H.M.: Verification of Workflow Task Structures:A Petri-Net-Based Approach. Information Systems 25(1), 43–69 (2000)
3. Manolescu, D.A., Johnson, R.E.: Dynamic Object Model and Adaptive Workflow. In: OOPSLA 1999 Metadata and Active Object-Model Pattern Mining Workshop, Denver, Colorado (November 1999)
4. Georgakopoulos, D., Hornick, M., Sheth, A.P.: An Overview of Workflow management: From Process Modelling to Workflow Automation Infrastructure. Diatributed and Parallel Databases 3(2), 119–152 (1995)
5. Kumar, A., Zhao, J.L.: Dynamic Routing and Operational Controls in Workflow Management Systems. Managemant Science 45(2), 253–272 (1999)
6. Muth, P., Wodtke, D., Werssenfels, J.: Enterprise-wide Workflow Management based on State and Activity Charts. In: Dogac, A., Kalinichenko, L. (eds.) Workflow management Systems and Interoperability, NATO Advanced Study Institute. Springer, Heidelberg (1998)
7. Workflow Management Coalition Workflow Standard Workflow Process Definition Interface– XML Process Definition Language. WFMC-TC-1025, V1.0 (2002)

The Framework of the Kernel Technology of Hibernate

Yutian Huang[1] and Xiaodong Feng[2]

[1] Taiyuan University of Technology, Taiyuan, P.R.C
[2] Shanxi Branch of China Unicom, Taiyuan, P.R.C
huangyutian@tyut.edu.cn, walkerxiaodong@sina.com

Abstract. This paper discuss the basic and yet kernel technology using implementing the hibernate which widely used in the j2ee project today. It shows how to use reflection technology to convert the two_demensions table struct to object which easily accessed in j2ee project. It also shows how to dynamically get the table column name and datatype using system query statements giving the example of oracle database. It also give the hibernate code and background java reflection code correspondingly.

Keywords: Hibernate, Reflection, Oracle, Implementation, Java.

1 Introduction

The hibernate technology is widely used today all over the world. It simply converts the two_demensions structure to the more familiar object in java. The frequently used getString(String column_name) methods are used instead of getUserName()(for example the column name is UserName). We will discuss the kernel technology with the database oracle as example. The kernel technology include java reflection technology, oracle column query statements, the initialization of hibernate factory, the xml file of hibernate, and so on.

2 Discussion of the Most Difficult Technology

The most difficult technology ----the reflection of java language. This involves ClassLoader ,Class<?>, Method of java.lang package. The method involved are listed below:

2.1 Public Static ClassLoader getSystemClassLoader() (java.lang.ClassLoader)

Returns the system class loader for delegation. This is the default delegation parent for new ClassLoader instances, and is typically the class loader used to start the application.

2.2 Public Class<?> loadClass(String name) (java.lang.ClassLoader)

Throws ClassNotFoundExceptionLoads the class with the specified binary name. This method searches for classes in the same manner as the loadClass(String, boolean)

G. Zhiguo et al. (Eds.): WISM 2011, CCIS 238, pp. 412–416, 2011.

method. It is invoked by the Java virtual machine to resolve class references. Invoking this method is equivalent to invoking loadClass(name, false).

2.3 Public Method getMethod(String name,Class<?>... parameterTypes) (java.lang.Class<T>)

throws NoSuchMethodException,

SecurityExceptionReturns a Method object that reflects the specified public member method of the class or interface represented by this Class object. The name parameter is a String specifying the simple name of the desired method. The parameterTypes parameter is an array of Class objects that identify the method's formal parameter types, in declared order. If parameterTypes is null, it is treated as if it were an empty array.

2.4 Public Object invoke(Object obj,Object... args) (java.lang.reflect.Method)

throws IllegalAccessException,

IllegalArgumentException,

InvocationTargetException Invokes the underlying method represented by this Method object, on the specified object with the specified parameters. Individual parameters are automatically unwrapped to match primitive formal parameters, and both primitive and reference parameters are subject to method invocation conversions as necessary.

If the underlying method is static, then the specified obj argument is ignored. It may be null.

The using of the the above method will be listed below:

getSystemClassLoader() ->loadClass(String name)->getMethod(String name,Class<?>... parameterTypes)->
invoke(Object obj,Object... args)

In the non programming language, it is first get a class loader, then load the class object with the class name, then get the method object from the class<?>,then using invoke method to call a method of the class<T > object.

The example code will be in the latter section.

3 How to Get the Column Properties of the Spcified Table in Oracle

Let us discuss how to get the column properties of the specified table in oracle. It will be the following SQL statements.

–to get table info, using

select * from dba_tables; select * from all_tables; select * from user_tables;

–to get column info using

select * from dba_tab_columns; select * from all_tab_columns; select * from user_tab_columns;

using desc user_tab_coumns, we will get

desc user_tab_columns;

```
Name                    Type               Nullable Default Comments
--------------------    -------------      ---------- ------ --------------------------
TABLE_NAME              VARCHAR2(30)                           Table, view or cluster name
COLUMN_NAME             VARCHAR2(30)                           Column name
DATA_TYPE               VARCHAR2(106) Y                        Datatype of the column
DATA_TYPE_MOD           VARCHAR2(3)   Y                        Datatype modifier of the column
DATA_TYPE_OWNER         VARCHAR2(30)  Y                        Owner of the datatype of the column
DATA_LENGTH             NUMBER                                 Length of the column in bytes
DATA_PRECISION          NUMBER        Y                        Length: decimal digits (NUMBER)
or binary digits (FLOAT)
DATA_SCALE              NUMBER        Y                        Digits to right of decimal point in a
number
...
```

Of course, the usefulcolumn will be

```
TABLE_NAME              VARCHAR2(30)                           Table, view or cluster name
COLUMN_NAME             VARCHAR2(30)                           Column name
DATA_TYPE               VARCHAR2(106) Y                        Datatype of the column
```

In Oracle database, data_type will be :Char ,varchar2 , nchar, nvarchar2,number, Date , Long ,Raw ,Long raw ,rowid, Blob ,Clob, nclob, Bfile, urowid ,float .

Then we could generate the corresponding get/set column class and the xml (column information, for example student.hbm.xml).It is easily to write a program to automatically generate the .java and .xml file.

4 Using the Reflection of Java Language to Convert Two_Demensions Table Struct to Object

But the most difficult problem that we faced is how to call the get/set column class dynamically without firstly knowing the method name when compiling. To solve this problem, we should know the reflection function. Using the class java.lang.ClassLoader,java.lang.Class<T> ,java.lang.reflect.Method.

For example, we want convert the one row(with the table name student, and with the column name and id) to one object a with the class name student.

```
try{
    ClassLoader
    a1=java.lang.ClassLoader.getSystemClassLoader();
    Class<?> c=a1.loadClass("com.fxd.student");
    Class[] parameterTypes = new Class[1];
    parameterTypes[0] = String.class;
    Method m=c.getMethod("setName",parameterTypes);
    String[] s=new String[1];
    s[0]=new String("fxd");
    Object a2=c.newInstance();
    m.invoke(a2,(Object[])s);//call setName("fxd");
}
catch(Exception e)
{
     e.printStackTrace();
}
```

Here we omit the code how to get the value of column name, it is easily solved using jdbc interface. We also omit the code how to parse the xml file, it is also easily parsed.

5 Example of Background Code of the Whole Procedure to Use Hibernate

Here is the hibernate.cfg.xml.

```
<?xml version='1.0' encoding='UTF-8'?>
<!DOCTYPE hibernate-configuration PUBLIC
 "-//Hibernate/Hibernate Configuration DTD 3.0//EN"
 "http://hibernate.sourceforge.net/hibernate-configuration-3.0.dtd">
<hibernate-configuration>
 <session-factory>
  <property name="show_sql">true</property>
  <property name="format_sql">true</property>
  <property
name="connection.driver_class">oracle.jdbc.driver.OracleDriver</property>
  <property
name="connection.url">jdbc:oracle:thin:@127.0.0.1:1521:hello</property>
  <property name="connection.username">lib</property>
  <property name="connection.password">lib123</property>
  <property name="connection.isolation">2</property>
  <property name="hbm2ddl.auto">none</property>
  <property name="current_session_context_class">thread</property>
  <property
name="cache.provider_class">org.hibernate.cache.NoCacheProvider</property>
  <property name="dialect">org.hibernate.dialect.Oracle9Dialect</property>
  <mapping resource="com/lib/xml/student.hbm.xml"/>
 </session-factory>
</hibernate-configuration>
```

When startup, the hibernate will parse the hibernate.cfg.xml file, to get the database information, for example url, username and password.

```
public static SessionFactory sessionFactory;// Data storage source
static {
   try {
       Configuration config = new Configuration().configure();
       sessionFactory = config.buildSessionFactory();

   } catch (Exception e) {
      e.printStackTrace();
   }
}
```

The above code will parse information xml hibernate.cfg.xml, connect to database and parse the mapping resource information.

The background code will be the following:

```
Class.forName("oracle.jdbc.driver.OracleDriver");
Connection conn =
```

```
DriverManager.getConnection( "jdbc:oracle:thin:127.0.0.1:1521:hello","lib","lib123");
      conn.setAutoCommit(false);
    When using the hibernate, it will be the following :
     Session session = sessionFactory.openSession();
      Transaction tx = null;
      try {
        tx = session.beginTransaction();
        List students = session.createQuery(
       "from student as c order by c.name asc").list();
                        //using reflecion method to save query
                       //info into list of object(student class)
        Iterator it = students.iterator();
        System.out.println("append:"+customers.size());
        while(it.hasNext())
        {
          Student c = (Student)it.next();
          System.out.println("ID:" + c.getId());
          System.out.println("Name:" + c.getName());
         }
        tx.commit();
      } catch (Exception e) {
        tx.rollback();
      } finally {
        session.close();
      }
```

We give the most important feature of the kernel technology of hibernate above. It is very easy to write the ungiven part with the jdbc driver api.

References

1. Java™ Platform, Standard Edition 6 API Specification
2. Hibernate api, jboss community
3. 《java语言程序设计》， 吴建平， 尹霞，冯晓冬，清华大学出版社，1997-8-1, ISBN: 9787302025375

Web Service Selection Based on Utility of Weighted Qos Attributes

Yanran Zhang and Minglun Ren

School of Management, Hefei University of Technology
Hefei, Anhui, P.R. China
ZhangYanran_88@163.com, hfutren@sina.com

Abstract. One of the service-oriented computing objectives is combining several simple services to form a powerful composite service. In recent years, an increasing number of online services with the same function but different quality attributes are provided, hence we have to conduct service selection based on Qos attributes. This paper analyzes the Qos attributes of web service, builds a service composition model according to users' Qos satisfaction, and extracts it with AGA. Further, an instance shows the feasibility of this approach.

Keywords: Web service composition, Web service selection, Qos, AGA.

1 Introduction

With the increasing number of web services with the same function on the Internet, a great amount of candidate services emerge. How to find a service according to the Qos requirements of users has become a hot issue in the area of the service computing study.

The problem of web service selection based on Qos is how to find a particular service from the candidate service set, making the entire Qos of composite service best to meet users' need. Numerous researchers have been studying this problem from different perspective. Yimei Mao and Jiajin Le [1] established a Web service composition optimization model, and performed by self-adapted GA. An optimum method based on GA is proposed, however the model just considers the sum of weighted property parameter, it is unable to solve the multi-objective optimization problem; In order to achieve the users' requirements of Qos, Meiling Cai, Maogui Li, and Jie Zhou [2] puts forward a pareto-based multi-object genetic algorithm for the web service selection with global optimal Qos. This method can produce a set of optimal Pareto services, and the most approving solution is selected as the final decision project according to users' requirements. Although this article proposes a multi-objective optimization problem, it fails to clearly express the method of attribute evaluation; Xiaoqin Fan, Changjun Jiang, and Zhijun Ding [3] design a discrete particle swarm optimization algorithm to facilitate the dynamic Web service selection, and prove its advantage by experiments. Nevertheless, its computing process is complex and not appropriate for large scale service composition problem. In this paper, we analyzed Qos attributes of Web service, defined a utility function to evaluate attributes. Then, we

G. Zhiguo et al. (Eds.): WISM 2011, CCIS 238, pp. 417–425, 2011.

develop a Web service composition model that can solve the multi-objective optimization problem, and present an AGA to solve the model. This paper has an advantage on the simple service composition problem.

2 Web Service Selection Model Based on Qos

2.1 Definitions

In this section, we define four important concepts concerning our following research: the web service description model, web service instance, candidate service, and service class as follows.

Definition 1. Web service description model:

$$W = \{F, Qos\}, \tag{1}$$

in which F represents the aggregation of services' function attributes; Qos denotes the aggregation of services' non- function attributes, that is the services' quality attributes. In the process of web service composition, we check whether a service can meet functional requirements based on its function attributes, and then choose the most suitable service from the service class based on its non- functional attributes.

Definition 2. Web service instance: it is a particular service issued by service providers. A service instance is composed of service name, service description, the service class it belonged to and semantics with respect to service quality. Different service instances may be similar in function attributes but not in quality attributes.

Definition 3. Candidate service: when function attributes of service instance W_{ij} are in accord with the task's function description in service process, we call this service the task's candidate service. Each candidate service has a Qos vector

$$Q_{ij} = \left(Q_{ij}^{\ 1}, Q_{ij}^{\ 2} .. Q_{ij}^{\ n}\right), \tag{2}$$

n is the number of Qos attributes.

Definition 4. Service Class: it means a set of service that has the same function attributes. A service class is composed of candidate services that have the same function attributes but different Qos attributes.

The problem of Web service selection based on Qos is how to choose Web service instance from different service classes to fulfill functional requirement, so that the Qos of composite service is the best under users' constraint conditions.

2.2 Web Service Composition Model Based on Qos

Qos is composed of some quality attributes. The main contents include as follows.

a) Response Time. it means the time taken to complete the assigned work. Its measure indexes mainly consist of response time, operation time and delay time.

b) Price. it means the expenses needed when a service requestor access services. Its measure indexes mainly consist of cost, system overhead, through put rate and so on.

c) Availability. it means the probability that a web service can be accessed.

d) Reliability. it means the probability that can successfully complete a web service. It represents the capability that a web service can be able to perform its regular function. Its measure indexes mainly contain successful rate.

e) Reputation. it means service users' assessment of web service. Its measure indexes mainly contain click rate, customer satisfaction and so on.

If the task proposed by service users can be done through m services' composition, then the service composition process is composed by m service classes, which are $S_1, S_2, ..., S_m$ and each service class contains j candidate services that is $S_{ij} (i = 1,2,...,m; j = 1,2,...,n)$.

Every candidate service includes above-mentioned quality attributes, namely,

$$Q_{ij} = \left(S_{ij}^{tim}, S_{ij}^{pri}, S_{ij}^{ava}, S_{ij}^{trel}, S_{ij}^{rep}\right)^T . \tag{3}$$

It can be denoted by

$$Q_{ij} = \left(S_{ij}^{1}, S_{ij}^{2}, S_{ij}^{3}, S_{ij}^{4}, S_{ij}^{5}\right)^T . \tag{4}$$

Then each service class S_i has a Qos matrix

$$Qos_i = (Q_{i1}, Q_{i2}, ..., Q_{in}) = \begin{pmatrix} S_{i1}^{1} & S_{i2}^{1} & \cdots & S_{in}^{1} \\ S_{i1}^{2} & S_{i2}^{2} & \cdots & S_{in}^{2} \\ S_{i1}^{3} & S_{i2}^{3} & \cdots & S_{in}^{3} \\ S_{i1}^{4} & S_{i2}^{4} & \cdots & S_{in}^{4} \\ S_{i1}^{5} & S_{i2}^{5} & \cdots & S_{in}^{5} \end{pmatrix} . \tag{5}$$

2.2.1 Standardization of Attributes

Each Qos attribute has its property: cost (the higher value, the lower quality), benefit (the higher value, the higher quality). Because of these attribute having different dimensional, we have to do the non-dimensional to dispel incommensurability coming with the different dimensional. At present, there are many methods to do the non-dimensional. Since proportion method uses the sum of raw data in the non-dimensional process and represents the result by the rate of raw data in the overall, it maintains the consistency of the original data. Therefore, we use proportion method to do the non-dimensional. The formula is as follows:

Benefit:

$$s_{ij}^{k*} = \frac{s_{ij}^{k}}{\sqrt{\sum_{j=1}^{n}(s_{ij}^{k})^2}} \quad (j = 1,2,\ldots,n; k = 1,\ldots,5) \tag{6}$$

Cost:

$$s_{ij}^{k*} = \frac{1}{s_{ij}^{k}} / \sqrt{\sum_{j=1}^{n}\left(\frac{1}{s_{ij}^{k}}\right)^2} \quad (j=1,2,\ldots,n; k=1,\ldots,5) . \tag{7}$$

After processing the data, we get the matrix of correction Qos_i*. Since users have different preferences on Qos attributes; we give every attribute a weight. Thus there is a weight set $W = \{w_1, w_2, w_3, w_4, w_5\}$, so we get weighted attribute matrix

$$Qo\tilde{s} = W * Qo\overset{*}{s} = (s_{ij}^{k\sim}) \quad (1 \le i \le m, 1 \le j \le n, 1 \le k \le 5) . \tag{8}$$

2.2.2 Attribute Evaluation

Since the utility function can well reflect the relationship between candidate service and target service, we use utility function to do attribute evaluation. By defining the ideal service and negative-ideal service in a service class, we can use the related distance to develop a utility function F_{ij}. The definitions are as follows.

Definition 5. Ideal Service:

$$s_i^{+} = \{s_{i1}^{+}, s_{i2}^{+}, s_{i3}^{+}, s_{i4}^{+}, s_{i5}^{+}\} \quad (1 \le i \le m) , \tag{9}$$

$s_{i1}^{+}, s_{i2}^{+}, s_{i3}^{+}, s_{i4}^{+}, s_{i5}^{+}$ is the maximum value of each attribute.

Definition 6. Negative-ideal Service:

$$s_i^{-} = \{s_{i1}^{-}, s_{i2}^{-}, s_{i3}^{-}, s_{i4}^{-}, s_{i5}^{-}\} \quad (1 \le i \le m) , \tag{10}$$

$s_{i1}^{-}, s_{i2}^{-}, s_{i3}^{-}, s_{i4}^{-}, s_{i5}^{-}$ is the minimum value of each attribute.

Definition 7. Ideal Distance:

$$u_{ij}^{+} = \sum_{k=1}^{5} w_k (s_{ij}^{k\sim} - s_{ik}^{+}) \quad (1 \le i \le m, 1 \le j \le n) . \tag{11}$$

It is the distance between target service and ideal point.

Definition 8. Negative-ideal Distance:

$$u_{ij}^{-}=\sum_{k=1}^{5} w_k \left(s_{ij}^{k\sim}-s_{ik}^{-}\right) \quad (1\le i\le m, 1\le j\le n) . \tag{12}$$

It is the distance between target service and negative-ideal point.

So we get utility function as follows:

Definition 9. Utility function:

$$F_{ij}=\frac{u_{ij}^{-}}{u_{ij}^{+}-u_{ij}^{-}} \qquad (1\le i\le m; 1\le j\le n) \cdot \tag{13}$$

It is used to show user's satisfaction with the target service.

2.2.3 Web Service Composition Model

On the basis of the additivity of the index (such as time, price, reputation) and the multipliability of the index (such as availability, reliability), we give a simplified model of Web service composition. It is very convenient for common Web service composition problem. So Web service composition problem can be described as optimization problem:

$$\left.\begin{array}{l}
\max \quad \sum_{i=1}^{m}\sum_{j\in S_i} F_{ij}x_{ij} \\
\sum_{i=1}^{m}\sum_{j\in S_i} S_{ij}^{1} * x_{ij} \le T_{\max} \\
\sum_{i=1}^{m}\sum_{j\in S_i} S_{ij}^{2} * x_{ij} \le P_{\max} \\
\prod_{i=1}^{m}\sum_{j\in S_i} S_{ij}^{3} * x_{ij} \ge Ava_{\min} \\
\prod_{i=1}^{m}\sum_{j\in S_i} S_{ij}^{4} * x_{ij} \ge \mathrm{Re}\, l_{\min} \\
\sum_{i=1}^{m}\sum_{j\in S_i} S_{ij}^{5} * x_{ij} \ge \mathrm{Re}\, p_{\min} \\
\sum_{j\in W_i} x_{ij} = 1 \\
x_{ij}\in\{0,1\} \qquad (i=1,2,\dots,m;\ j\in W_i)
\end{array}\right\}$$

In a service class W_i ,if candidate service W_{ij} is selected, then x_{ij} is 1,otherwise the result is 0; $T_{\max}$, $P_{\max}$, $Ava_{\min}$, $\mathrm{Re}\, l_{\min}$, $\mathrm{Re}\, p_{\min}$ denote respectively the service composition's maximum response time, maximum price, minimum availability requirement, minimum reliability requirement, minimum reputation requirement.

3 Model Solving with Aga

3.1 AGA

The problem of multi-objective optimization under certain constraint condition is NP-hard. As a kind of new heuristic optimization method, Adaptive Genetic Algorithms (AGA) has several characteristics such as flexible, efficient and robust. Also it has a strong ability to find the optimal solution in solving composition optimization problems. Therefore, we use AGA to get the solution, the steps are:

a) Use binary coding to get the population, initialize parameters including: Group size (SIZE), Number of generations (Gmax), Gap of generations (GAP);

b) Generate initial population: generate SIZE individuals randomly, check the satisfaction rate, form the initial population, in the given feasible region of variable;

c) Select and copy: calculate the fitness of individuals according to the fitness function, use roulette method to select operators and select SIZE times to replicate SIZE individuals.

d) Crossover and mutation: adaptive crossover and mutation probability are as follows:

$$p_c = \begin{cases} p_{c1} - \dfrac{(p_{c1} - p_{c2})(f' - \bar{f})}{f_{max} - \bar{f}}, f' \geq \bar{f} \\ p_{c1}, f' \geq \bar{f} \end{cases}$$

$$p_m = \begin{cases} p_{m1} - \dfrac{(p_{m1} - p_{m2})(f - \bar{f})}{f_{max} - \bar{f}}, f \geq \bar{f} \\ p_{m1}, f \geq \bar{f} \end{cases}$$

p_{c1}, p_{c2} are control parameters of crossover; p_{m1}, p_{m2} are control parameters of mutation; f_{max} is the largest fitness in the population; $\bar{f}$ is the average fitness in the population; f is the fitness of an individual to mutation; f' is the larger fitness of two individuals to cross.

If the fitness of an individual is equal to the largest fitness, the individual is retained; otherwise, calculate the crossover probability and do t crossover action, then calculate the mutation probability and do mutation action.

e) Determine the termination condition: judge whether the number of generations is Gmax. If not, go to step 3), otherwise output the value of each variable and the final function value.

3.2 Experiment and Analysis

Experiment adopts a tourism service composition scene. It is composed by three services performed in the sequence; tourist spots reservation service, airplane ticket reservation service, hotel reservation service. Each service class has several candidate services, the number of candidate services and Qos parameters are generated by random, then solve the problem by AGA. The purpose of this experiment is to verify the feasibility of this algorithm to find the global optimum of the model based on Qos.

In the case of each service class which has eight candidate services, we give the modified Qos matrix:

$$Qos_1 = \begin{bmatrix} 0.3486 & 0.3753 & 0.3411 & 0.3449 & 0.3281 \\ 0.3822 & 0.2949 & 0.3411 & 0.3406 & 0.3486 \\ 0.3426 & 0.3753 & 0.3625 & 0.3322 & 0.3179 \\ 0.3312 & 0.4128 & 0.3454 & 0.3449 & 0.3896 \\ 0.3680 & 0.3176 & 0.3710 & 0.3785 & 0.3691 \\ 0.3613 & 0.3753 & 0.3539 & 0.3575 & 0.3896 \\ 0.3549 & 0.3176 & 0.3625 & 0.3659 & 0.3589 \\ 0.3368 & 0.3440 & 0.3497 & 0.3617 & 0.3179 \end{bmatrix} \quad Qos_2 = \begin{bmatrix} 0.3717 & 0.2558 & 0.2438 & 0.2493 & 0.2770 \\ 0.3097 & 0.2164 & 0.2679 & 0.2581 & 0.2260 \\ 0.2859 & 0.3126 & 0.2348 & 0.2434 & 0.2551 \\ 0.3379 & 0.2814 & 0.2528 & 0.2522 & 0.2697 \\ 0.3717 & 0.2345 & 0.2468 & 0.2640 & 0.2187 \\ 0.2859 & 0.2558 & 0.2559 & 0.2552 & 0.2406 \\ 0.3379 & 0.2164 & 0.2619 & 0.2464 & 0.2624 \\ 0.3097 & 0.2814 & 0.2378 & 0.2581 & 0.23335 \end{bmatrix}$$

$$Qos_3 = \begin{bmatrix} 0.2127 & 0.2145 & 0.1909 & 0.2020 & 0.2207 \\ 0.2304 & 0.1815 & 0.2082 & 0.2091 & 0.1789 \\ 0.2212 & 0.1967 & 0.2033 & 0.1901 & 0.2207 \\ 0.2765 & 0.1475 & 0.1934 & 0.2186 & 0.1909 \\ 0.2514 & 0.1686 & 0.2033 & 0.2139 & 0.2088 \\ 0.2404 & 0.2360 & 0.1983 & 0.2067 & 0.2028 \\ 0.2304 & 0.2360 & 0.2057 & 0.2020 & 0.2028 \\ 0.2212 & 0.1311 & 0.2008 & 0.2139 & 0.19685 \end{bmatrix}$$

The operational results of F_{ij} are:

$$F_{ij} = \begin{bmatrix} 3.2858 & 3.5423 & 3.1302 & 3.8690 & 3.4447 & 3.8130 & 3.3117 & 3.1093 \\ 5.2776 & 3.7310 & 4.1626 & 4.7126 & 4.4191 & 3.7948 & 4.4240 & 4.0859 \\ 3.6516 & 3.1058 & 3.4002 & 3.0197 & 3.1281 & 3.8134 & 3.6973 & 2.5505 \end{bmatrix}$$

The optimum solution under constraint condition is $W_{16} \rightarrow W_{21} \rightarrow W_{36}$.Its number of generation is as shown in Figure 1:

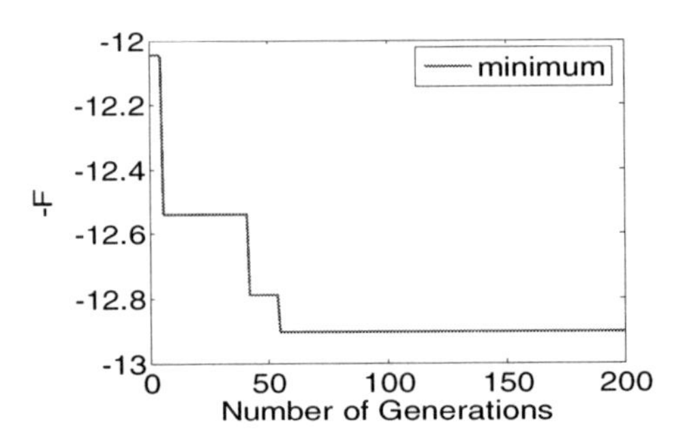

Fig. 1.

In reality, there are always a lot of candidate services, the change of running time when the number of candidate services increase is shown in Figure 2:

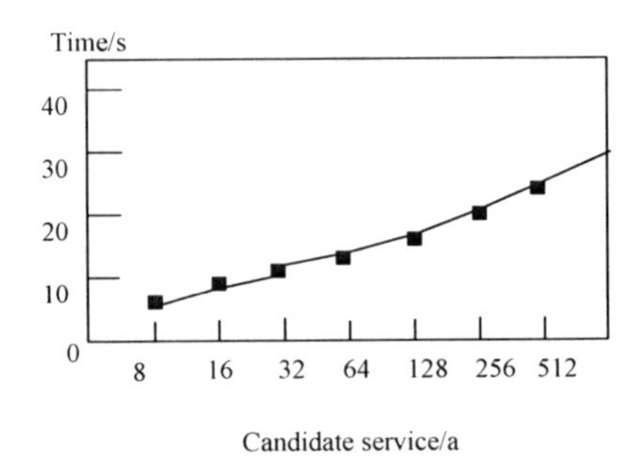

Fig. 2.

We can find from the figure above that: running time has a positive correlation with the number of candidate services. With the increase in the number of candidate services, running time increases linearly, and the growth rate is flat.

Although this experiment is still in the prototype stage, but its feasibility has been shown.

4 Conclusion

This article aims to solve the problem of service selection in the process of service composition. From the start of meeting the users' quality requirements, this paper propose a service selection model based on utility of weighted Qos attribute, and solve it by using AGA. At last, we show the feasibility of the model and algorithm by an experiment. Futher study should be to consider the internal structure of the service composition.

Acknowledgments. The authors gratefully acknowledge the support of NSFC (grant NO.70871032), and the National High Technology Research and Development Program of China (grant NO.2009AA043403).

References

1. Mao, Y., Le, J.: Research on genetic algorithm based web service composition optimization. Computer Applications and Software 25, 199–201 (2008)
2. Cai, M., Li, M., Zhou, J.: Multi-choice Web services composition based on multi-object genetic algorithm. Computer Engineering and Applications 46, 202–205 (2010)
3. Fan, X., Jiang, C., Fang, X., Ding, Z.: Dynamic Web service Selection Based on Discrete Particle Swarm Optimization. Journal of Computer Research and Development 47, 147–156 (2010)
4. Li, J., Xia, J., Tang, W., Zeng, J., Wang, X., Wu, L.: Survey on Web services selection algorithms based on Qos. Application Research of Computers 27(10), 3622–3627 (2010)
5. Li, J., Liu, Z., Liu, Z., Han, L.: Research on multi-constraint Web service composition. Computer Engineering and Application 26, 110–114 (2010)
6. Tao, C., Feng, Z.: Qos-aware Web service Composition Based on Probabilistic Approach. Journal of Tianjin University 43, 860–865 (2010)
7. Yun, B., Yan, J., Liu, M.: Method optimize Web service composition based on Bayes trust model. Computer Integrated Manufacturing Systems 16, 1103–1110 (2010)
8. Cao, H., Nie, G., Chen, D.: Web services composition method based on functional attribute semantics. Journal of Computer Applications 30, 2344–2347 (2010)
9. Yin, K., Zhou, B., Zhang, S., Xu, B., Chen, Y., Jiang, D.: QoS-based bottom-up service replacement for Web service composition. Journal of Zhejiang University 44 (2010)
10. Li, S., Zuo, M., Feng, M.: Web services Composition Based on Ontology and QoS. Computer & Digital Engineering 37 (2009)
11. Qiu, L., Shi, Z., Lin, F., Chang, L.: Agent-Based Automatic Composition of Semantic Web Services. Journal of Computer Research and Development 44, 643–650 (2007)

A New Architecture of Geospatial Information Service for Cloud Computing

Liu Rong*, Huang Ruiyang, and Xie Geng

Zhengzhou Institute of Surveying and Mapping, Zhengzhou, China
xq050105@sohu.com, chxycy@126.com

Abstract. Cloud computing is the extension of parallel computing technology that provide robust storage and computing capacity for geospatial information services. The author introduces the classifications, architecture and characteristic of cloud computing, analyzes the relationship among the SaaS model, traditional service model and SOA, proposes a new geospatial information system architecture that fits to the 'software-and-service' model of cloud computing, at last studies the key technologies for the architecture.

Keywords: Cloud Computing; SaaS; Geospatial Information Service; Software and Service.

Nowadays, Cloud Computing is one of hot words in the IT research field. Since first proposed in 2007, Cloud Computing has been catching more and more eyes and developing rapidly. Quite different from Grid Computing., the development of Cloud Computing technology is not propelled by science institutes but by commercial giants such as Google, Amazon, IBM and Microsoft. In these years after 2007, Cloud Computing has already walked out from laboratories to practical application and Some Cloud Computing platforms, such as Google API, Amazon EC2, Windows Azure, have been built and come into actions. The design theory, architecture and realize technology of Cloud Computing shed a light to Geospatial Information Services.

1 Cloud Computing Platform

Cloud Computing is the development of Parallel Computing, Distributed Computing and Grid Computing. Although the goal of Cloud Computing still lies in sharing resources, However, Shared resources is not restricted within software and data, but is so extensive that including computing resources, storage resources, knowledge resources and so on. With the help of commercial model of Cloud Computing platform, service providers can get benefits from sharing scalable resources.

* Supported by 863 project(2009AA12Z228).

G. Zhiguo et al. (Eds.): WISM 2011, CCIS 238, pp. 426–432, 2011.

1.1 Classification of Cloud Computing

There is not a legalized definition for Cloud Computing now, and the classification of Cloud Computing platform is usually in term of offering service types and service modes.According to service types, Cloud Computing Platform can involve three components: infrastructure cloud, platform cloud and application cloud. According to service mode, Cloud Computing Platform can involve public cloud, private cloud and hybrid cloud.

1.2 Cloud Architecture

According to service types offered, a typical cloud architecture is usually a 3-tier architecture and involves three types of service:

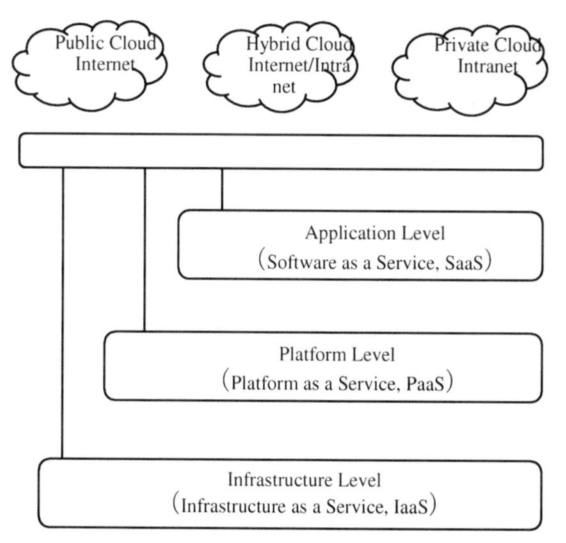

Fig. 1. Cloud Computing Architecture

- Infrastructure

Cloud infrastructure services, also known as Infrastructure as a Service (IaaS), deliver computer infrastructure as a service based on internet protocols, and it involves provisioning of dynamically scalable and often virtualized resources.

- Platform

Cloud Platform Services, also known as Platform as a Service (PaaS), deliver a computing platform as a service and supply end-users, especially software developer, with "cloud platform middleware", such as databases, application servers, message handlings.

- Application

Cloud application Services, also known as Software as a Service (SaaS), deliver software as a service over the internet, It can eliminate the need to install and run the application on the customer's own computers and simplifying maintenance and support as long as Customers can access to applications via the web.

1.3 Key Characteristics of Cloud Platform

As a new distributed computing platform, the characteristics of Cloud Computing can be summed up as follows:

1) Hardware and software are resources in the cloud computing platform and seen as service.

2) The whole resources and be highly scalable that can dynamically extended and deployed according to the requirement of users.
3) Resources in the cloud computing platform are separated physically but united logically and can be seen a whole.
4) Pricing on a utility computing basis is fine-grained with usage-based options, so the cost is always greatly reduces.

Cloud computing is a model that enables convenient, on-demand network access to a shared pool of configurable computing resources, which seems like public utilities such as electricity, gas, and water.

2 App Cloud—SaaS

We know that run geographic information service in the cloud computing environment is developing App cloud service, which is called SaaS. The SaaS means software as a service. When a consumer uses the software, in fact, he is consuming the service, and the provider of software is the producer and provider of service.

2.1 SaaS and Traditional Service Mode

SaaS is a new software service mode. There are many differences between SaaS and traditional service mode, which are in table 1.

Table 1. SaaS and traditional service mode

	Traditional service mode	SaaS
Develop cost	High	low
equipment	Consumer purchase	Consumer need not purchase
Connect cost	high	low
Response time	A lot of time	Little time
maintenance	difficult	easy
customization	unflexible	flexible

2.2 SaaS and SOA

The core of SOA(Service Oriented Architecture) and SaaS are web service. SOA aimed to software design and structure mode. Broadly speaking, any technique that provide service can realize SOA, for example, Web Service, RMI, CORBA, JMS, MQ, JSP and SERVLET. On the other hand, SaaS emphasize software how to provide applications. It is the mode that software as a service that can access through Internet. In general, SOA is a method and SaaS is transfer model. It is very clear that there is little difference between SaaS and SOA. They can be called "SOA gets SaaSy", which means utilizing SaaS to realize SOA.

2.3 SaaS and Cloud Computing

There are some difference between SaaS and cloud computing. SaaS appeared in 2003, which is earlier than cloud computing. But cloud computing bring strength to SaaS. SaaS providers can solve many problems through cloud computing, for example, computing capability, storage, and bandwidth and so on. Simultaneity, the scope of cloud computing is widened to software, and have more consumers including enterprises and common consumers.

3 Structure Design of Geographic Information Service Platform

3.1 Platform Description

Geographic information service platform is a geographic information application service (SaaS) based on cloud compute platform (PaaS). To satisfy the application requirements on the cloud computation platform, the platform should have two feathers:

1) Service architecture. The architecture of platform should follow SOA, and its elementary functional components adopt two models: .Net and Web Service. .Net component is used for the tightly-coupled integration inside the platforms and the formulation of desktop edition. And web service component is used for the loose-coupled integration of service form and the formulation of server edition.
2) Service network. It decompounds traditional GIS to loose GIS, form service network by much deploy of platform or by mix deploy of SOA platform, actualize virtualization management to the service resource in network, actualize "controllable state" to the resource, and compose geographic information service grid.

3.2 Architecture

To consider the non-state characteristic and the implementation efficiency of Web Service, and ensure the efficiency of geographic information service, platform design adopts "S+S"(Soft+ Service) structure (Fig. 2).

The platform data consists of three parts mainly:

1) Massive spatial data or non-spatial data in data server cluster. It can implement data's comical parallel access, which is the main storage mode of spatial data. The platform implements parallel access to dispersive database cluster by GFS technology.
2) Spatial data document. It is the main storage and access mode of spatial data in traditional GIS.
3) Spatial data service. It mainly includes spatial data services that accord with OGC standard. Service can be provided by exterior commercial data server as well as by the platform via issuing all the data that can be accessed.

The platform function includes mainly four parts:

1) Platform self-provided function. Such as basic graphic display, graphic operate, statistical inquiry, spatial analysis, plot, output of printing, which can be implemented by the tightly-coupled integration of .Net component.
2) Other Platform functions. Embedding plug-ins from other platforms into the system via secondary-development.
3) OGC Web processing service. WPS is a more and more important web processing service standard, which can be the expansion of system function.
4) Nodes and service management. This function mainly refers to the management of virtual organization (adding nodes, deleting nodes, grouping nodes), which packing the disparate nodes into relative-tight-coupled service cluster to provide pertinent service. The service management mainly refers to the inserting, invoking and compounding of service.

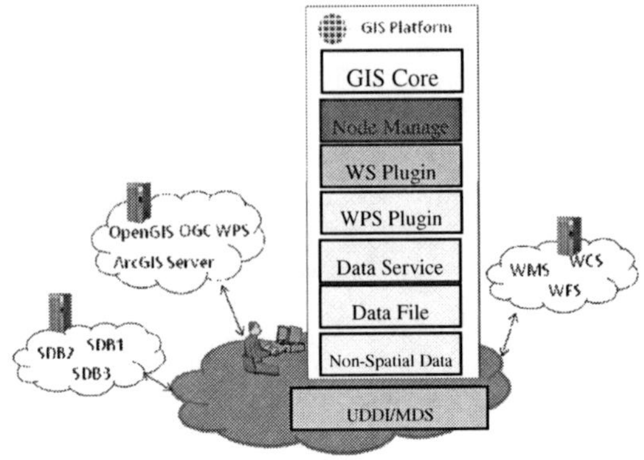

Fig. 2. Architecture of GIS for Cloud Computing

Registry is the key part that manages service discovery, validation, registering and grouping; it's implemented mainly based on UDDI or MDS standard. Furthermore, service accounting system, authentication system and some other systems are also implemented by cloud computation platform.

4 Key Technology Points of the Geographic Information Service Platform

4.1 Open Architecture

GIS based on cloud computing platform is not only using software no longer, but cooperates with net. Traditional net model made by a database server, a net server and innumerable clients would become a complex service net made by innumerable server, server group and innumerable clients. GIS platform might be server or client which depended on the role of service progress. It is a high demand on the open degree and standardization of GIS software. For more effective and open geographic information service, GIS software is divided into kernel components and service components. Kernel components use tightly coupled structure.

4.2 Dynamic Evolution Technology of the Platform

Evolution is the basic property of the software. Especially for cloud computing software, it needs to deal with the trouble of any node at any moment. While a part of components does not work normally or meet need of users, it needs to find out the common components at other node into self system, which become a part of the software. So that it is a core problem that it should insure the system work usefully at steady, robust state and providing 24hs service. The technology should pay attention to the evolution model, service similar match, service transmit and so on.

4.3 Distributed Files Saving Technology

The files of the geographic information service platform might be divided into more than 100M files and less data quantity and mass little files such as label files. There are different strategies for them. For larger files distributed files system(HDFS) uses open source cloud computing platform provided by Hadoop, which uses distributed saving, multi-back copy technology architecture by which larger files are divided into block that location different computer. For less files, the technology needs to change HDFS, relationship between files and saving.

4.4 Distributed Computing Technology

Similar to the theory of the distributed files saving, distributed computing technology (Map Reduce) of cloud computing divides into lots of little block from an application. Each work cell deals with access data copy, which is named Map. And the computing results are gathered the final results to caller, which is named Reduce. This is a technology which was promoted by Google firstly, and also realized by Hadoop. For mass data and complex application, distributed computing technology based on files download is necessary.

5 Discuss and Expectation

The development of GIS and computer technology is tightly related, and research on distributed technology such as cloud computing, grid computing and so on supports GIS. Compared with net computing, cloud computing is recognized application version of grid computing. No matter what, the target is providing a robust distributed computing environment. ESRI promotes the style ArcGIS 10 which has been released now. So the effect has formed primarily. But cloud computing is still in development time, the standard and security still need to be solved and GIS would be changed with it.

References

1. LiuPeng: Cloud Computing. Publishing House of Electronics Industry, Beijing (2010)
2. Cloud Computing Group, Virtualize and Cloud Computing. Publishing House of Electronics Industry, Beijing (January 2010)

3. Wang, L., JinagWei, ZhouLa.: SaaS: Update for Software Service Providing Mode. World Communiction (July 2009)
4. Ye, w.: The Software Revolution of Internet Times- SaaS Architecture Design, vol. 33. Publishing House of Electronics Industry, Beijing (2009)
5. Yang, F.-q.: Research on Development of Software Engineering. Journal of Software 16(1), 1–7 (2005)
6. Wang, J.-y., Cheng, Y., Wu, M.-g., Sun, Q.-h.: Development and Developing of Geographic Information System. Journal of Geomatics Science and Technology 25(4), 235–240 (2008)
7. Wang, J.: Development of geographic information system and developing geographic information system. Engineering Science 11(1), 10–16 (2008)
8. Wu, s.: Research and Practice on Distributed GIS Component Platform. Institute of Surveying and Mapping, Information Engineering University (2001)
9. Bi, J.-t.: Research on Web-oriented Sharing Geospatial Information Service. Chinese Academy of Sciences (2005)
10. Wu, G.-h.: Studies and Implementations of Distributed Geographic Information Services. Institute of Surveying and Mapping, Information Engineering University (2006)
11. Di, L., Chen, A., Yang, W., et al.: The Development of a Geospatial Data Grid by Integrating OGC Wev Services With Globus-based Grid Technology. Concurrency And Computation: Practice and Experience. Pract.Esper. (2008)
12. OpenGIS Consortium, Open GIS Web Services Architecture (Version0.3) (2003), http://www.OpenGIS.org/docs/03-025.pdf

The Personalized E-Learning System Based on Portal Technology

Xu Cui and Shaotao Zhang

Shandong Urban Construction Vocational College, China
Shandong Guosheng Investment & Guaranty company, China
cuixu20021981@163.com, zst1981@sohu.com

Abstract. The paper proposed an intelligent Personalized E-Learning System as a way to make the E-learning more personalized. Portal technology serves an essential role to the development of the personalized learning, and offers convenient tools for users. Although most E-learning system offer complete learning information, they simply offer functions of view and search and do not provide a personalized learning environment. This study aims to develop a Personalized E-Learning System Based on Portal Technology that unlike the traditional E-learning system use a number of intelligent methods, including information extraction, information retrieval, and some heuristic algorithm and provide more meaningful and personalized information to users.

Keywords: Portal technology, Information extraction, Information retrieval, Personalized E-Learning.

1 Introduction

The portal is an environment through which a user can access Web-based information and tools from a single Internet location. The portal is designed to present only the information and tools that each user needs, without the clutter of information and tools he/she doesn't use. The portal is not just another homepage — it is a personalized view of Web-based information. Depending on the user, the portal will include access to institutional data, personal data, productivity tools, and other information of interest [1].

Nowadays the commonly accepted way is to divide web portal into two categories which are "Horizontal Portal" and "Vertical Portal" [2]. To be brief, a horizontal portal focuses on a breadth-oriented integration of web content information while a vertical portal emphasizes depth-oriented integration. That is to say, a vertical portal focuses on managing certain aspect content such as education, industry and so on. Based on the advantage of the Vertical Portal and its important property of configuring and personalizing the portal, we mainly use Vertical Portal technology in E-learning to realize the personalization.

This article will give an overview of the Personalized Learning System Based on Portal Technology. Section 2 is about the disadvantages of current E-learning and the solving method. Section 3 gives an introduction of the technology related to this article. Section 4 describes how to design the system of the personalized learning

G. Zhiguo et al. (Eds.): WISM 2011, CCIS 238, pp. 433–439, 2011.

based on portal technology. The application of the system has been described in section 5. Section 6 concludes the paper.

2 The Problem Exiting in Current E-Learning and Solving Method

E-learning is a new education form which is different from the traditional education form with the development of the Internet and the multi-media technology [3]. Learners do not be limited by time and space and can finish their learning plans at any time and any place. Currently, E-learning has provided an ideal environment for users who are fond of learning, but they can not provide a personalized learning environment which has much relation to the learning results. This has been a burning problem in learning process because without it most learners' latent capabilities can not be stimulated.

On the other hand, in most system exited the center of the learning is the site itself which is not convenient to users. Ed Lightfoot and Weldon Ihrig indicate in their article "The Next-Generation Infrastructure" [4] that "higher education needs a next-generation infrastructure that will allow our institution to be user centered, to establish and maintain life-long relationships with individuals, and to provide personalized, secure and seamless connections with all constituents". In order to reach the personalized learning, organizations has turned to the use of portal technology.

Referring to the application that has been done in enterprise information portal such as "the Jinan talent Internet", we also adopt portal technology in E-learning. In "the Jinan talent Internet", visitors can select favorite color, design and layout of pages according to one's fondness. By doing so most of them can be absorbed by the portal interface, then the visit efficiency can be improved. The same to E-learning, learners can personalize their interface and improved the learning efficiency. What's more in the new system, there will be some other information that related to visitors can appear in the reticulation automatically accompany with the passion of users which can improve visit efficiency.

3 Related Work

To make an E-learning system more personalized, it is necessary to integrate a number of intelligent methods. This section will explain the methodology and conception for developing a personalized E-learning system in this study which are information extraction [5], information retrieval [6] and cluster algorithm [7].

3.1 Information Extraction

In order to make the E-learning system capable of meeting different users, it is necessary to collect information related to users. In a web page, there are only two parts information that related to users which are personalized configure information and visited information. These two parts information are stored in structural or nonstructural data in an out-of-order way, so extracting useful data is a pivotal task must be solved first. There has been much kind of way to extract data and change them into a consistent and strict form such as [8], [9]. We can get uniform format data after using these technologies.

3.2 Information Retrieval

Once we can extract useful data from personalized configure information and visited information and change them into a consistent and strict form, next we should match the extracted information with all web pages of the system. By calculating the similarity of all the web pages and the extracted information, we can select web pages related to the topics users interested. In calculating the similarity we can refer to the information retrieval method [10].

3.3 Cluster Algorithm

In this paper, we adopt cluster algorithm to help E-learning system integrate web page contents. Since cluster algorithm has the features of fast problem-solving ability and convenient using property, it is proper using it here. In the framework of this study, cluster algorithm is a computational module and can be replaced by any other suited algorithm.

4 Personalized Learning System

This study developed a system called Personalized E-Learning System Based on Portal Technology, which consist several methods. They are information extraction, information retrieval and cluster algorithm. In this section, we will discuss the system's design and implementation.

4.1 The Frame of the System

Unlike the traditional E-learning system, in this paper we propose a new frame related to portal technology. At present, most E-learning systems only have the function of view or search, but they don not have personalized learning environment. From the experiences, if a website can support different face to users according to users' fondness and can support information users interest, then it can absorb more users and can raised the learning efficiency. The different between traditional learning system and our personalized e-learning system is show in Fig 1. Fig 1 illustrates the process of the system service to users and the Personalized E-Learning System Based on Portal Technology can support a personalized learning environment which can center by users. Define abbreviations and acronyms the first time they are used in the text, even after they have been defined in the abstract. Abbreviations such as IEEE, SI, MKS, CGS, sc, dc, and rms do not have to be defined. Do not use abbreviations in the title or heads unless they are unavoidable.

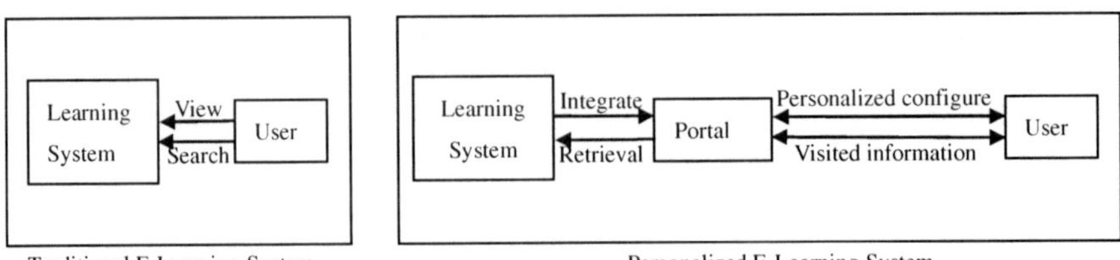

Fig. 1. Comparison of the Traditional E-Learning System and the personalized E-Learning System

4.2 The Design of the System

In the framework of the system we investigate that the Portal plays a middleware role, through which the function of the system has been personalized. At the same time, to reach this goal, the proposed system combines some of the conception of information extraction, information retrieval and cluster algorithm etc. In this section we will illustrate the procedure of the proposed system.

Procedure 1: Fetch the personalized configure information and visited information, parse them in order to extract useful information which can be saved by uniform format and we call these information "extracted web information (EWI)". The personalized configure information and visited information should be managed separately, because the former is used to design the personalized faces and the latter is used to retrieving the related web pages. The system can filter out the unusable content (like picture information) through portal in addition to retrieval information.

Procedure 2: The main object of this procedure is to retrieve related web pages by referring to EWI. Matching technique is used to finish retrieving the suited web pages. But the number of the web page is large in a system, so we should condense the number of the pages first. Because a system is made up of many web pages and all the web pages can form a tree and the root of the tree is the home page. When retrieving web pages, the elements that effect retrieving efficiency are the number of the web pages. So to condense the number of the web page has been an important problem. Fig 2 introduces a method how to solve the problem has been proposed.

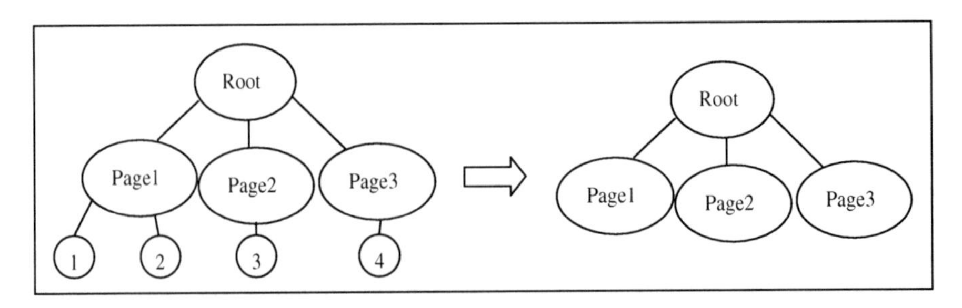

Fig. 2. Condense the Number of the Web Pages

In fig 2, we condense the number of the web page tree by deleting all the leaves. Why we do it that? On one hand, in a web page tree, the leaves are pages without super link which support information related to EWI. And its super link has been defined by its father or grandfather web page. On the other hand, the EWI only includes the super link information which may be related to the main information collected from the web pages. So by using the method figure 2 proposed can raise the retrieving efficiency and can not leave out information.

Procedure 3: When retrieving web pages, the matching criterion is to compute the similarity of web pages and EWI and the order of retrieving web page tree start from the tree leave. That is because if in a layer of the tree, all super links are selected, then we can stop retrieving the rest of the tree leaves and we can get proper results. After doing that, all matched pages are related to some EWI; all pages have been divided into several parts according to the EWI after matching. We have finished the automatically cluster at the same time.

Because the extracted web information is super link information, we adopt double-gene inverted-index technology [11] [12] to extract the feature subsets and store them. The following explains the ways of calculating the degree of similarity.

$$W_{ab} = \sum_{i=1}^{n} X(a_i, b) \tag{1}$$

In the formula, " a " is a feature of the feature subsets; " b " is a page related to feature " a "; " n " is the number of features in feature subsets; $X(a_i, b)$ is the similarity of the feature and the web page. And all $X(a_i, b)$ sums up can get W_{ab}.

$$X(a_i, b) = sw(a_i) \tag{2}$$

This is the formula of the $X(a_i, b)$, $sw(a_i)$ is the number of a_i appear in b. Select all from W_{ab} where W_{ab} >min which can be given by real problem as the results. From (1) and (2) we can get all the web pages may be contribute to the final result integration.

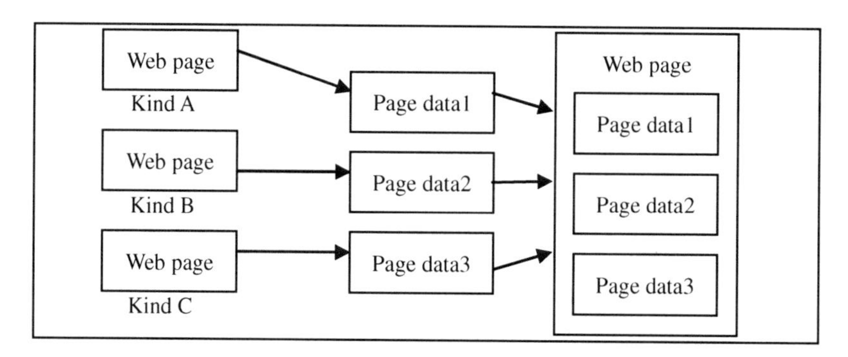

Fig. 3. Integrate Cluster Kind

In this part, we deal with every web page according to its extracted web information, and then integrate the extracted information. Making it is easy to show in one web page.

Procedure 4: Finally, in this procedure, the system produces an XML file and an HTML file [14], including java script procedures in the HTML file which is responsible for presenting the XML file, as well as the XML file preserved merger information which was obtained from procedure 1 to procedure 4.

5 Application

An example is taken to show the application of the Personalized E-Learning System Based on Portal Technology. In the example, we use the Personalized E-Learning System Based on Portal Technology developed by the study to experiment different personalized configure information and visited information by different users. First, the system collect the different personalized configure information and visited information and extract useful information from them, then select proper web pages that suitable to the user, at last integrate the related information that may be interested by users.

In using the system, we design a counter and a time table to measure the use of the system. Comparing two kinds system traditional and personalized, the number of click count has doubled and users tested by the same problems, the accuracy has doubled nearly. It is clear that the personalized system can meet users better. The comparison result as follow in Table1.

The system proposed in this paper still exit disadvantages such as something proposed by system to users but can not meet users completely. In the further work, we will modify the personalized system to make it more power and efficiency.

Table 1. The comparison of the two system

	Click counts(all super links)	Accuracy count(a number of problems)
Traditional system(1 month)	**5019**	**321**
Personalized system(1 month)	**11532**	**534**

6 Conclusions

The paper proposed a framework and the rudiments of a Personalized E-Learning System Based on Portal Technology. Unlike the traditional system it can raise the click counts and accuracy counts, and can reach the personalized request. This design has been divided into five procedures that use information extraction, information retrieval, and some heuristic algorithm. The system can make users interesting information that the system has supported and can raise visit efficiency.

References

1. http://www.wsu.edu/portal-project/background.html
2. Ed Lightfoot and Weldon Ihrig. The Next-Generation Infrastructure. EDUCAUSE Review (November 2002)
3. Buttler, D., Liu, L., Pu, C.: A Fully Automated Object Extraction System for the World Wide Web. In: 21st International Conference on Distributed Computing Systems, pp. 361–370 (April 2001)
4. Chang, C.-H., Liu, S.-C., Wu, Y.-C.: Applying pattern mining to Web information extraction. In: 5th Pacific Asia Conference on Knowledge Discovery and Data Mining (PAKDD 2000), pp. 4–6 (April 2001)
5. Baeze-Yates, R., Ribeiro-Neto, B.A.: And Artificial System. MIT Press, Boston (1922)
6. Moffat, A., Zobel, J.: Compression and Fast Indexing for Multi-Gigabit Text Databases. Australian Compute 26(1), 19 (1994)
7. Hao, X.W., Ma, J.: A platform for distance education. In: The 3th Asia and Pacific Conference on Web Computing, pp. 117–123. Xian Kluwer Academic Publisher, Dordrecht (2000)
8. Tsai, C.-W., Ho, J.-H., Liang, T.-W., Yang, C.-S.: An intelligent Web portal system for Web information region integration, pp. 3878–3883 (October 2005)
9. Adomavicius, G., Tuzhilin, A.: User profiling in personalization applications through rule discovery and validation. In: Proceeding of the 5th International Conference on Data Mining and Knowledge Discovery, pp. 377–381 (1999)
10. Guarino, N.: Formal Ontology and Information System. In: Proc. of FIOS 1998 Formal Ontology in Information Systems,Trento, Italy (June 1998)
11. Christ, M., Krishnan, R., Nagin, D., Gunther, O.: Measuring Web portal utilization. In: Proceedings of the 35th Annual Hawaii International Conference System Sciences, HICSS, pp. 2647–2653 (2002)
12. Guarion Semantic Matching:Formal Ontological Distinctions for Information Organization,Extraction and Integration. In: Pazienza, M.T. (ed.) SCIE 1997. LNCS, vol. 1299, pp. 139–170. Springer, Heidelberg (1997)
13. Wu, Y.-H., Chen, Y.-C., Chen, A.L.P.: Enabling personalized recommendation on the Web based on user interests and behaviors. In: Proceedings. Eleventh International Workshop Research Issues in Data Engineering, pp. 17–24 (2001)
14. Kotsakis, E., Bohm, K.: XML Schema Directory: a data structure for XML data processing. Web Information Systems Engineering (2000); Smith, T.F., Waterman, M.S.: Identification of Common Molecular Subsequences. J. Mol. Biol. 147, 195–197 (1981)

A Novel Approach to Cluster Web Traversal Patterns Based on Edit Distance

Xiaoqiu Tan and Miaojun Xu

School of mathematics, physics and information
Zhejiang Ocean University, Zhoushan China
{tanxq,girl_xu}@zjou.edu.cn

Abstract. Edit distance, as a similarity measure between user traversal patterns, satisfies the need of varying-length of user traversal sequences very well because it can be computed between different-length symbol strings which needs lower time and storage expense. Moreover, web topology is skillfully used to compute the relationship between pages which is used as a measure of cost of an edit operation. Finally, two-threshold sequential clustering method (TTSCM) is used to cluster user traversal patterns avoiding specifying the number of cluster in advance, and reducing the dependency between the clustering results and the clustering order of traversal patterns. Experimental results test and verify the effectiveness and flexibility of our proposed methods.

Keywords: Edit distance, Clustering, Traversal Pattern, Web Topology.

1 Introduction

User traversal behaviors are influenced by their characters, interests and performances. It is crucial to mine traversal patterns from traversal history database. It is helpful for Webmasters to supply better services, such as personalized recommendation and improvement of web topology structure.

Clustering is a practical choice to analyze user common behaviors, so clustering has been becoming a hot research, and many related literatures are published. Y. Fu etc. al in [1] organized all sessions in a hierarchical structure in a partial order, and a hierarchical clustering algorithm, BIRCH, was proposed to analyze useful traversal patterns based on conception generalization. In [2], each session is treated as a Markov chain, and *Kullback-Liebler* distance between Markov chains is used to measure the dissimilarity of user traversal behaviors, and then Access Interest Perception is put forward to reflect user further performance to cluster user traversal patterns. In [3], artificial immune system is applied to find common interests and performance to clustering users. In [4], a weighted bipartite graph is introduced to improve Hamming distance calculation so that the similarity between patterns can be described accurately to cluster user traversal patterns well.

Even if most pervious researches brought in good clustering effect, they possessed one or more drawbacks as follow:

G. Zhiguo et al. (Eds.): WISM 2011, CCIS 238, pp. 440–447, 2011.

(1)The representation of traversal pattern is required to be uniform length, such as [1] and [2]. However, in practice, various user sessions usually comprise of different-length pages. A vector or a matrix is usually used to represent user traversal patterns of high time and storage expense.

(2)The accessed orders of pages are not been take into consideration such as [4].

(3)Sparse problem. Sparse problem is a recurrent issue in pattern recognition, such as [1], [2] and [3]. A session usually consists of a small fraction of pages. A number of elements are meaningless except for causing storage pressure and extra time consuming if a vector or matrix is used to represent a traversal pattern.

Considering these problems, we introduce edit distance as the measure of dissimilarity among user traversal patterns creatively because it can be computed between different-length patterns. Moreover, the page relationship referring to web topology is used as costs of edit operations revealing further similarity between pages. Finally, a two-threshold sequential clustering method (TTSCM) is chosen to cluster traversal patterns avoiding specifying the number of cluster in advance.

2 Relationships between Pages Referring to Web Topology

Webmasters always make all effort to arrange the web structure for users to navigate web site conveniently. The most common web topology is organized hierarchically. Every hierarchy comprises of various themes which contain several sub themes or pages. Sub themes or pages under the same themes are related conceptually, and granularities of themes in the higher hierarchy are bigger than in lower hierarchy, namely more generalized.

In Fig. 1, homepage, /, contains two themes *news*, and *courses*, and there are two themes, *Language* and *Computer* in *courses*. In comparison, the granularities of *news* and *courses* is bigger than the granularities of *Language* and *Computer*, i.e. the former two themes are more generalized.

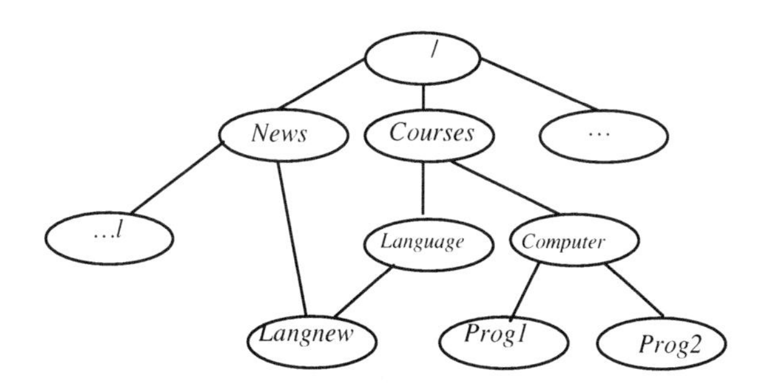

Fig. 1. An illustration of web topology

The conceptual relationship between pages reflects user common interests to some extent, i.e. if two users access some pages which are related intensely, it can be implied that they possess similar interests. In previous works, only the same pages are contributed to the similarity. The relationships between pages can further reveal user behaviors. Therefore, a novel method is proposed to compute the relationship between pages referring to web topology.

Definition 1 (Web topology). A web site is a directed graph G=(V,E), where $V=\{url_1,url_2,\ldots url_n\}$ is a vertex set, and each vertex corresponds to a page, $E=\{e \mid 1\le i,j\le n\}$ is a directed edge set, and $e=<url_i,url_j>$ indicates there is a hyperlink from url_i to url_j.

Definition 2 (Page Level). Assuming that there is no less than one path from homepage to any other page, and the level of homepage is assigned 0. If page *A* points to *B* with a hyperlink, then the level of *B* is increment of the level of *A*.

Definition 3 (Page Distance). The sum of the edges of the paths from two pages to their shortest common ancestor is called the distance of the two pages.

On account of their different levels the relationships of page pairs are maybe different even if the distance between various page pairs are equal. In order to reflect this fact, normalized page distance is computed as a measure of the relationship between pages. Detail algorithm is depicted as follows:

```
Algorithm 1: Computing Distance between Two Pages
int Page_distance(a,b)
  { if a=b then   return 0;
  while (true)
  { P= all parents of a;
  for each c belongs to P do
    {if c is an ancestor of b then
         append c into Candi;
      append all parents of c into P1; }
    if Candi is not empty then
      break;
    P=P1; }
   k = min_p{p.level | p ∈ Candi} //return a normalized page-dis
return 1-2*p.level/(a.level+b.level);}
```

Algorithm 1 searches the nearest common ancestor of *a* and *b*. Since there is one path from homepage to others at least, one common ancestor exists between any page pair at least, which ensures that algorithm 1 will stop after finite steps.

3 Edit Distance between Traversal Patterns

3.1 Edit Distance

The conception of edit distance is pioneered by Levenstein in [5], which is used to represent the lowest edit cost to transform one symbol string into another symbol string. Bellman in [6] firstly advanced a dynamic programming to compute edit distance. Marzal and Vidal [7] proposed normalized edit distance (NED) to avoid the influence from the different length of patterns. Edit distance is widely used various pattern recognition, such as handwritten text recognition [8], biological sequence analysis [9]. However, edit distance is used to analyze user traversal patterns are seldom published, so applying edit distance as a measure of dissimilarity between traversal patterns is creative in some sense.

Definition 4 (Edit distance). Suppose that *A,B* are two symbol string, the edit distance between *A* and *B* referred to the lowest cost of all edit operation sequences including *Substitution*, *Insertion* and *Deletion*, to transform *A* into *B*, denoted by

$$D(A,B) = \min_j[Substitution(j) + Insertion(j) + Deletion(j)] \tag{1}$$

where j comprises of all possible path from A to B, obviously $D(A,B) = D(B,A)$.

3.2 Edit Distance between User Traversal Patterns

Suppose that $S_1 = \{t_1, t_2, \ldots, t_d\}$, $S_2 = \{r_1, r_2, \ldots, r_k\} \in TDB$ (Traversal Database) are treated as two symbol strings, each page in S_1, S_2 as a symbol. A dynamic programming algorithm is used to compute edit distance between S_1 and S_2. Detailed algorithm is as follows:

```
Algorithm 2: Edit Distance of two Traversal Patterns
double Edit_dis(S1,S2)
{    d=size(S1);k=size(S2);
          if d=0 and k=0 then
             return 0;
     if d=0 and k>0 then
            return Edit_dis(S1,S2)+1;
  if k=0  and d>0
           return Edit_dis(S1,S2,d-1,k)+1;    //try
diagonal transforming
c1=Edit_dis(S1,S2,d-1,k-1)+Relationship[S1(d)][S2(k)];
//try horizontal transforming
c2=Edit_dis(S1,S2,d,k-1)+1;// try vertical transforming
c3=Edit_dis(S1,S2,d-1,k)+1;
return Min(c1,c2,c3);}
```

where *RelationShip*, a distance matrix of pages, is erected by calling *Algorithm 1* before executing *Algorithm 2*. The time complexity and space complexity are all *O(dk)* which is farther lower than $O(n^2)$,the lowest time expense of similarity computation in many algorithms.

Edit distances between all patterns can be computed by calling *Algorithm 2*:

```
Algorithm 3: Edit Distances of all Traversal Patterns
double Pattern_edit_distance(TDB)
{  m= sizeof(TDB);
   for i=1 To m do
   { d=sizeof(Si);
```

```
    for j=1to i-1 do
    {k=sizeof(Sj);
     result[i,j]= (2*Edit_dis(S1,S2,d,k))/max(d+k);}
   return result;}
```

Algorithm 3 costs a large fraction of time on calling *Algorithm 2*, *m(m-1)/2* times in all, where m is the size of transaction database. Now that the time complexity of *Algorithm 2* is *O(dk)*, the integrated time complexity is $O(m^2dk)$.

4 Clustering User Traversal Pattern Based on Edit Distance

As one of dynamic clustering algorithms, sequential clustering method clusters every pattern only though a threshold θ avoiding a specified number of cluster. The value of θ is crucial for final results. An inappropriate threshold results in meaningless clustering results, so two thresholds $\theta_1, \theta_2 (\theta_1 \le \theta_2)$ are set to weaken the influence from a single threshold. If the dissimilarity, $d(x,C) < \theta_1$ between x and the nearest cluster C is lower than θ_1, assign x into cluster C, and if $d(x,C) > \theta_2$, create a new cluster *N* and assign x into cluster N, otherwise the cluster of *x* is postponed to determine. Detailed procedure is as follows:

```
Algorithm 4: Two-threshold Sequential Clustering Method
int TTSCM(TDB,Thr1,Thr2)//return the number of clusters
{  m= size(TDB);
   N=0;  //a counter of cluster
flag[1:m]=0;//if session i is clustered, flag[i]=0
changed=0; //if no session is clustered, change=0
k=find(flag=0);
kc=size(k);
while kc>0 do //there are unclustered patterns
{  if change=0 then       (*)
     {i=k(1);//index of the first unclustered pattern
     N=N+1;
         flag[i]=1;//set the clustered  flag of patter i
       assign session i into cluster C[N];}
   changed=0;
   for i=2 to kc Do
      {  j=k(i);
   [d,index]=min(j,C);
  if d<Thr1 then//the distance lower than threshold 1
     {  assign session j into cluster C[index];
            flag[j]=1; changed=1; }
       //the distance is greater than threshold 2
 if d>Thr2 then
  {  N=N+1; //create a new cluster
     assign session i into cluster C[N];
     flag[j]=1; changed=1; }
   }
k=find(flag=0); kc=size(k);}
return N;}
```

In algorithm 4, if-statement in the line marked with (*) ensures that at least one session is clustered per cycle so that outer loop will stop after *m* cycles at most. Inner loop is executed *m* cycles per outer loop. Synthesizing above analysis, the time complexity of algorithm 4 is $O(m^2)$, where m is the number of traversal patterns.

5 Experiments and Evaluation

5.1 Experimental Methods

All these algorithms given earlier are complemented based on Windows XP professional using Visual C++ and Matlab 2008a. Transaction database is erected based on web log of DePaul CTI Web server in April 2002[10], and comprised of 683 pages, 13745 sessions. In experiments, 8574 sessions are selected by removing the sessions whose lengths are lower than 4.

5.2 Results of Experiments

First, we extract more than about 50 sessions from original transaction database randomly to construct a simulated transaction database, *Exper*, and then *Exper* is enlarged by 10 times, namely the number of session exceeds 500, finally each session in *Exper* is executed random disturbance, including substitution, insertion and deletion of pages to ensure that these sessions from a same original session are more similar than these sessions from different original sessions, but are completely not identical. If sessions from a same original session can be clustered into a same cluster, the clustering method is effective, so an operator is defined as equation (2) to verify our proposed methods.

$$Accuracy = \frac{\text{the number of session clustered correctly}}{\text{the number of all sessions}} \tag{2}$$

In experiments, θ_2 is assigned 0.9, and the variance of accuracy with θ_1 is illustrated in Fig 2. Fig 2 shows the fact that the beginning accuracy is becoming greater with θ_1 until a certain peak. However, if θ_1 continues to increase, accuracy will decrease instead. The reason is analyzed as follows. At the beginning, more and more sessions can be clustered correctly because the dissimilarity between sessions is lower than θ_1, even from a same original session. When θ_1 exceed a certain peak and continues to increase, the dissimilarity of most sessions is lower than θ_1 so that they are clustered into a same cluster even if they don't originate from a same session, so accuracy will decrease.

Moreover, in order to reflect the efficiency of the proposed approach, we compared it with the method, *IHD-based*, proposed in [4]. Firstly, original sessions were repeatedly extracted independently and randomly, and then enlarged by 10 times and disturbed with similar operations mentioned in above paragraph to build eight data sets which consist of 100, 200, 400, 500, 800, 1000, 1500, 2000 user sessions respectively. Secondly, each data set was preprocessed according to corresponding method to build Hamming-distance matrix in the method proposed in [4] and ED-distance matrix in our proposed method. At the end, both clustering algorithms were executed respectively. In method proposed in [4] the threshold Λ is specified by the average value of all Hamming-distances between users' session, while in our proposed method, the two thresholds θ_1 and θ_2 are specified by 0.65 and 0.90 respectively. The precisions of two methods are compared as showed in Table 1.

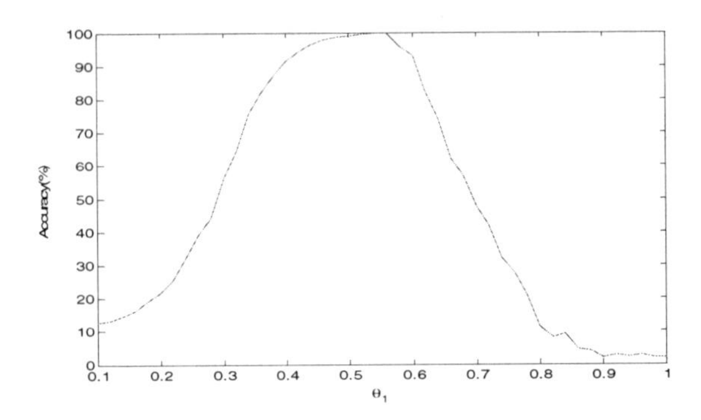

Fig. 2. The variance of clustering accuracy with thresholds

Table 1. Comparison of precisions between IHD-based and ED-based

	1	2	4	5	6	7	8	9
IHD-based	0.94	0.92	0.82	0.86	0.83	0.75	0.72	0.734
ED-based	0.91	0.885	0.87	0.868	0.858	0.794	0.78	0.79

From Table 1, we can know the precisions in two methods were high relatively, and our proposed method is of a higher average precision and stability, especially when the size of data set becomes greater. The possible causes are ED-based method take into consideration the accessed order of page and the structure of web site is introduced to measure similarity between pages.

Finally, the process of experiments is executed repeatedly to examine whether the results are sensible to the clustering order of traversal patterns. In each repeated experiments, the transaction database is rearranged randomly, and the final results are very close which manifests our proposed methods are strongly independent of the clustering order of patterns

6 Conclusions

User traversal pattern clustering is very important to analyze user intentions and performance. In this paper, since edit distance related to the relationship between pages referring to web topology is used to measure the dissimilarity between traversal patterns, rather clustering effectiveness is obtained. Moreover, Two-threshold sequential clustering can avoid specifying the number of cluster in advance, and reduce the dependency between the clustering effectiveness and the clustering order of traversal patterns.

Acknowledgment. Our research is supported by Project of Zhejiang Provincial Natural Science Foundation of China (Grant No. Y5100054) and Project supported by the Open Foundation from Ocean Fishery Science and Technology in the Most Important Subjects of Zhejiang (20100212).

References

[1] Fu, Y., Sandhu, K., Shih, M.-Y.: A Generalization-Based Approach to Clustering of Web Usage Sessions. In: The International Workshop on Web Usage Analysis and User Profiling, pp. 21–38 (1999)
[2] Lin, W., Liu, Y., Zhu, Q., Xi, D.: Web User Clustering and Personalized Recommendation Based on Mixture of Hidden Markov Chain Models. Journal of the China Society for Scientific and Technical Information 28(4), 557–564 (2009)
[3] Jia, L.v.: Web Log Mining Based upon Immune Clustering. Journel of Chongqing Normal University (Natural Science Edition) 24(3), 32–35 (2007)
[4] Li, S.S., Fang, S.H.: Improved clustering algorithm based on web usage mining technology. Computer Engineering and Design 30(22), 5182–5184 (2009)
[5] Levenshtein, V.I.: Binary Codes Capable of Correcting Deletions, Insertions and Reversals. Doklady Akademii Nauk SSSR 163(4), 845–848 (1965)
[6] Bellman, R.: Dynamic Programming. Princeton Univ. Press, Princeton (1957)
[7] Marzal, A., Vidal, E.: Computation of Normalized Edit Distance and Applications. IEEE Trans. Pattern Analysis and Machine Intelligence 15(9), 926–932 (1993)
[8] Oncina, J., Sebban, M.: Learning stochastic edit distance: Application in handwritten character recognition. Pattern Recognition 39, 1575–1587 (2006)
[9] Saigo, H., Vert, J.-p., Ueda, N., Akutsu, T.: Protein homology detection using string alignment kernelsEddy. Bioinformatics (February 2004)
[10] Depaul CTI Web Usage Mining Data, `http://maya.cs.depaul.edu/~classes/etc584/resource.html`

Evaluation of Web Search Engines

Luo XiaoLing and Xue he ru

Computer Science and Information Engineering College of Inner Mongolia
Agriculture University, Hohhot, Inner Mongolia, P.R. China
{luoxl,xuehr}@imau.edu.cn

Abstract. Using the proper search engine is crucial for efficient and effective web search. The objective of this paper is to develop methodologies to evaluate search engines in a systematic and reliable manner. A new model for evaluation and comparison of search engines is proposed. This hierarchical model classifies the most common features found in search engines and search results into groups and subgroups.

Keywords: engine, hierarchical model, groups.

1 Introduction

People can find information very quickly using any one of the existing search engines by simply entering the desired keywords. The usefulness of the returned hits is open to question, and is left to be judged by the user. The 'goodness' of the results depends on the choice of the search words as well as the effectiveness of the search engine. There are many search engines available these days, though only a few dominate. Each search engine has its own characteristics and effectiveness depending on the user's keywords and search criteria. It is not easy for a user to choose the most appropriate search engine for his/her particular use. The most common criteria for evaluating search engines, such as precision, recall, user effort, and coverage. Most search engine evaluation works do not include many criteria into their evaluation model. After analyses of existing evaluation models and search engine characteristics, we proposed seventy evaluation criteria, including some new ones, are categorized into two groups, features and performance.

The seventy parameters are classified into two major groups based on their functionality. The "Performance Group" includes features and capabilities that enhance the usability of the engine. The "Features group" includes various metrics for evaluating search results. Elements in each group can be further subdivided into subgroups and subsubgroups, resulting in a hierarchy of evaluation parameters. This provides additional flexibility to the users. Because of the clear hierarchical structure it is easy for users to focus on a specific group of evaluation parameters of interest. Users can also compare several search engines based on the selected features.

G. Zhiguo et al. (Eds.): WISM 2011, CCIS 238, pp. 448–454, 2011.

2 A Search Engine Evaluation Model

2.1 Weighed Parameters and Summary Core

A hierarchical structure allows us to rate the search engines at various abstraction levels of details. In general, the mathematical model or score of a collection of parameters can be expressed as

$$\text{Score} = \sum_{i=1}^{x} WiPi \quad (1)$$

where wi is the weight assigned to the ith parameter of a group with X parameters. Considering the two major groups of evaluation parameters, the total score of a search engine is represented as

$$\text{Score}=\text{WfeaturePfeature}+\text{WperformancePperformance} \quad (2)$$

A very important part of developing the evaluation model is the assignment of weights to different parameters. The sign for the weight is rather objective. If the presence of a feature makes the search engine more useful, the sign is positive, otherwise it is negative. A negative weight can be used to indicate the undesirable impact of a parameter on the total score. For example, the higher the number of dead links in the returned results, the worse the search engine is compared to others. Therefore the parameter indicating the number of dead links should have a negative weight. On the contrary, the magnitude of a weight is subjective. If one feels that performance is more important than the features of a search engine, the weights assigned may be 0.6 and 0.4, respectively.

The scores for features and performance, Pfeature and Pperformance, in turn, are derived from subsequent scoring equations at lower levels of the evaluation hierarchy. The value of a parameter is either a 0 or 1 to indicate whether a feature or capability exist in that search engine. A range between 0 and 1 is assigned to parameters that have various degrees of quality. The sum of the weights assigned to all parameters within a group must be equal to 1. This ensures the consistency of weight distribution among different groups.

This flexibility of tailoring the scoring system to individual needs, by changing the weights of the parameters or deleting an unwanted feature or adding a desirable feature, makes the proposed evaluation model very attractive to search engine users, subscribers, and providers. As pointed out in a workshop position paper [1], specific web search engines are effective only for some types of queries in certain contexts. Using our model and making the proper adjustment as described, a user will be able to find the particular search engine that suits his/her needs.

2.2 Feature Parameters

We further classified feature parameters into six major categories as shown in Figure 1.This hierarchical structure pools collections of related parameters into subgroups and sub-subgroups.

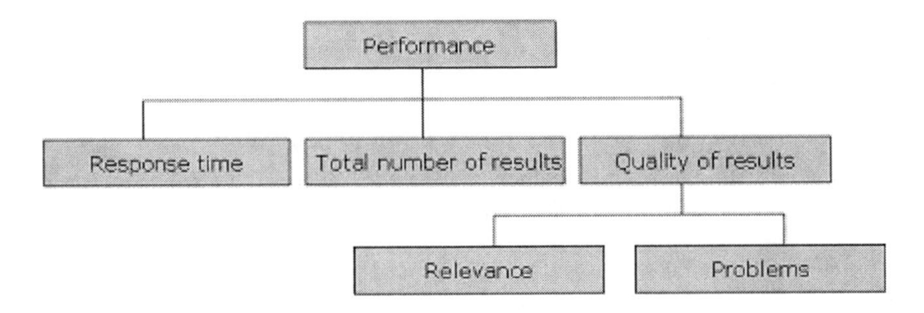

Fig. 1. Feature parameters considered

1. Home Page Features: This category indicates how user friendly the home page is regarding various help and user selection menus. This feature includes a subjective user evaluation, the availability and visibility of help links, result language selection, topic directory selection, and advanced search selection.
2. User Preferences: This category includes a choice of the home page language, the availability of safe search filtering, the control of the number of results per page, the choice of displaying the results in a new window, intelligent input correction, search default setting, search options within the result page, and news search.
3. Search Options: This category is further divided into subgroups of search modifier (at least one, case sensitive, etc.), search field (title, url, links, etc.), search focus selection (web site, web page, directory, etc.), search constraint selection (language, file format, publication date, etc.), and search meta words for focused search (specified sites only, similar pages, geographic regions, etc.).
4. Keyword Entry Options: This category considers the capability of the search engine in stop word interpretation, case sensitivity, exact phrase specification, wildcard allowance, search by pronunciation, and Boolean operators.
5. Database: This category indicates the number of groupings arranged in directories and the total number of indexed pages of the corresponding search engine.

Result Features: This category reviews display features such as, whether there is indication for the total number of hits, the number of pages, and search time; the capability to search within results; whether the results are ordered and numbered; whether different file formats are allowed in the returned items; whether a pay listing is allowed (a negative weight); web page snap; further search for related pages and the presence of hits' date, size and summarization.

The following tables 1 show the detailed information for each feature group including the description of each parameter, the possible value of the parameter, whether the evaluation is subjective or not (if users evaluate according to their own judgment, the evaluation is subjective; it is objective if the evaluation is based on facts.), and the sign of the assigned weight.

2.3 Performance Parameters

There are three major groups of performance metrics as shown in Figure 2: the response time, the total number of hits as indicated on the result page, and the quality of results.

Table 1. Feature parameters: user preferences

Description	Parameter's value	Subjective	Sign of Weight
User preferences	0 到 1		
Homepage interface language selection	yes=1, no=0	no	"+"
Safe search filtering	yes=1, no=0	no	"+"
Number of results (number of results per page)	yes=1, no=0	no	"+"
Results window (in a new window)	yes=1, no=0	no	"+"
Intelligent input correction	yes=1, no=0	no	"+"
Set the search default homepage	yes=1, no=0	no	"+"
Search option on the result page	yes=1, no=0	no	"+"
News display capability on search result page	yes=1, no=0	no	"+"

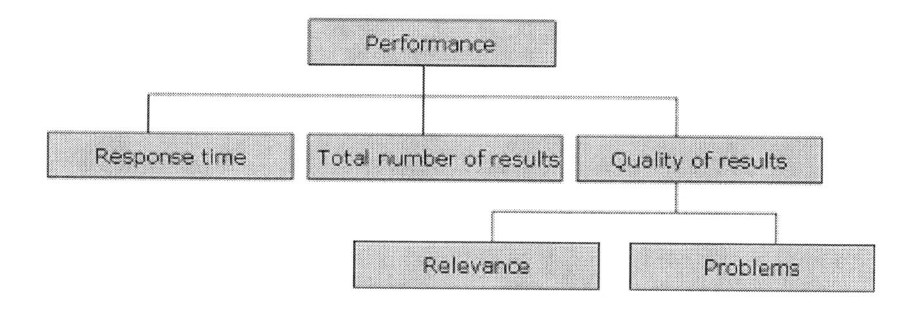

Fig. 2. Performance parameters

The Quality of Results group consists of subgroups Problems and Relevance. The Problems subgroup indicates the severity and frequency of problems encountered when a user tries to access the search engine or the returned hits. This includes the number of times that the search site is down during trial and experimentation, the number of broken links such as host not found and connection time out, and duplicates. All these parameters within the Problems group carry negative weights.

In order to obtain the Relevance score, one can solicit the assistance of humans to examine the relevance of the returned items with respect to the keywords. The scores are averaged as an attempt to eliminate any inherent potential bias and subjectivity in human interpretation. On the other hand, one can eliminate the subjectivity of humans by using a common list approach as described in Section 2.4. In such case, the Relevance subgroup includes the parameters precision @ N and recall @ M.

2.4 Evaluating Relevance Using a Common List

In order to eliminate subjectivity and the labor intensive process of using humans to provide the relevance score, an algorithmic approach is desirable. Comparing individual search engine's results against a common baseline is a reasonable means to evaluate relevance. The critical issue is how to generate such a common list. Since all search engines return what they regard as high ranking items in an ordered list, therefore it makes sense to combine the individual lists into a single list. This single list can be considered as the most accurate list as it consolidates the expertise of all the search engines. Matching items to this common list gives a sense of quality as well as quantity in terms of relevance.

Traditionally, there are three metrics commonly used to measure the relevance of the matched results in information retrieval of a bounded database:

Precision: the proportion of the returned items that are deemed relevant.
Precision @ N: precision evaluated from the top N highest ranked items.
Recall: the proportion of relevant items returned.

In the context of web search, it is impractical to examine all the returned items of a search engine, which is in the order of thousands and even hundreds of thousands. Also, it is impossible to measure the total of number of relevant items exists on the web, not to mention that this number is changing constantly. Therefore the first and the third metrics listed above are impossible to determine. We need to modify the definition of the above metrics to fit our current context with some assumptions.

First, an item in a search engine's list is deemed relevant if it also appears in the common list. This makes sense as it can be assumed that all the search engines are experts and the consensual common list derived collectively has a high probability of holding the truly high-ranked items. Second, extending this argument, we can treat the number of items in the common list as the total relevant items, since we are interested only in highly relevant items in most cases. Third, most search engine users are interested only in the top ten or so items as appear on the first result page, and therefore considering only the top ten ranked items is sufficient for our purpose.

For the Relevance factor, the following two revised metrics are used:

Precision @ N = the number of items that also appear in the common list over N.
Recall @ M = the number of items that also appear in the common list over the number of items M in the common list.

In many other relevance evaluations, precision @ N is taken as the ratio of relevant items as interpreted by humans over that of the top N items, while recall is not measured at all [2]. Our revised precision @ N concept eliminates human effort and subjectivity by matching items to a common list. These revised metrics actually measure relevance in terms of both quantity and quality. Two commonly employed metrics, coverage and overlap, are not used in our study because these measures have already been indirectly incorporated into the common list.

3 Experiments

We have performed some experiments in assigning different weights to the various parameters. Indeed, the evaluation model is sensitive to the weights as well as the values of the parameters. Google is not always the clear winner as the specific experiment shown in Figure 3.

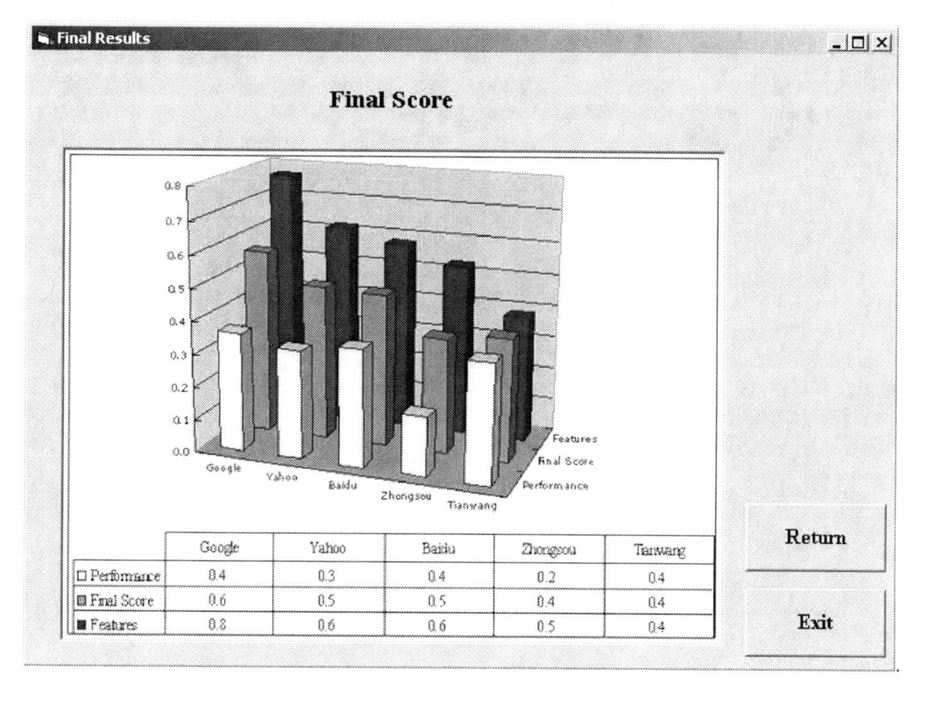

	Google	Yahoo	Baidu	Zhongsou	Tianwang
Performance	0.4	0.3	0.4	0.2	0.4
Final Score	0.6	0.5	0.5	0.4	0.4
Features	0.8	0.6	0.6	0.5	0.4

Fig. 3. Comparison of Search Engines

4 Discussions

In this paper, we introduce our search engine evaluation model, and describe the selected parameters in details. The advantage of this model is that it can be tailored to an individual's needs by changing the weight assignment, deleting parameters, and adding parameters. A user can also focus on the evaluation of a particular aspect easily due to the model's hierarchical structure. We introduce the common list as the benchmark in our model. Currently, we use a simple method to generate the common list, but further investigations on the concept of the common list and the exploration of more sophisticated algorithms are necessary.

Acknowledgement. This work is supported by the NDPYTD2010-9.

References

1. Allan, J., et al.: Challenges in Information Retrieval and Language Modeling. Report of a Workshop held at the Center for Intelligent Information Retrieval, University of Massachusetts Amherst (September 2008)
2. Chu, H., Rosenthal, M.: Search Engines for the World Wide Web: A Comparative Study and Evaluation Methodology. In: ASIS 1996 Annual Conference, October 19-24 (1996)
3. Lewandowski, D., Wahlig, H., Meyer-Bautor, G.: The Freshness of Web search engines' databases. Journal of Information Science (2006)

4. Voorhees, E.M., Harman, D.K.: Text Retrieval Conferences, March 9 (2002), http://trec.nist.gov/
5. Ma, L.: An Introduction and Comparison of Chinese Search Site. eSAS World, 139–146 (July 1998) (in Chinese), http://www.mypcera.com/softxue/txt/s35.htm (March 9, 2008)
6. Luk, R.W.P., Kwok, K.L.: A Comparison of Chinese Document Indexing Strategies and Retrieval Models. ACM Transactions on Asian Language Information Processing 1(3), 225–268 (2007)
7. Cresssie, N.: Spatial Prediction and Ordinary Kriging. Math. Geol. 20(4), 405–421 (1988)
8. Lophaven, S.N., Nielsen, H.B., Søndergaard, J.: DACE - A MATLAB Kriging Toolbox. Technical Report IMM-TR-2002-12, Informatics and Mathematical Modeling, Technical University of Denmark (2002)
9. Hardy, R.L.: Multiquadratic Equations of Topography and Other Irregular Surfaces. J. Geophus. Res. 76, 1905–1915 (1971)
10. Buhmann, M.D.: Radial Basis Functions. Cambridge University Press, Cambridge (2006)

Author Index